158
Advances in Polymer Science

Editorial Board:
A. Abe · A.-C. Albertsson · H.-J. Cantow
K. Dušek · S. Edwards · H. Höcker
J. F. Joanny · H.-H. Kausch · K.-S. Lee
J. E. McGrath · L. Monnerie · S. I. Stupp
U. W. Suter · G. Wegner · R. J. Young

Springer

*Berlin
Heidelberg
New York
Barcelona
Hong Kong
London
Milan
Paris
Tokyo*

Polymers for Photonics Applications I

Volume Editor: K.-S. Lee

With contributions by
C. Bosshard, M. Canva, L. Dalton,
U. Gubler, J.-I. Jin, H.-K. Shim,
G. I. Stegeman

Springer

This series presents critical reviews of the present and future trends in polymer and biopolymer science including chemistry, physical chemistry, physics and materials science. It is addressed to all scientists at universities and in industry who wish to keep abreast of advances in the topics covered.

As a rule, contributions are specially commissioned. The editors and publishers will, however, always be pleased to receive suggestions and supplementary information. Papers are accepted for „Advances in Polymer Science" in English.

In references Advances in Polymer Science is abbreviated Adv Polym Sci and is cited as a journal.

Springer APS home page: http://link.springer.de/series/aps/ or
http://link.springer-ny.com/series/aps/
Springer-Verlag home page: http://www.springer.de

ISSN 0065-3195
ISBN 3-540-42384-2
Springer-Verlag Berlin Heidelberg New York

Library of Congress Catalog Card Number 61642

This work is subject to copyright. All rights are reserved, whether the whole or part of the material is concerned, specifically the rights of translation, reprinting, re-use of illustrations, recitation, broadcasting, reproduction on microfilms or in other ways, and storage in data banks. Duplication of this publication or parts thereof is only permitted under the provisions of the German Copyright Law of September 9, 1965, in its current version, and permission for use must always be obtained from Springer-Verlag. Violations are liable for prosecution under the German Copyright Law.

Springer-Verlag Berlin Heidelberg New York
a member of BertelsmannSpringer Science+Business Media GmbH
http://www.springer.de

© Springer-Verlag Berlin Heidelberg 2002
Printed in Germany

The use of registered names, trademarks, etc. in this publication does not imply, even in the absence of a specific statement, that such names are exempt from the relevant protective laws and regulations and therefore free for general use.

Typesetting: Data conversion by MEDIO, Berlin
Cover: MEDIO, Berlin
Printed on acid-free paper   SPIN: 10741860   02/3020hu - 5 4 3 2 1 0

Volume Editor

Prof. Kwang-Sup Lee
Dept. of Polymer Science & Engineering
Hannam University
133 Ojung-Dong
Taejon 306-791, Korea
E-mail: kslee@mail.hannam.ac.kr

Editorial Board

Prof. Akihiro Abe
Department of Industrial Chemistry
Tokyo Institute of Polytechnics
1583 Iiyama, Atsugi-shi 243-02, Japan
E-mail: aabe@chem.t-kougei.ac.jp

Prof. Ann-Christine Albertsson
Department of Polymer Technology
The Royal Institute of Technolgy
S-10044 Stockholm, Sweden
E-mail: aila@polymer.kth.se

Prof. Hans-Joachim Cantow
Freiburger Materialforschungszentrum
Stefan Meier-Str. 21
D-79104 Freiburg i. Br., FRG
E-mail: cantow@fmf.uni-freiburg.de

Prof. Karel Dušek
Institute of Macromolecular Chemistry, Czech
Academy of Sciences of the Czech Republic
Heyrovský Sq. 2
16206 Prague 6, Czech Republic
E-mail: dusek@imc.cas.cz

Prof. Sam Edwards
Department of Physics
Cavendish Laboratory
University of Cambridge
Madingley Road
Cambridge CB3 OHE, UK
E-mail: sfe11@phy.cam.ac.uk

Prof. Hartwig Höcker
Lehrstuhl für Textilchemie
und Makromolekulare Chemie
RWTH Aachen
Veltmanplatz 8
D-52062 Aachen, FRG
E-mail: hoecker@dwi.rwth-aachen.de

Prof. Jean-François Joanny
Institute Charles Sadron
6, rue Boussingault
F-67083 Strasbourg Cedex, France
E-mail: joanny@europe.u-strasbg.fr

Prof. Hans-Henning Kausch
c/o IGC I, Lab. of Polyelectrolytes
and Biomacromolecules
EPFL-Ecublens
CH-1015 Lausanne, Switzerland
E-mail: kausch.cully@bluewin.ch

Prof. Kwang-Sup Lee
Department of Polymer Science & Engineering
Hannam University
133 Ojung-Dong
Teajon 300-791, Korea
E-mail: kslee@mail.hannam.ac.kr

Prof. James E. McGrath
Polymer Materials and Interfaces Laboratories
Virginia Polytechnic and State University
2111 Hahn Hall
Blacksbourg
Virginia 24061-0344, USA
E-mail: jmcgrath@chemserver.chem.vt.edu

Prof. Lucien Monnerie
École Supérieure de Physique et de Chimie
Industrielles
Laboratoire de Physico-Chimie
Structurale et Macromoléculaire
10, rue Vauquelin
75231 Paris Cedex 05, France
E-mail: lucien.monnerie@espci.fr

Prof. Samuel I. Stupp
Department of Measurement Materials Science
and Engineering
Northwestern University
2225 North Campus Drive
Evanston, IL 60208-3113, USA
E-mail: s-stupp@nwu.edu

Prof. Ulrich W. Suter
Department of Materials
Institute of Polymers
ETZ,CNB E92
CH-8092 Zürich, Switzerland
E-mail: suter@ifp.mat.ethz.ch

Prof. Gerhard Wegner
Max-Planck-Institut für Polymerforschung
Ackermannweg 10
Postfach 3148
D-55128 Mainz, FRG
E-mail: wegner@mpip-mainz.mpg.de

Prof. Robert J. Young
Manchester Materials Science Centre
University of Manchester and UMIST
Grosvenor Street
Manchester M1 7HS, UK
E-mail: robert.young@umist.ac.uk

Preface

The future of information technology requires ultra-high speed processing and large data storage capacity. Since the electronics technology using semi-conductors and inorganic materials is about to reach its limits, much current research is focused on utilizing much faster photons than electrons, namely photonics. To achieve any significant effect on the actual use of the science of photonics, developments of more efficient photonics materials, better optical property evaluations, manufacture of devices for system applications, etc. are the subjects which need to be explored. In particular, the development of photonics materials stands in the forefront of research as this constitutes the most pertinent factor with regard to the development of ultra-high speed and large capacity information processing. In this respect, there has been continuous research on photo-responsive materials through molecular structure design and architecture and the results so far are very promising as functions and performances are beginning to realize their high expectations.

The two special volumes "Polymers for Photonics Applications" give authoritative and critical reviews on up-to-date activities in various fields of photonic polymers including their promising applications. Seven articles have been contributed by internationally recognized and they deal with, polymers for second- and third-order nonlinear optics, quadratic parametric interactions in polymer waveguides, electroluminescent polymers as light sources, photoreflective polymers for holographic information storage, and highly efficient two-photon absorbing organics and polymers. This review should provide individuals working in the field of photonics polymers with invaluable scientific knowledge on the state of the art while giving directions for future research to those deeply interested.

Finally, I would certainly like to give my heartfelt thanks to the contributors to this two volume special issue and although the publication has been delayed, I am very grateful that this issue can be dedicated to Professor Gerhard Wegner in celebration of his 60th birthday.

Taejon, August 2001   Kwang-Sup Lee

In Recognition
Professor Gerhard Wegner

Polymer scientists do not hesitate in naming Professor Gerhard Wegner as one of the most influential scholars in the field of polymer science today. During the dynamic growth period of polymer science, its conception by Professor H. Staudinger, Professor Wegner has made large contributions in the areas of polymer synthesis, new synthetic methodology for macromolecules, polymer structure analysis, and characterizations. His achievements have been particularly prominent in self-organization of macromolecules, supramolecular architectures, solid polyelectrolytes, electrical and optical polymers, and functional nano-structured films based on polymeric materials, where many valuable fundamental results were established through years of creative research in advanced polymer science. His efforts are still continuing today with great energy and concentration.

Professor Wegner was one of the founders of the Max-Planck-Institute for Polymer Research in 1983. He was instrumental in making the Institute an internationally recognized organization in polymer science research, largely by his excellent academic and administrative efforts. He is now the vice-president of the Max-Planck Society.

Professor Wegner received his Ph.D degree in polymer chemistry for a thesis on his work at the University of Mainz with Professor W. Kern and Professor R. C. Schulz in 1965. Between 1966 and 1968, he worked at Yale University (USA) as a research staff chemist for Professor H. G. Cassidy. After returning to Germany, he worked for his professorship in the group led by Professor E. W. Fisher at the University of Mainz. It was during this period that he opened for the first time the area of topological polymerization of diacetylene. Since then there has been a large number of scientific papers related to the chemistry and physics of polydiacetylenes. From 1974 to 1984 he worked as a co-director of the Institute of Macromolecular Chemistry (Herman Staudinger Haus) at the Freiburg University, after which he moved to the Max-Planck-Institute for Polymer Research where he is still the director.

Professor Wegner has published over 450 publications related to various areas of polymer science and his works have and are still being very widely cited. For the recognition of his achievements in polymer science, he has received the Adolph-Bayer-Award, the Philip-Morris-Prize, the Herman Staudinger Medal of the German Chemical Society, American Chemical Society Award in Polymer Chemistry and other numerous awards and honors. Professor Wegner is currently serving as an editorial board member for many international scientific journals, and the contributions he has made to Advances in Polymer Science during the last ten years have been invaluable.

Apart from his outstanding scientific achievements, his concern for polymer education, cooperative global activities at individual, group and institutional levels have produced excellent results and many have benefited from his efforts.

We hope that Professor Wegner continues to benefit us with the far-reaching insight into the interesting world of polymer science and we truly look forward to many more years of his contributions. On the occasion of his sixtieth birthday, we wish him all the best in all the facets of his scientific and personal life.

Kwang-Sup Lee

Advances in Polymer Science
Now Also Available Electronically

For all customers with a standing order for Advances in Polymer Science we offer the electronic form via LINK free of charge. Please contact your librarian who can receive a password for free access to the full articles. By registration at:

http://link.springer.de/series/aps/reg_form.htm

If you do not have a standing order you can nevertheless browse through the table of contents of the volumes and the abstracts of each article at:

http://link.springer.de/series/aps/
http://link.springer-ny.com/series/aps/

There you will find also information about the

– Editorial Board
– Aims and Scope
– Instructions for Authors

Contents

Nonlinear Optical Polymeric Materials: From Chromophore Design to Commercial Applications
L. Dalton 1

Quadratic Parametric Interactions in Organic Waveguides
M. Canva, G. I. Stegeman 87

Molecular Design for Third-Order Nonlinear Optics
U. Gubler, C. Bosshard 123

Light-Emitting Characteristics of Conjugated Polymers
H.-K. Shim, J.-I. Jin 193

Author Index Volumes 101–158 245

Subject Index 259

Contents of Volume 159

Polymers for Photonics Applications II
Nonlinear Optical, Photorefractive and Two-Photon Absorption Polymers

Volume Editor: K.-S. Lee

Organics and Polymers with High Two-Photon Activities and Their Applications
T.-C. Lin, S.-J. Chung, K.-S. Kim, X. Wang, G.S. He, J. Swiatkiewicz, H.E. Pudavar, P.N. Prasad

Photorefractive Polymers and Their Applications
B. Kippelen, N. Peyghambarian

Polymeric Materials and Their Orientation Techniques for Second-Order Nonlinear Optics
F. Kajzar

Nonlinear Optical Polymeric Materials: From Chromophore Design to Commercial Applications

Larry Dalton

Loker Hydrocarbon Research Institute, University of Southern California,
Los Angeles, CA 90089–1661 and Department of Chemistry, University of Washington, Seattle, WA 98195–1700, USA
e-mail (USC): dalton@usc.edu; e-mail (UW): dalton@chem.washington.edu

Abstract. Polymeric electro-optic materials have recently been developed that, when fabricated into devices such as Mach-Zehnder interferometers, permit drive (V_π) voltages of less than 1 V to be realized at the telecommunications wavelength of 1.3 and 1.55 microns. Operation of polymeric electro-optic modulators to frequencies (bandwidths) of greater than 100 GHz has been demonstrated. The total insertion loss of polymeric electro-optic modulators has been reduced to values as low as 5 dB, which is competitive with values obtained for lithium niobate modulators and is much lower than that obtained for gallium arsenide electro-absorptive modulators. Polymeric electro-optic modulators can be operated for long periods of time at temperatures on the order of 100 °C. Techniques have been developed for seamlessly integrating polymeric electro-optic circuitry with passive low loss optical circuitry (e.g., silica long-haul transmission fiber and medium-range fluoropolymer fibers) and with very large scale integration (VLSI) semiconductor electronics. These advances have created a considerable interest in the commercialization of polymeric electro-optic materials. Polymeric electro-optic materials are now being evaluated for applications such as phased array radar, satellite and fiber telecommunications, cable television (CATV), optical gyroscopes, electronic counter measure (ECM) systems, backplane interconnects for high-speed computers, ultrafast (100 Gbit/s) analog-to-digital (A/D) converters, land mine detection, radio frequency (rf) photonics, and spatial light modulators. This review discusses the structure-function relationships that had to be defined and the synthesis and processing advances achieved before materials appropriate for commercialization could be produced. Topics discussed include the design and synthesis of chromophores that simultaneously exhibit large molecular hyperpolarizability, low optical absorption, processability (e.g., solubility in various processing media), and the prerequisite (thermal, chemical, electrochemical, and photochemical) stability to survive conditions encountered in the fabrication and operation of polymeric electro-optic devices. Chromophore-chromophore intermolecular electrostatic interactions have been shown to be the most serious problem impeding the optimization of electro-optic activity in organic materials. The quantitative theoretical treatment of such interactions by equilibrium and Monte Carlo statistical mechanical methods is discussed. Rules for designing chromophores with shapes leading to maximum obtainable electro-optic activity are discussed and the synthetic realization of such structures is reviewed. The role of electrostatic interactions in influencing the choice of processing conditions is also discussed. A number of processing steps, including spin casting, electric-field poling, lattice hardening, fabrication of buried channel waveguides (including deposition of cladding layers), electrode deposition and connection with electronic circuitry, and integration of active and passive optical circuitry, are required. Each of these steps can affect device performance (e.g., influence optical loss, electro-optic activity, and stability). The systematic optimization of each of these steps is reviewed. Finally, device design and operation are reviewed and speculation on the future of the field is expressed.

Keywords. Electro-optic materials, Electro-optic devices, Electric field poling, Hyperpolarizability, Intermolecular electrostatic interactions

1	**Introduction – Concepts, Materials, and Applications**	4
2	**Materials Requirements**	11
3	**Materials Characterization Methods**	15
4	**Chromophores**	22
4.1	Optimization of Molecular Hyperpolarizability	22
4.2	Auxiliary Properties of Chromophores	23
5	**The Problem of Translating Microscopic to Macroscopic Optical Nonlinearity**	29
5.1	Theoretical Treatment of Large Intermolecular Electrostatic Interactions	30
5.1.1	The Single Phase Region	30
5.1.2	Phase Separation and Phase Diagrams	39
6	**Processing Steps**	41
6.1	Spin Casting, Electric Field Poling, and Lattice Hardening	41
6.2	Fabrication of Buried Channel Waveguides	50
6.2.1	Reactive Ion Etching Techniques	51
6.2.2	Photolithographic Techniques	51
6.3	The Unique Problem of Fabricating Tapered Transitions	52
6.4	Electrodes and Claddings	53
7	**Sophisticated Packaging**	54
7.1	Vertical and Horizontal Integration with VLSI Semiconductor Chips	54
7.2	Integration with Passive Optical Circuitry and Diode Lasers . . .	55
7.3	Special Electrode Structures for High (100 GHz and Above) Frequencies	56
8	**Evaluation of Device and System Performance**	60
8.1	Bandwidth	60
8.2	Drive Voltage	60
8.3	Optical Loss	62
8.4	Stability under Harsh Conditions	63
9	**Applications and Commercialization**	65

10	Future Prognosis	69
11	Appendix: Conversion Factors and Useful Relationships	76
References		76

Abbreviations

ac	alternating current
ATR	attenuated total reflection
APC	amorphous polycarbonate [poly(bisphenol A carbonate-*co*-4,4'-(3,3,5-trimethylcyclohexylidene)diphenol]
dc	direct current
α	molecular polarizability
β	molecular (first) hyperpolarizability
β_0	molecular hyperpolarizability at zero frequency
β_{EO}	molecular hyperpolarizability at operational wavelength
ε	electrical permittivity
CF	3-cyano-5,5-dibutyl-2-dicyanomethylene-4-methyl-2,5-dihydrofuran
$f(\omega)$	local field (Onsager) factor
k	Boltzmann constant
D	Debye
DSC	differential scanning calorimetry
DEC	double end crosslinkable
$\Delta\phi$	phase shift of light caused by application of an electric field
E	allied electric poling field strength
esu	electrostatic units
EFISH	electric field induced second harmonic generation
F	electric field at the chromophore
FT	Fourier transform
EO	electro-optic
γ	molecular second hyperpolarizability
Hz	hertz (1 Hz=s^{-1})
MHz	megahertz (10^6 Hz)
GHz	gigahertz (10^9 Hz)
HRS	hyper-Rayleigh scattering
IR	infrared
L	Langevin function
μ	dipole moment
M	chromophore molecular mass (weight)
n	index of refraction
NLO	nonlinear optical
N	chromophore number density (number of molecules per cubic centimeter) in polymer host

P	macroscopic polarization
p	molecular polarization
PDS	photothermal deflection spectroscopy
PMMA	poly methyl methacrylate
PU	poly urethane
r	electro-optic coefficient
r_{33}	electro-optic coefficient in the direction of the applied electric field
rf	radio frequency
θ	angle between the poling field and chromophore principal axis
T	Kelvin temperature
T_d	chromophore decomposition temperature
T_g	glass transition temperature of polymer material
TBDMS	tert-butyldimethylsilyl
TGA	thermal gravimetric analysis
TMA	thermal mechanical analysis
V_π	voltage required to produce a phase shift of π
$<\cos^3\theta>$	order parameter relevant to electro-optic activity
$\chi^{(1)}$	linear optical susceptibility
$\chi^{(2)}$	second-order nonlinear optical susceptibility
$\chi^{(3)}$	third-order nonlinear optical susceptibility
UV	ultraviolet
w	weight fraction of chromophore in a polymer matrix
wt	weight
ω	angular frequency
λ	wavelength

1
Introduction – Concepts, Materials and Applications

In this review, the focus is on electro-optic materials. Such materials are members of the more general class of second-order nonlinear optical materials, which also includes materials used for second harmonic generation (frequency doubling). The term second-order derives from the fact that the magnitude of these effects is defined by the second term of the power series expansion of optical polarization as a function of applied electric fields. The power series expansion of polarization with electric field can be expressed either in terms of molecular polarization (p; Eq. 1) or macroscopic polarization (P; Eq. 2)

$$P_i = \alpha_{ij} E_j + \beta_{ijk} E_j E_k + \gamma_{ijkl} E_j E_k E_l + ... \qquad (1)$$

$$P_I = \chi_{IJ}^{(1)} E_J + \chi_{IJK}^{(2)} E_J E_K + \chi_{IJKL}^{(3)} E_J E_K E_L + ... \qquad (2)$$

Equation (2) can be expanded to a form that more directly shows the existence of high frequency (second harmonic generation or frequency doubling – f=2ω) and low frequency (electro-optic – f=0) phenomena.

$$P=\chi^{(1)}E_0\cos(\omega t-kz)+(1/2)\chi^{(2)}E_0^2[1+\cos(2\omega t-2kz)]+\chi^{(3)}E_0^3[(3/4)\cos(\omega t-kz)+(1/4)\cos(3\omega t-kz)] \qquad (3)$$

Actually, the "low frequency" (f=0) term typically involves frequencies in the range 10^3–10^{11} Hz, i.e., from 1 kilohertz to 100 gigahertz. Such radio, microwave, and millimeter wave frequencies are low frequency only with respect to much higher optical frequencies (e.g., f=2.3×10^{14} Hz, λ=1.3 µm; f=1.9×10^{14} Hz, λ=1.55 µm). Unlike the first- and third-order terms of the above equations, there is a symmetry requirement for the second-order terms; namely, that of noncentrosymmetric or acentric symmetry. The molecules or materials cannot have a center of inversion symmetry. Although second-order nonlinear optical activity can arise from higher order octupolar symmetry [1], such materials have yet to be used in the fabrication of devices. Thus, for practical purposes, organic second-order nonlinear optical materials can be considered to be dipolar molecules organized into noncentrosymmetric (acentric or "ferroelectric") chromophore lattices. It will shortly become clear to the reader that this requirement of ordered chromophore lattices is the most daunting materials requirement to be satisfied in the development of electro-optic materials.

Electro-optic activity can be thought of as control of the index of refraction of a material by application of a finite dc or ac voltage. As noted above, ac voltages can be characterized by frequencies from 1 Hz to greater than 100 GHz. Index of refraction relates to the speed of light in a material so another way of thinking about electro-optic activity is to view it as a voltage-controlled phase shift of light. The application of an electric field acts to change the charge distribution in a second-order nonlinear optical chromophore. If the chromophore is of the common charge-transfer type, the electric field can be viewed as changing the mixing of neutral and charge-separated states. This change in charge distribution will, in turn, alter the velocity of light propagating through the material. If the chromophores are randomly ordered, the influence of individual chromophores will cancel out as light propagates through the material. If all chromophores are pointing in the direction of the applied field, then the effect of individual chromophores will simply be additive and the electro-optic activity of the material will be given by $r_{33}=2N\beta f(\omega)/n^4$ where N is the chromophore number density, $f(\omega)$ is a product of local field factors taking into account that applied fields are attenuated by the local environment of the chromophore, and n is the index of refraction. If the noncentrosymmetric order is not perfect, which will certainly be the case if order is induced by electric-field poling, then the macroscopic electro-optic coefficient will be given by Eq. (4).

$$r_{33}=|2Nf(\omega)\beta<\cos^3\theta>/n^4| \qquad (4)$$

The order parameter, $<\cos^3\theta>$, will thus play a critical role in defining the maximum achievable electro-optic activity that can be realized, particularly if this quantity is dependent on chromophore concentration.

From the above introduction to the basic phenomenon of electro-optic activity, several points are clear. First, a conceptually attractive class of organic electro-optic materials involve charge-transfer molecules of the general form, D-(π-electron connective segment)-A, where D is an electron-donating group such as an amine and A is an electron-accepting group such as a cyano or nitro group.

Second, the effect of applied electric field changing the mixing of neutral and charge-separated forms of the charge-transfer chromophores will define the magnitude of electric field required to achieve a given phase shift. Molecular first hyperpolarizability β can be understood within the framework of the parameters of this two-state charge-transfer process, as shown in Eq. (5) (see also the Appendix):

$$\beta = (\mu_{ee} - \mu_{gg})(\mu_{ge})^2/(\Delta E_{ge})^2 \tag{5}$$

where the critical parameters are the difference between the dipole moments of the ground and excited states, the HOMO-LUMO energy gap, and the transition matrix element between the ground and excited state.

Third, achieving the largest possible order parameter will be important for optimizing electro-optic activity. Moreover, this order parameter must be stable over the operational lifetime of a device utilizing the electro-optic material.

Polymeric electro-optic materials are used in a variety of device configurations to achieve a number of functions including electrical-to-optical signal transduction, switching at nodes in optical networks, control of the phase of radio frequency and optical signals, electromagnetic field sensing, etc. Three of the most common device configurations are shown in Fig. 1

The phase retardation of light transiting a material to which an electric field has been applied is given by Eq. (6):

$$\Delta\phi = 2\pi\Delta nL/\lambda = \pi n^3 rEL/\lambda \tag{6}$$

Consider the Mach-Zehnder (MZ) interferometer device configuration (top, Fig. 1). Application of an electric field to one arm of the MZ results in a phase retardation relative to the signal traversing the second arm and in destructive interference at the output of the device. The consequence of this voltage-controlled destructive interference is that the applied electrical signal is transduced onto the optical beam as an amplitude modulation. The birefringent modulator shown in Fig. 1 also acts as an electrical-to-optical signal transducer. However, in this modulator, both TM and TE optical modes travel through the electro-optic material and the application of an electrical field produces a voltage-dependent birefringence. This birefringence is turned into an amplitude modulation by positioning a polarizer at the output of the device. The voltage, V_π, required to achieve full-wave modulation with a birefringent modulator is 1.5 times that required for a Mach-Zehnder modulator. This is a good reminder to us that material requirements will vary from one device to another and we will need to keep the details of devices in mind when discussing applications of electro-optic materials. The directional coupler, shown at the bottom of Fig. 1, consists of two side-by-side waveguides separated by only a few micrometers. The overlap of the guided waves in the two waveguides couples energy back and forth between the waveguides. When an electric voltage is applied, both the phase difference be-

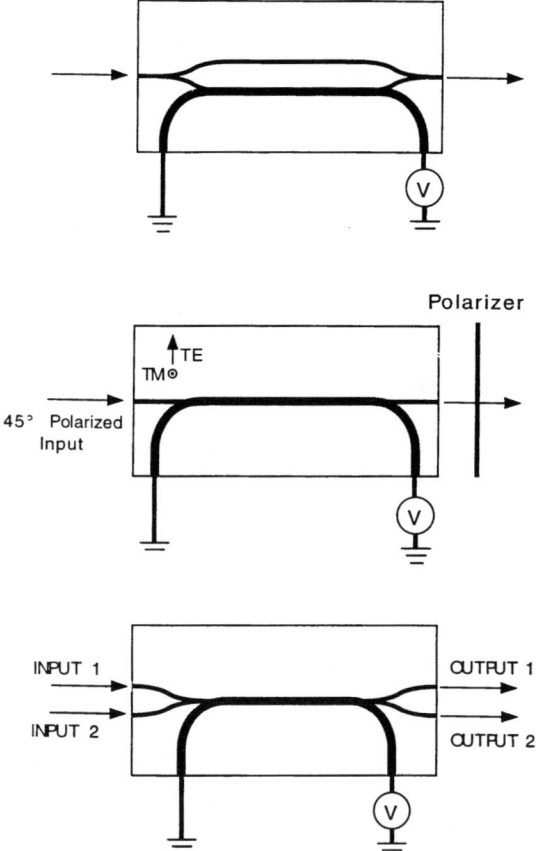

Fig. 1. Three common electro-optic device configurations: *Top* Mach-Zehnder interferometer; *middle* birefringent modulator; *bottom* directional coupler

tween modes in the two waveguides and the coupling coefficient are affected with the result that the output is toggled between outputs 1 and 2. The smallest voltage needed for switching an optical signal between the two output ports is 1.7 times larger than for a Mach-Zehnder modulator. A more detailed discussion of the performance of these three device structures is given elsewhere [2]. It is readily seen that these simple device configurations can be used to effect electric-to-optical signal transduction and switching of optical signals. The Mach-Zehnder and birefringent modulator configurations can also be used for voltage sensing. The fundamental device configurations, discussed above, can be used to fabricate more sophisticated devices such as ultrafast analog-to-digital signal converters [3].

If electro-optic modulators and directional couplers are to be driven by the electrical outputs from very large-scale integration (VLSI) semiconductor chips, then the required drive or V_π voltages of EO devices must be less than 6 V (for

TTL chips) and on the order of 1 V or less for the fastest semiconductor chips. If EO modulators are to be used in fiber optic communication links, it can be noted that the link gain is inversely proportional to $(V_\pi)^2$ and the noise figure is directly proportional to $(V_\pi)^2$. The higher the electro-optic activity of materials (and thus the lower the V_π of the EO device) the better! Drive voltages in the order of 1 V or less are required for lossless (transparent) communication links.

If electro-optic devices are to improve information processing and communication speeds, then the bandwidths of such devices must be greater than 10 GHz and ideally should approach 100 GHz. Also, devices must have low insertion loss. Total insertion losses should be significantly below 10 dB and the per-unit-length (of electro-optic waveguide) optical loss should be 1 dB/cm or less. Finally, the device must exhibit sufficient longevity under in-field operating conditions. Not only must modulators survive temperature variations of considerable magnitude, but they must also exhibit significant photochemical stability. If EO modulators are to be used in space applications they must be capable of withstanding bombardment from high-energy particles and electromagnetic radiation.

Modulation can be accomplished in a variety of ways. For example, the output of a laser can be modulated directly or external modulation can be applied using electro-optic or electro-absorptive modulators. Currently, commercial electro-optic modulators are fabricated from lithium niobate. The performance characteristics of one family of lithium niobate modulators, produced by Lucent Technologies (Breinigsville, PA), are given in Table 1.

Lithium niobate is a crystalline material and, as such, modulators fabricated from lithium niobate cannot be straightforwardly integrated with VLSI semiconductor electronics. A lithium niobate modulator must typically be connected to its electrical signal source by wires and techniques such as flip-chip bonding. Lithium niobate has a high dielectric constant, which can strongly perturb the incoming electromagnetic field in radar applications [5] and is a detriment both to modulation efficiency and bandwidth. The other alternative to electro-optic modulation is electro-absorptive modulation employing gallium arsenide multiple quantum well materials. Electro-absorptive devices currently suffer from high optical loss but hold the promise of low drive voltages. Like polymeric EO materials, gallium arsenide electro-absorptive materials can still be considered to be in a developmental stage. In Table 2, we summarize the properties of lithium niobate, gallium arsenide and polymeric EO materials.

A brief comment on the importance of the relationship of the dielectric constant, ε, to the index of refraction, n, is useful. With polymeric materials, the π-

Table 1. Performance characteristics of the Lucent lithium niobate electro-optic modulator

V_π	6 V @ 1550 nm
Bandwidth	30 GHz
Maximum operational temperature	70 °C
Insertion loss	6 dB

Table 2. Comparison of lithium niobate, gallium arsenide and polymeric materials

Property	Gallium arsenide	Lithium niobate	EO polymers
EO coefficient (pm/V at 1.3 microns)	1.5	31	>70
Dielectric constant, ε	10–12	28	2.5–4
Refractive index, n	3.5	2.2	1.6–1.7
Bandwidth-length product, GHz/cm	>100	10	>100
Voltage-length product, V/cm	1–5	5	1–2
Figure of merit	6	10	100
Optical loss (dB/cm at 1.3 microns)	2	0.2	0.2–1.1
Thermal stability (°C)	80	90	80–125*
Maximum optical power (mW)	30	250	250*

*Obtained for a hardened polymeric material

electrons play the dominant role in defining both ε and n. Thus, $\varepsilon = n^2$. Moreover, for polymeric materials, these quantities are essentially frequency independent. Because of the above equality, optical and electric waves will co-propagate in polymeric EO materials with the same velocity. Thus, they will not de-phase and can interact (through the EO material) over significant interaction lengths. This feature of polymeric materials permits large bandwidth-length products to be realized. Modulation efficiency can be increased by taking advantage of longer active waveguide lengths. For example, 3-cm modulators have recently been employed by IPITEK (formerly TACAN) Corporation to obtain V_π values of 0.8 V [6]. Before leaving the discussion of types of electro-optic materials, we should also draw a distinction between crystalline organic electro-optic materials and polymeric electro-optic materials. In the 1980s, Professor Anthony Garito (University of Pennsylvania) was able to prepare large crystals of methylnitroaniline (MNA). These materials were subsequently investigated by corporate laboratories such as Celanese (which later became Hoechst-Celanese). As might be expected, the molecular polarizability of the small methylnitroaniline moieties was simply not large enough to translate to useable macroscopic electro-optic activity. Thus, device-oriented research on this material was ultimately abandoned. Considerable excitement was generated by the discovery by Marder, Perry, and co-workers (then at California Institute of Technology and now at University of Arizona) of electro-optic crystals of 4'-dimethylamino-*N*-methyl-4-stilbazolium tosylate (DAST) [7]. It appeared initially as if these crystals might exhibit electro-optic coefficients in the range of several hundred picometers per volt [7–9]. However, DAST crystals suffer from severe growth anisotropy and thus have, until recently, been obtained only as long needles. Device development with such needle-like crystals is very difficult indeed. Thus, continuing research efforts on DAST have focused on obtaining thin films of the material or at least rod-like crystals with reduced growth anisotropy [8, 10]. Collaborative research efforts include those of Marder and Perry with Professor Stephen Forrest at Princeton using molecular beam quasi-epitaxy in the presence of an electric poling field. Efforts that show the greatest promise of yielding device quality materials currently appear to be those of Professor Peter Gunter [10] at the ETH, Zurich, Professor Thukar at Auburn [8], and those of various research groups at Tohoku University [10] in Japan. However, it is unlikely that DAST materials can be modified to yield the magnitude of electro-optic activity obtained with chromophore-containing, electrically poled, polymer materials. Before leaving our introductory discussion, it is worth noting that electro-optic activity is a tensorial property. For example, growth anisotropy (crystal shape) will determine which tensor components of the material electro-optic tensor can be utilized in device operation. For crystals, Zyss and Oudar [11] have shown that macroscopic nonlinear optical susceptibility can be related to molecular hyperpolarizability by the following equations:

$$\chi^{(2)} = N\beta f(\omega)\Gamma(\Omega) \tag{7}$$

$$\Gamma(\Omega) = (1/N_c)\Sigma_{ijk}\Sigma_s(\cos\theta^s)_{Ii}(\cos\theta^s)_{Jj}(\cos\theta^s)_{Kk} \tag{8}$$

The orientation factor provides the projection of the molecular β tensor onto the crystal or laboratory axis (note that the crystal will be oriented intentionally with respect to the applied electric field by the act of depositing electrodes on the crystal). In Eq. (8), the indices I, J, K and i, j, k refer, respectively, to the principal directions of the crystal and molecular coordinate systems. N_c is the number of equivalent positions in the unit cell.

For dipolar chromophores that are the subject of this chapter, only one component of the molecular hyperpolarizability tensor, β_{zzz}, is important. Thus, the summation in Eq. (8) disappears. Electric field poling induces C_v cylindrical polar symmetry. Assuming Kleinman [12] symmetry, only two independent components of the macroscopic second-order nonlinear optical susceptibility tensor survive.

$$\chi^{(2)}_{zzz} = N\beta f(\omega)<\cos^3\theta> \tag{9}$$

$$\chi^{(2)}_{zxx} = N\beta f(\omega)<\cos\theta><\cos^2\theta> \tag{10}$$

Equation (9) defines r_{33} while Eq. (10) defines r_{13}. For electrically poled polymer materials, it is common to assume that $r_{33} = 3(n_o/n_e)^4 r_{13}$ [3]. However, the reader should note that this is not always the case [13, 14] and is seldom the case with use of laser-assisted (photochemical) poling [15, 16]. Thus, we need only consider Eq. (4) in our further discussion. However, it should be noted that devices such as the birefringent modulator depend on the difference between r_{33} and r_{13} so that both components can enter the equation for device performance.

Clearly, a requirement for device quality second-order nonlinear optical materials is a noncentrosymmetric dipolar chromophore lattice. There are several ways by which such lattices have been achieved. With all methods, a force must be identified that will act to align chromophores so that they point in the same direction. Such forces can be surface electrostatic forces, as in the case of Langmuir-Blodgett methods, covalent bonds in the case of sequential synthesis methods, electric poling field/chromophore dipole interaction with electric field poling, and host/chromophore electrostatic interactions in the case of inclusion compounds [3, 5]. Such forces must overcome intermolecular chromophore-chromophore electrostatic interactions that will favor centrosymmetric ordering of the chromophores. In the rare case of natural noncentrosymmetric crystal growth, such as in the case of DAST, the chromophore must experience some ionic or steric interaction that strongly favors noncentrosymmetric order. The DAST chromophore is positively charged so that interaction of the chromophore with negatively charged counter ions strongly influences the ordering of chromophores. We have reviewed elsewhere [3, 5] the use of sequential synthesis, Langmuir-Blodgett fabrication of thin films, crystal growth methods, growth of chromophore-containing inclusion compounds, and films prepared by vapor phase quasi-epitaxy. In the remainder of this review, we will focus our attention on electric field poling of chromophore-containing polymer (and dendrimer) materials. Of alternative methods to electric field poling, only the sequential synthesis methods of Professor Tobin Marks (Northwestern) show continued improvement and remain the subject of active development [17–24].

The reader is also referred to other literature for a more comprehensive introduction to various aspects of optical nonlinearity and for more in-depth reviews of certain topics than can be covered in this review [2, 3, 5, 24–87]. It is also suggested that the reader use these cited reviews to access literature not cited here. The literature on the topic of organic electro-optic materials now stands at many thousands of papers. It is unrealistic to attempt to even cite this entire body of literature here let alone discuss in depth all of the salient features of research into organic electro-optic materials. It is hoped that a readable overview of poled polymer electro-optic materials is provided in the following sections and that insight is provided both into the technological potential and problems associated with these materials.

2
Materials Requirements

For fabrication of electro-optic devices, one must consider three different types of materials that are critical to device function: (1) Electro-optic chromophores, (2) chromophore-containing polymeric active EO materials; and (3) polymer cladding materials. A cladding material is indispensable for preventing the optical wave propagating in the active EO polymer waveguide from seeing the optically lossy metal electrodes used to deliver the electrical driving field. The best electro-optic material is of little use without a compatible cladding material. The reader should keep in mind that an electro-optic device typically consists of a stacked architecture consisting of a substrate, lower electrode, lower cladding layer, active electro-optic layer, upper cladding layer, and upper electrode.

Let us first focus on the materials requirements that must be satisfied for a chromophore to be useful. We have already noted that devices characterized by drive voltages of less than 6 V are required to be compatible with semiconductor electronics. There are, of course, applications (such as cable TV, CATV and voltage sensing in the electric power industry [88]) that do not require small V_π voltages and hence do not require large material electro-optic coefficients; however, the majority of applications will require digital-level drive voltage capability. To achieve a drive voltage of 1 V, a material electro-optic coefficient on the order of 100 pm/V is required. Realization of digital level drive voltages and matching the drive voltage performance of lithium niobate requires electro-optic coefficients on the order of 30 pm/V. To achieve such electro-optic coefficient values, in turn, requires molecular $\mu\beta$ product values in the order of 2000×10^{-48} esu or greater. For comparison, aminonitrostilbene is characterized by a value of 482×10^{-48} esu.

For a chromophore to be useful for electro-optic applications, it must not exhibit significant optical absorption at anticipated operating wavelengths (e.g., telecommunication wavelengths of 1.3 or 1.55 microns). Normally, vibrational absorption from C-H overtones will dominate absorption at telecommunication wavelengths. Chromophore vibrational absorption will not pose a problem for device operation although it can contribute a fraction of a dB/cm optical loss at very high chromophore loading. Normally, the polymer host material will contain a higher density of protons than the chromophore and will occupy a greater weight fraction of the final material. Thus vibrational absorption from the polymer host will be a more significant problem. Optical loss from such absorption can typically be in the order of 1 dB/cm at 1.3 microns and even somewhat higher at 1.55 microns. The most serious problem with respect to optical absorption from chromophores is that associated with interband (HOMO-LUMO) electronic excitation. This typically is not a serious problem for operation at telecommunication wavelengths if the λ_{max} of the lowest energy chromophore electronic absorption lies below 700 nm. For the type of molecules considered here, the lowest energy electronic absorption will be the charge-transfer excitation that also dominates molecular hyperpolarizability. As might be expected, for a given class structural class of chromophores, β and λ_{max} will be related (see Appendix). However, it is important to keep in mind that the host polymer matrix can influence the HUMO-LUMO absorption spectrum of the chromophore both through dielectric constant (local field) effects and due to exciton effects if chromophore aggregation occurs due to chromophore/polymer incompatibility.

Chromophores must be thermally robust enough to withstand temperatures encountered in electric field poling and subsequent processing of chromophore/polymer materials. Chromophore decomposition temperatures can be assessed by techniques such as thermal gravimetric analysis (TGA) and differential scanning calorimetry (DSC). TGA and DSC measurements on neat chromophore samples in air will tend to yield decomposition temperatures lower than those for the same chromophores in hardened polymer lattices. Typically, to be useful for development of device quality materials, a chromophore must exhibit thermal stability of 250 °C or higher (with thermal stability defined as the temperature at which decomposition is first observed). Chromophores must also exhibit chemical, electrochemical, and photochemical stability [2, 3, 5, 50, 63, 64, 87, 89–92]. The exact definition of the required stability depends on the application for which the material is applied and is a subject that we will defer to later in this review.

Chromophores and host polymer must exhibit appropriate and comparable solubility in spin-casting solvents. The preparation of optical quality films frequently requires a delicate balance of solution viscosity and solvent volatility. Finally, the chromophore must have a shape and segmental flexibility appropriate for efficient poling. As we shall shortly discuss at some length, the choice of a shape that minimizes unwanted intermolecular electrostatic interactions is extremely important for maximizing macroscopic electro-optic activity. The chromophore must also be compatible with the polymer host in which it is to exist. Phase separation quickly leads to unacceptably high optical loss due to scattering from aggregates or microdomains.

The host polymer must exhibit good thermal stability, low optical absorption, and good solubility in spin-casting solvents. The glass transition temperature of the host polymer should be sufficiently high to impede chromophore relaxation after the poling field is turned off and the material is returned to ambient (or device-operating) temperatures [2, 3, 5, 50, 63, 64, 93–131]. The presence of a chromophore will produce some plastization of the host polymer, particularly if the chromophore is not covalently coupled to the polymer main chain but is rather simply physically incorporated into the polymer to form a composite material. Thus, one typically looks for polymers with glass transition temperatures in the range of 150–250 °C. Frequently, optimum poling temperatures lie in the range 90–140 °C. For higher temperatures, sublimation of chromophores can be a problem for chromophore/polymer composite materials. To avoid electrical conductivity that will attenuate the poling voltage felt by chromophores, both chromophores and polymers must be free of ionic impurities [2, 3, 5, 63, 64]. An example of the diffusion of ionic impurities reducing effective poling voltage and second-order optical nonlinearity is shown in Fig. 2.

If the host polymer does not have a high glass transition temperature then the structures of polymer and chromophore must be such that post-poling lattice hardening can be carried out to assure thermally stable electro-optic activity [2, 5, 50, 63, 64, 110, 132–150].

Ideally, the cladding material has a larger electrical conductivity than that of the active polymer layer [151–153]. Such conductivity permits the electric field to be dropped across the cladding layer yielding a greater field felt by the chromophores in the active layer. This can be useful for realizing improved electric field poling efficiency. If a cladding material with high ac conductivity could be realized, it is conceivable that reduced drive voltage (improved modulation efficiency) could also be achieved. Unfortunately, microwave conductivity is unknown at this time for optically transparent polymer materials. The cladding material should be characterized by low optical loss and its index of refraction should be lower than that of the active polymer layer to support mode confine-

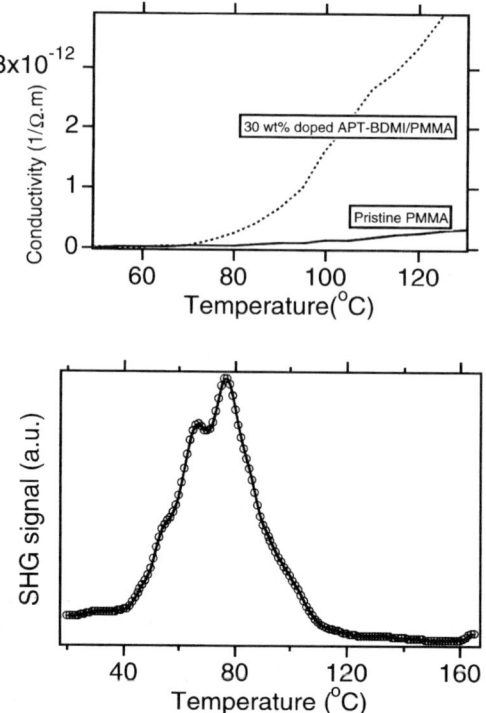

Fig. 2. Correlation of ionic conductivity with poling efficiency (second-order nonlinear optical activity). The upper graph shows conductivity measured as a function of temperature while the lower graph shows second-order nonlinear optical activity (measured by second harmonic generation, SHG) as a function of temperature. Note that second-order NLO activity starts to decrease with the onset of conductivity. Conductivity in this case was shown to arise from ionic impurities

ment in the active layer. The cladding material must be compatible with and exhibit good adhesion to the active polymer material. Spin-casting solvents used in the deposition of the cladding layer cannot be permitted to dissolve the active layer. Such solvent damage invariably leads to unacceptably high optical loss. The glass transition temperatures of the active and cladding polymers must be compatible to avoid problems during electric field poling. If the cladding polymer is a UV-curable epoxy, the active layer material must be capable of withstanding UV radiation. The cladding material must be thermally and photochemically stable. Photochemical degradation of the cladding material can result in mode expansion and the evolution of the waveguide from single mode to multi-mode. Surprisingly, we have traced the photo-degradation in the performance of some devices to photochemically induced changes in the cladding material and not the active material.

It should be clear to the reader from the preceding sections that a large number of often-conflicting materials requirements must be simultaneously satisfied. For a device to work well, all of its component materials must function well. Clearly, the topic of organic electro-optic materials is much more complex than envisioned in the early research and development efforts of the 1980s. The fact that organic electro-optic materials are essentially modular in nature is both an advantage and a curse. Probably all problems are ultimately solvable by systematic variation of component materials and processing protocols; however, such variation can be a very time-consuming process. The initial failure of quick commercialization of organic electro-optic materials in the late 1980s and early 1990s was not due to lack of effective research efforts but rather simply reflected the enormous number of issues that had to be addressed and the number of property/structure relationships that had to be defined. Any future commercialization will owe a great debt to the research efforts of the late 1980s and early 1990s. The reader checking the literature will note a decline in the number of papers published in this field since 1996. This, of course, reflects the departure of major industrial research groups (at IBM, Hoechst-Celanese, 3M, Eastman Kodak, etc.). With recent advances in materials and device performance, this trend appears to be reversing.

3
Materials Characterization Methods

In addition to characterization of the structure of chromophores and chromophore-containing polymer materials by standard techniques [such as FT-NMR, FT-IR, mass spectrometry, UV-Visible spectrometry, elemental analysis, thermal methods (including TGA, DSC, and TMA), and electrochemical (e.g., cyclic voltammetry) methods], characterization of relevant optical and nonlinear optical properties must be carried out for both chromophores and chromophore-containing polymer materials.

Molecular hyperpolarizability, β, must be defined and this is typically accomplished employing either electric field induced second harmonic generation (EFISH) [154–158] or hyper-Rayleigh scattering (HRS) [159–168]. EFISH measures the product of dipole moment (μ) and β while HRS measures β directly; moreover, HRS can be applied to octupolar and ionic molecules as well as to dipolar molecules while EFISH measurements are restricted to dipolar molecules. The applied electric field of EFISH breaks the macroscopic centrosymmetric symmetry of the solution permitting second harmonic generation. Extraction of β from EFISH measurements requires knowledge of both μ and γ. Both EFISH and HRS techniques are scientifically challenging and several factors can contribute to erroneous results. Although hyper-Rayleigh scattering has the advantage of permitting measurement of β for both dipolar and nondipolar molecules, the second harmonic generated intensity is weak since it arises only from fluctuations of density and orientation of the dissolved chromophores. Thus, hyper-Rayleigh scattering is sensitive to disruptive processes such as multiphoton-induced fluorescence [162, 163]. An elegant solution to this problem is the use of time-resolved HRS [164]; unlike fluorescence, SHG in HRS is an instantaneous process, enabling discrimination between hyper-Rayleigh scattered light and fluorescence emission. Based on the lessons learned from time-resolved experiments, an improved technique was developed recently for suppressing multiphoton-induced fluorescence by high-frequency modulation of the fundamental laser beam [165, 166]. The modulation method circumvents the disadvantage of time-resolved experiments of requiring costly instrumentation. Another route to discriminating between HRS and multiphoton fluorescence involves tuning the fundamental wavelength such that the fluorescence does not overlap the generated second harmonic signal [162, 167]. Yet another approach involves addition of a suitable fluorescence quencher [168].

Since β will vary with wavelength, measurement over a range of wavelengths is desirable to ascertain the relevant β value for a particular device operating wavelength. Unfortunately, reporting the full spectral dependence of β is not very practical. For the sake of comparing hyperpolarizabilities among chromophores, values of $\mu\beta$ are frequently reported at the "off-resonance" wavelength of 1.9 microns (e.g., see Table 3).

Frequently, values of β for wavelengths where experimental data do not exist are estimated by extrapolation using a two-level model description of the resonance enhancement of β (see Appendix). Levine and co-workers [170] have also shown how to estimate the wavelength (frequency) dispersion of two-photon contributions to β. Because of the potential of significant errors associated with each measurement method, it is important to compare results from different measurement techniques. Perhaps the ultimate test of the characterization of the product of $\mu\beta$ is the slope of electro-optic coefficient versus chromophore number density at low chromophore loading. It is, after all, optimization of the electro-optic coefficient of the macroscopic material that is our ultimate objective.

A number of different techniques have been employed to measure electro-optic coefficients; these include ellipsometry [171–174], attenuated total reflection (ATR) [175–177], and two-slit interference modulation [178, 179]. Second harmonic generation is also used, particularly for in situ monitoring of the induction of second-order optical nonlinearity by electric field poling [121]. Care must be exercised in relating second harmonic generation and electro-optic data as the linear electronic interband absorption (resonance enhancement) influences each of these measurements to different degrees. Of course, the V_π values from different devices can be related to material electro-optic coefficients [2, 35, 63, 64, 180]. A particularly attractive measurement configuration for accessing macroscopic electro-optic coefficients independent of chromophore rotational relaxation has recently been described [176, 181–183] and is shown in Fig. 3.

As with the measurement of β, it is important to compare electro-optic coefficients determined by different methods. It is particularly important to establish consistency between electro-optic coefficients obtained from initially poled thin films and those existing in functional prototype devices. Such consistency not only adds credibility to reported material electro-optic activity, but also as-

Table 3. Measured µβ values for representative second-order NLO chromophores

Structure		µβ (10^{-48} esu)
	DR	580
	APII	926
	ISX	2000
	FCN	3300
	APTEI	4000
	TCI	6100
		9800
		13,000
	SDS	15,000
	FTC	18000

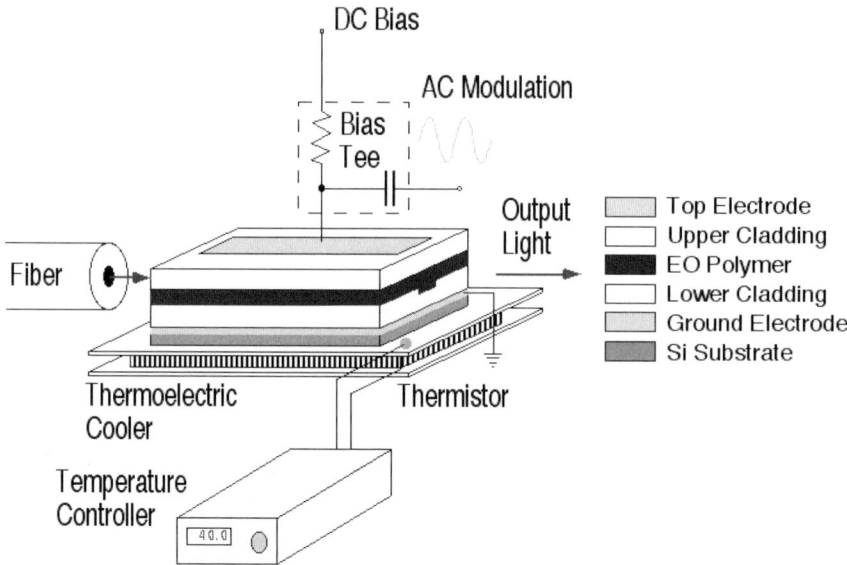

Fig. 3. Device for measuring electro-optic activity of chromophore/polymer materials near the glass transition temperature of the material matrix. This configuration avoids rotational relaxation of chromophores in the absence of an applied poling field

sures that device fabrication has not altered the initially obtained electro-optic activity. In this review, we will focus only on materials that are suitable for prototype device fabrication and for which characterization of electro-optic properties has been effected by several techniques. It should be noted that electro-optic coefficient values, like molecular hyperpolarizability values, vary with measurement wavelength. Values reported in this review typically correspond to the telecommunications wavelength of 1.3 microns. Values at other wavelengths can be estimated from experimentally defined values at a particular wavelength by extrapolation using the two-level model for estimating resonance enhancement (see Appendix).

In addition to characterization of molecular and macroscopic electro-optic activity, it is important to define optical loss. Optical loss can be influenced both by absorption and by scattering effects. In order to minimize overall loss, it is important to understand the independent contributions made by scattering and absorption. To separate these effects, we need to determine the contributions made by both chromophore and polymer host to the optical absorption at device operating wavelengths. Chromophore interband electronic absorption can be measured on resonance by traditional UV-Visible spectrometry; however, we will typically be concerned with optical absorption at telecommunication wavelengths of 1.3 and 1.55 microns where such techniques do not provide accurate information. Total optical absorption at 1.3 microns is occasionally determined by both the interband electronic absorption of the chromophore and by C-H vi-

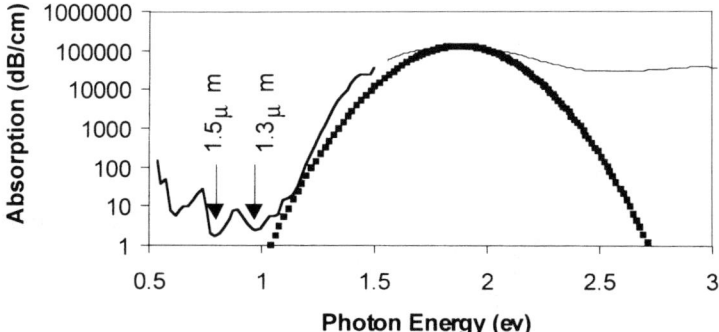

Fig. 4. Photothermal deflection spectroscopy (PDS) is used to measure absorption from 0.5 to 3 eV for a sample of CLD-1 in PMMA. Also shown in the figure (*squares*) is the fitting of the HOMO-LUMO charge-transfer absorption by a simple two-state model. This figure shows that interband electronic transitions do not contribute significantly to optical loss at 1.3 and 1.55 microns. At these wavelengths loss is clearly dominated by C-H vibrational overtone absorption. Note that PDS is not good for accurately measuring optical loss but is excellent for assessing contributions from various absorption mechanisms at particular wavelengths

brational absorption from the chromophore or from the polymer host. These contributions are best distinguished by performing photothermal deflection spectroscopy (PDS) measurements over a range of wavelengths. While PDS is not very reliable for absolute absorption measurements, it is very useful in defining the relative contributions from various sources, as is evident from the example shown in Fig. 4.

For the chromophore loading levels (number densities or weight/weight fractions) that are typically utilized, interband electronic absorption will not contribute significantly to total absorption at 1.3 microns if the λ_{max} of the chromophore lowest energy interband electronic absorption lies below 700 nm. Note that the charge-transfer absorptions of most electro-optic chromophores will lie in the range 400–700 nm. Absorption at 1.3 microns will typically be dominated by C-H vibrational overtone absorption coming largely from the host polymer. For operation at 1.55 microns, it can be safely assumed that absorption loss will be dominated by C-H vibrations of the polymer host. This is easily understood in terms of the host polymer occupying a larger wt/wt fraction and in terms of the host polymer having a larger proton density. Absorption from C-H vibrations can be reduced by substitution of halogens and deuterium atoms for protons. However, replacement of H by D does not always result in a reduction of optical loss for the two telecommunication wavelengths; the exact absorption will depend on precisely how the operating wavelength relates to the λ_{max} values for the various vibrational overtones. Moreover, replacement of H by halogens such as fluorine can change the solubility of chromophores and polymers in traditional spin-casting solvents requiring the development of new spin-casting protocols to obtain optical quality films.

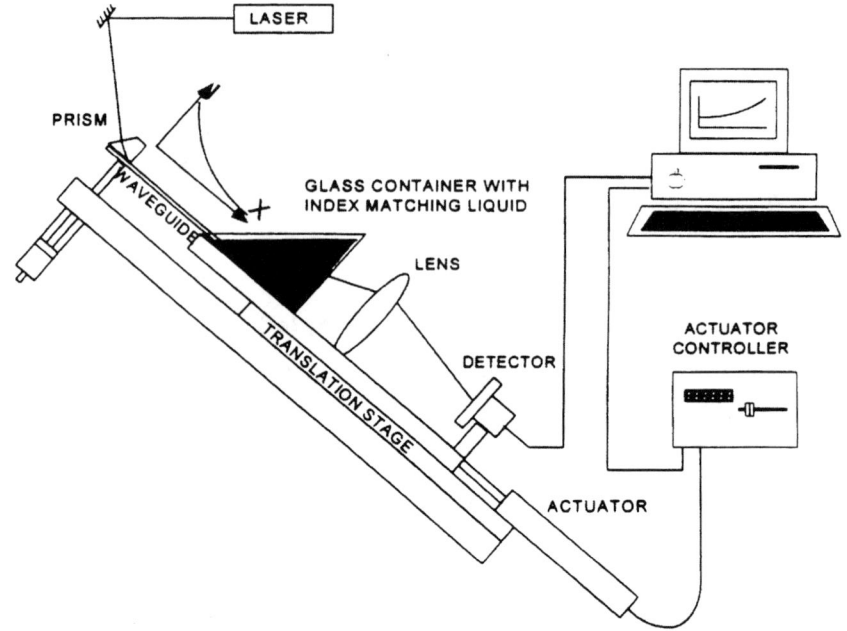

Fig. 5. Apparatus used to measure optical loss by the nondestructive immersion method of Teng [184]

Optical loss can arise from scattering introduced by dust particles or from microdomains introduced during processing (spin casting, poling, lattice hardening). It can also arise from poling-induced surface damage and from pitting during reactive ion etching. Optical loss from scattering typically is detected in measurements of light propagation through slab or channel waveguides. The two most common methods are the cut-off method and the immersion method of Teng [184]. The cut-off method is a destructive method that simply involves measuring optical loss as sections of the waveguide are systematically cut off. A schematic of the immersion method is given in Fig. 5.

Electron microscopy and atomic imaging methods are frequently used to assess surface roughness after processing steps such as reactive ion etching [5]. The insight provided by these methods can be useful in understanding observed optical loss.

Assessing thermal and photochemical stability is important. Thermal stability can be readily measured by measuring properties such as second harmonic generation as a function of heating at a constant rate (e.g., 4–10 °C/min) [121]. The temperature at which second-order optical nonlinearity is first observed to decrease is taken as defining the thermal stability of the material [2, 3, 5, 63, 63]. It is important to understand that the loss of second-order nonlinear optical activity measured in such experiments is not due to chemical decomposition of the electro-optic material but rather is due to relaxation of poling-induced acentric

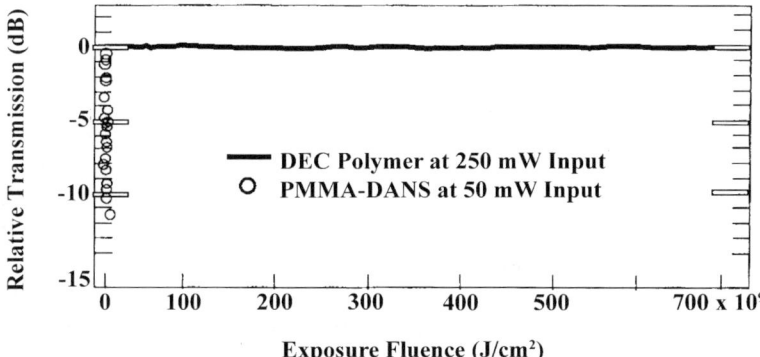

Fig. 6. Measurement of photostability of two polymeric electro-optic materials as carried out by researchers at IPITEK (TACAN) Corporation. The data represented by the *solid line* correspond to the LRD-3 DEC material of Dalton and co-workers [138] while the data represented by *open circles* correspond to a diaminonitrostilbene chromophore/poly(methyl methacrylate) guest/host material produced by IBM Almaden Laboratories. The dramatic improvement observed for the DEC material can be associated with increased lattice hardness from the chromophore coupling to adjacent polymer chains

chromophore order by rotational diffusion of the chromophores within the polymer matrix.

Photochemical stability can be evaluated by examining electro-optic activity or second harmonic generation as a function of increased optical power. Another commonly employed means of assessing photochemical stability is to measure changes in linear absorption of the interband electronic transition as a function of increasing optical power at the operating wavelength. Care must be exercised in such measurements to keep the power of the observing light sufficiently low that it does not induce photochemical reactions; this is not so easily accomplished, as measurements are frequently made over long periods of time. Unfortunately, the erroneous assumption that the decomposition of chromophores is independent of polymer lattice has frequently been made. In point of fact, the photochemical stability of a chromophore will depend strongly on the hardness of the polymer lattice to which it is attached. This point is illustrated by data obtained by the TACAN Corporation, which is shown in Fig. 6.

For operation at telecommunication wavelengths, one must be concerned with two-photon excitation processes leading to photochemical decomposition mechanisms. Femtosecond pulse techniques, such as those described by Strohkendl et al. [185], prove useful in evaluating two-photon excitation cross sections and excited state energy transfer processes relevant to photochemical decay pathways.

Dipole moment, dielectric permittivity, optical absorption, and index of refraction are determined for chromophores by standard methods.

4
Chromophores

4.1
Optimization of Molecular Hyperpolarizability

Quantum mechanical calculations have provided the necessary guidance for designing chromophores with ever improving molecular hyperpolarizability [51, 186–208]. Even simple particle-in-box calculations advance the qualitatively correct idea that molecular hyperpolarizability will increase with increasing length of the π-electron system. Unfortunately, chemical reactivity and optical absorption in the infrared can also increase so that optimization of molecular optical nonlinearity by simply increasing the length of chromophores is not a practical option. However, the chromophores that we are considering here are

Table 4. Theoretically calculated dipole moments for representative chromophores

	Chromophore (R=H)	Dipole moment (μ)
ISX		8.6021
JH		10.2698
FTC		12.1898
CLD		13.4668
GLD		13.8811

charge-transfer molecules and a two-level quantum mechanical model [see Eq. (5)] provides considerable insight in designing chromophores. From a consideration of Eq. (5), it is clear the strength of the donor and acceptor moieties and bond alternation in the π-electron bridge region will directly affect hyperpolarizability. The HOMO-LUMO gap, ΔE_{ge}, of Eq. (5) will go to zero as bond length alternation decreases and the sign of β will be defined by the sense of bond alternation. The strength of donor and acceptor moieties will influence the ground, μ_{gg}, and excited, μ_{ee}, state dipole moments and the transition moment μ_{ge}. The overall effect is to lead to a sinusoidal variation of β with bond length alternation or E_{ge}. Variation of donor and acceptor strengths can be viewed as changing the mixing of the neutral and charge-separated forms of the charge-transfer chromophore molecule. If the neutral form dominates the ground state of the chromophore, then the sign of β will be positive with the converse true if the charge-separated form dominates the ground state. One can systematically move along the theoretical curve by changing the strength of the donor and acceptor moieties and by changing the nature of the π-electron bridge (e.g., by varying polyene, thiophene, phenyl, etc. segments). Other correlations (based on two-level theory) of the structural features of organic chromophores with molecular first hyperpolarizability have been discussed by Harper [209].

Quantum mechanical calculations are also useful in predicting minimum energy conformations of chromophores and for predicting properties such as dipole moments. Calculated dipole moments are shown for several chromophores in Table 4. Such calculations are particularly useful in understanding how chromophore molecular conformation changes with addition of substituents.

4.2
Auxiliary Properties of Chromophores

Although quantum chemical calculations are a great aid in optimizing molecular hyperpolarizability, many other factors must be taken into account in the design of chromophores. Unless the chromophores have the requisite chemical stability to withstand materials processing and device operating conditions, they are of no value. Moreover, chromophores must possess adequate solubility in spin-casting solvents such that optical quality films can be prepared. An electro-optic chromophore must have a structure that can reorient under the influence of the electric poling field. Thus, although record molecular hyperpolarizabilities have been obtained with chromophores containing long and unprotected polyene bridges, such materials have no potential for practical application because of their chemical instability. Extremely stiff rigid-rod chromophores have likewise been found to be unusable for device fabrication because they cannot be effectively poled. To the best of my knowledge this has also been the case with phthalocyanines and related macrocyclic materials [86]. This does not mean these materials cannot be used to generate electro-optic materials, it simply means that they are not suitable for generation of electro-optic activity by electric field poling of chromophores in condensed phase media. The exceptional chemical and thermal stability of such chromophores continues to make them attractive candidates for continued exploration.

As might be expected from simple kinetic considerations, chromophore decomposition typically involves bimolecular processes and the rate of decomposition usually depends on the ability of species to diffuse in condensed media. The denser the polymer medium (e.g., the more heavily crosslinked), the slower the rate diffusion and the slower the rate of chromophore decomposition. Decomposition also often depends on attack at certain reactive positions of the chromophore consistent with well-known principals of organic chemistry. Steric protection of these reactive sites can dramatically improve chemical stability including at elevated temperatures [183, 210–212].

As already noted, electro-optic chromophores are modular in nature consisting of electron-donating, electron-accepting, and π-electron connective segments. Most chromophore development to date has focused on exploring various types of acceptor moieties and various types of π-electron bridges. Essentially all chromophores used for the fabrication of prototype electro-optic devices have involved the use of amine donor groups. It has been noted by researchers at IBM [50] that arylamines lead to better thermal stability than alkylamines. Interestingly, for the materials studied by the IBM researchers, thermally induced decomposition was defined by the amine donor rather than by the π-electron bridge structure.

A large number of acceptor moieties, including nitro, isoxazolone, thiobarbituric acid, cyanovinyl, etc. groups, have been explored [2, 3, 5, 63, 64]. Tricyanovinyl groups lead to some of the largest observed electro-optic coefficients. Unfortunately, chromophores containing unprotected tricyanovinyl functionalities do not exhibit good chemical stability. Considerable research effort has been undertaken to attempt to achieve the optical nonlinearity of such groups while imparting improved chemical stability by protecting reactive sites. Among the most successful efforts to date is the use of 3-cyano-5,5-dibutyl-2-dicyanomethylene-4-methyl-2,5-dihydrofuran (cyanofuran, CF) moieties such as in the FTC, CLD, and GLD chromophores shown in Table 4. Such chromophores exhibit thermal stability of approximately 300 °C while also yielding $\mu\beta$ values in excess of $10,000 \times 10^{-48}$ esu. Values of electro-optic coefficients for these chromophores dissolved in poly(methyl methacrylate), PMMA, are shown in Fig. 7. Data for various forms of the FTC and CLD chromophores are shown in Figs. 8 and 9.

As is evident from a consideration of Figs. 7–9, each of these chromophores has exhibited electro-optic activity exceeding that of lithium niobate while at the same time exhibiting auxiliary properties of chemical stability (T_d >300 °C) and solubility that permits preparation of device quality materials [183, 210–212]. These materials also illustrate another major direction in the preparation of electro-optic materials; namely, the development of bridging segments that lead to improved chemical stability, improved solubility in spin-casting solvents, improved compatibility with polymer host materials, and which inhibit unwanted intermolecular electrostatic interactions (we shall discuss such interactions

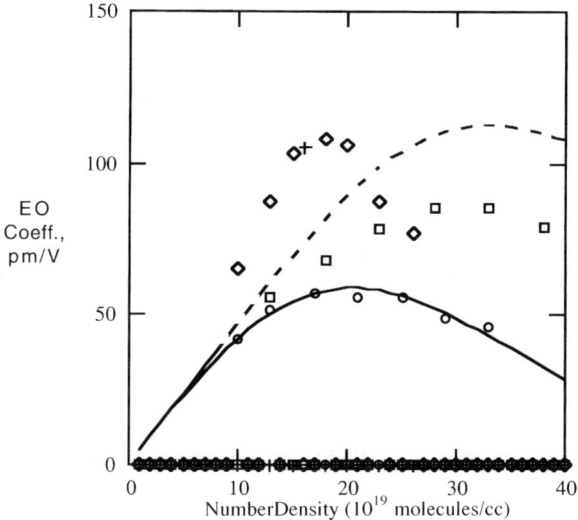

Fig. 7. EO coefficient data, as a function of chromophore number density, for FTC (*circles*), CLD (*squares*), GLD (*diamonds*), and CWC (*cross*) chromophores (see text and synthetic schemes) in PMMA. Also shown are theoretical curves computed treating FTC as an ellipse (*solid line*) and as a sphere (*dashed line*). The slope of the various graphs at low number density is determined by $\mu\beta F$. Consistent with EFISH and HRS measurements, GLD and CWC appear to have larger $\mu\beta$ values than FTC and CLD

shortly). It is clear from the preceding sections that a number of factors must be kept in mind when designing bridge segments; these include flexibility adequate for electric field poling. Quantum mechanical and molecular dynamical calculations are very useful in understanding both molecular hyperpolarizability and poling dynamics.

To be useful for commercial devices, chromophores must be synthesized by high-yield, cost-effective chemical pathways. Fortunately, the modular nature of chromophores permits a variety of structures to be explored and workhorse reaction schemes to be optimized. In Schemes 1–4 we illustrate the synthesis of the chromophores of Fig. 7.

Although the synthesis of each chromophore material is to some extent unique (and hence requires individual description which is beyond the scope of this review), most chromophores used for device prototyping involve amine donor moieties. Thus, once a large quantity of aldehyde-terminated amine donor/partial bridge material has been synthesized, it can be reacted via Knoevenagel condensation with a variety of acceptor groups to form different series of chromophores. The acceptor in Schemes 1–4 is 3-cyano-5,5-dibutyl-2-dicyanomethylene-4-methyl-2,5-dihydrofuran (which we refer to as the cyanofuran, CF, acceptor). This is synthesized by a literature method [213].

Note that chromophores are typically synthesized with reactive functionalities (e.g., hydroxyl groups) at one or both ends of the chromophores. Such func-

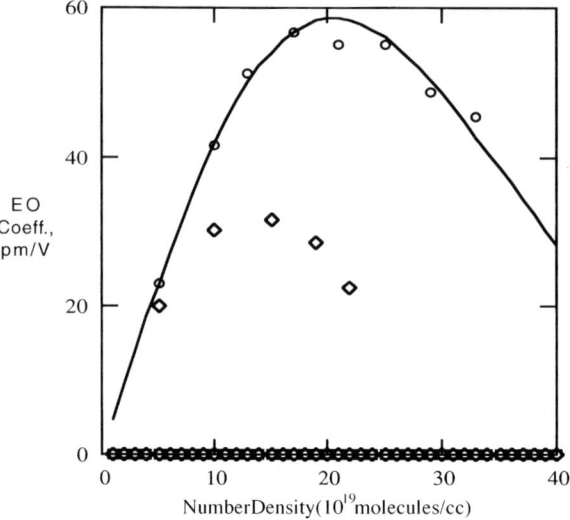

Fig. 8. EO coefficient data, as a function of chromophore number density, for FTC (*circles*) and FTC-2H (*diamonds*) chromophores in PMMA. Also shown is the theoretical curve computed for FTC. Note that for FTC-2H, the two butyl groups (attached to the thiophene ring) are replaced by protons. The more ellipsoidal FTC-2H exhibits a smaller maximum electro-optic activity and the position of the maximum is shifted to lower number density. Consistent with EFISH, HRS, and other measurements, the dipole moments and molecular first hyperpolarizabilities of these two chromophores are comparable (The values for FTC-2H may be slightly larger)

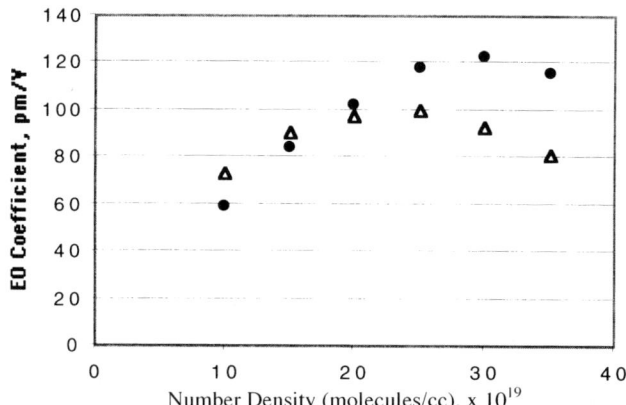

Fig. 9. EO coefficient data, as a function of chromophore number density, for CLD-type chromophores with (*solid circles*) and without (*open triangles*) isophorone protection of the polyene bridge. The maximum achievable electro-optic activity is smaller for the naked polyene bridge structure and the maximum of the curve is shifted to lower number density. The dipole moment and molecular first hyperpolarizability values are comparable (The unprotected polyene bridge variation may exhibit slightly higher values of $\mu\beta$)

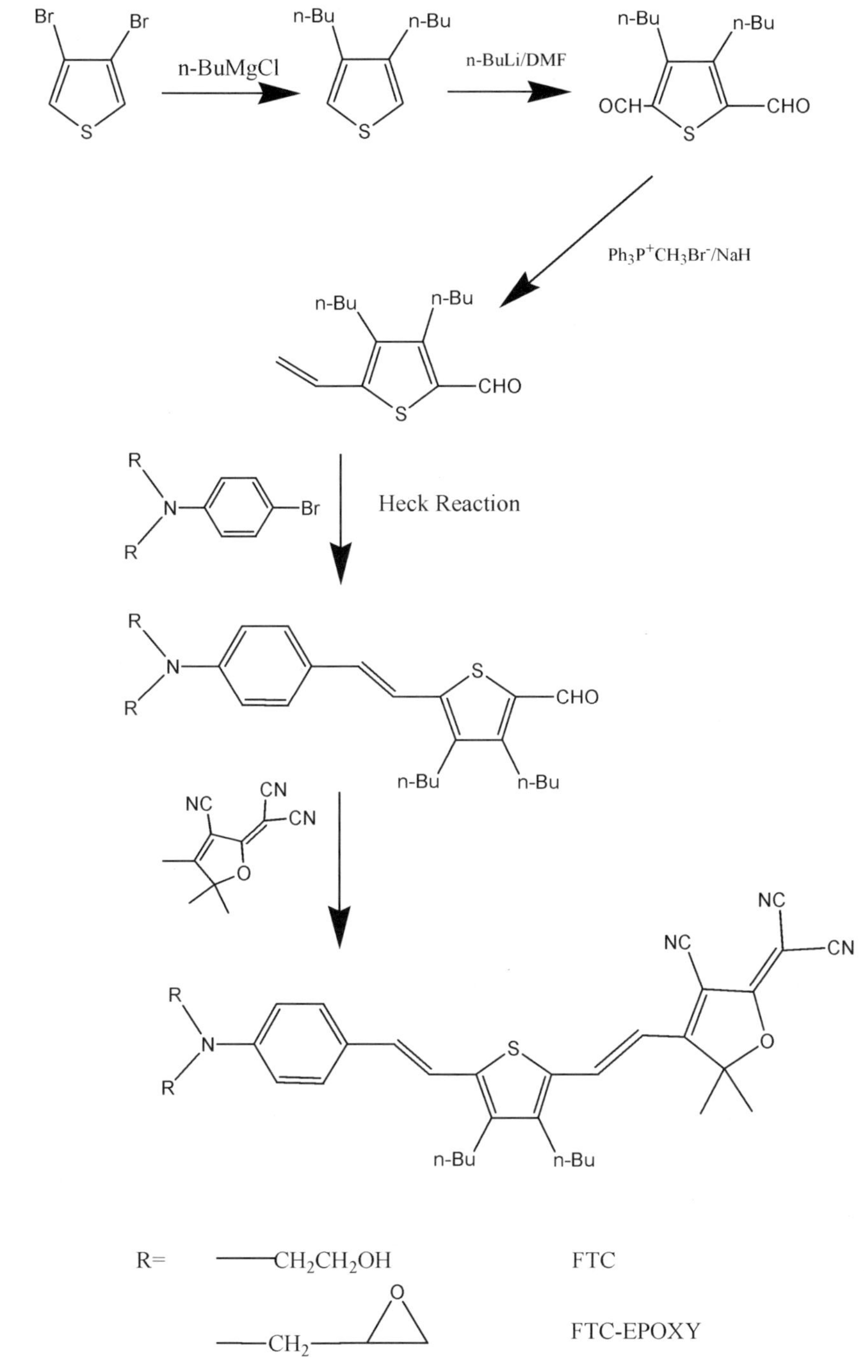

Scheme 1. Synthesis of the FTC chromophore

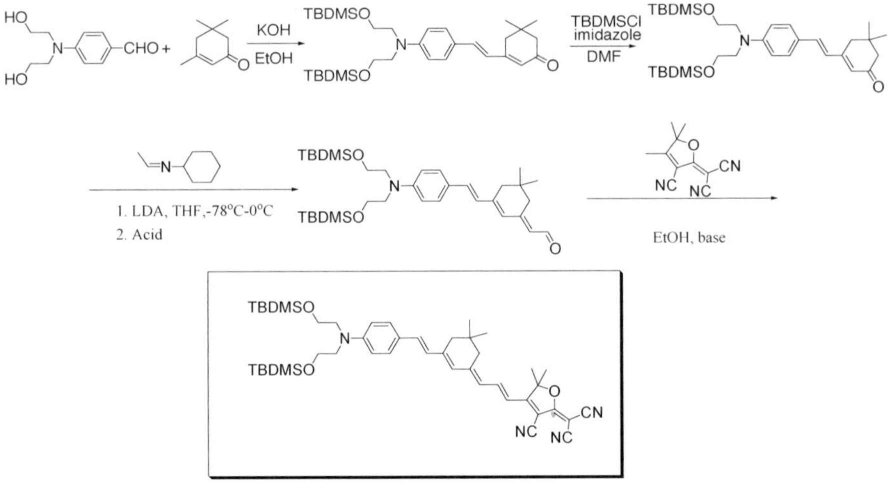

Scheme 2. Synthesis of the CLD-1 chromophore

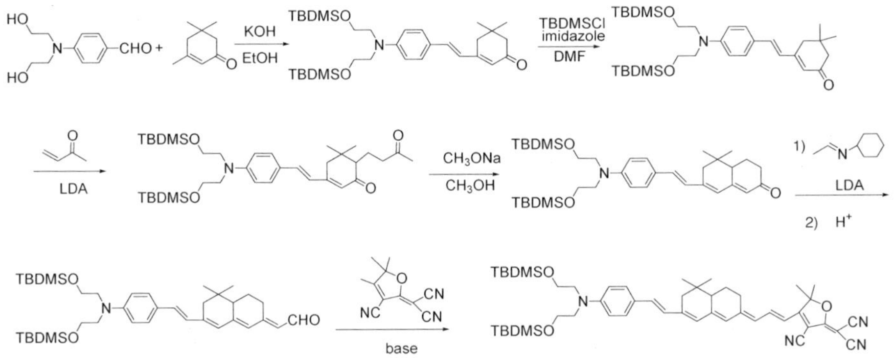

Scheme 3. Synthesis of the GLD chromophore

tionalities are used to covalently couple chromophores to a polymer lattice. When double-end crosslinkable (DEC) chromophores are used, the chromophore acts as a crosslinking agent to harden the polymer lattice [138]. Such reactive groups are protected (until needed) by protecting groups such as TBDMS (tert-butyldimethylsilyl). Hydrogen bonding involving amine donor groups can significantly influence chromophore hyperpolarizability, which also varies with chromophore molecular conformation. Thus, it is not safe to assume that the fundamental molecular optical nonlinearity is the same for a chromophore in its various forms. Hydroxyl groups can also chemically react with certain acceptors so care must be exercised with each chromophore synthesis.

With all of the above comments in mind, let us turn our attention to the most dramatic issue affecting material performance – the role of intermolecular elec-

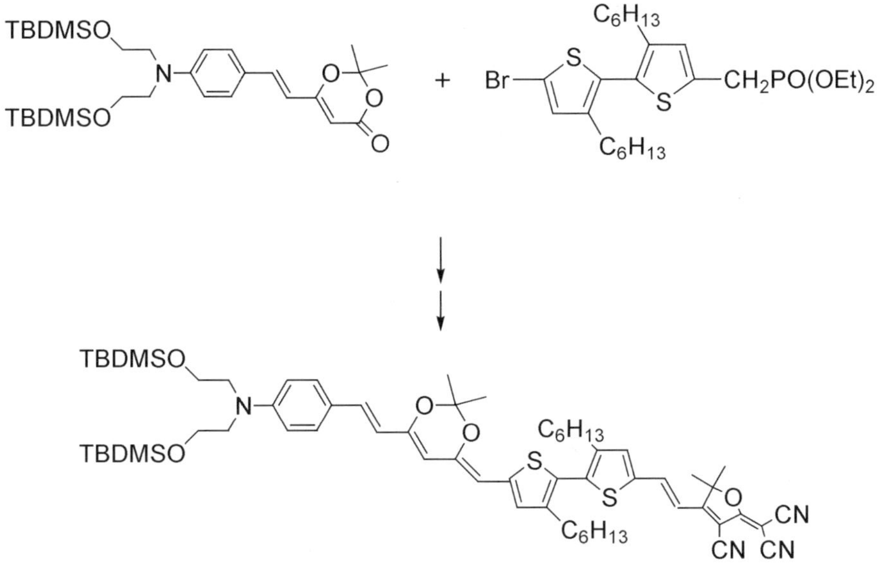

Scheme 4. Synthesis of the CWC chromophore

trostatic interactions in influencing chromophore organization into acentric (noncentrosymmetric) lattices with electric field poling.

5
The Problem of Translating Microscopic to Macroscopic Optical Nonlinearity

From a consideration of Eq. (4), it is clear that once we have optimized β, it remains for us to optimize $<\cos^3\theta>$, the acentric order parameter, and to achieve as high a chromophore number density, N, as possible. In the late 1980s, it was assumed that N and $<\cos^3\theta>$ were independent. Maximizing N was simply a matter of adding as much chromophore as could be tolerated by the host medium. As we shall shortly show by mathematical derivation, in the limit of non-interacting chromophores, $<\cos^3\theta>=L_3(\mu F/kT)=\mu F/5kT$. Thus, it was assumed that the order parameter (or poling efficiency) was limited only by electric field strength F (i.e., by material dielectric breakdown), dipole moment, and poling temperature. The larger the dipole moment, the better. Indeed, efforts were even made to synthesize head-to-tail coupled chromophores in hopes of exploiting the giant dipole moments associated with such macromolecular structures. These efforts did not yield the expected large electro-optic activity.

From 1990 to 1997, a number of chromophores with ever improving $\mu\beta$ values were synthesized by research groups led by Alex Jen, Seth Marder, Tobin Marks, Tony Garito and many others [2, 3, 5, 63]. Unfortunately, these chromophores did not translate to materials with ever-increasing electro-optic activity. Thus, as late as 1997, it was not clear that any organic material had surpassed the electro-optic activity of lithium niobate. Our experience was the same as that of other laboratories. To understand why our ever-improving chromophores were not yielding improved electro-optic materials, we launched a systematic investigation into the poling of chromophores characterized by high μβ values. Analogous to the data shown in Figs. 7–9, a maximum was always observed in plots of electro-optic activity versus chromophore number density. As has been discussed in the literature [2, 3, 5, 64–66, 76, 214–218], this is now known to be true for more than a dozen classes of chromophores and has been the case for every high μβ chromophore studied to the present time.

In looking for an explanation of the variation in electro-optic activity with chromophore loading, we attempted to systematically rule out possibilities. As was shown in Fig. 2, ionic impurities can lead to a concentration dependence of poling-efficiency and electro-optic activity. However, such effects are readily detected by a detailed investigation of the variation of electro-optic activity with both temperature and chromophore concentration. Moreover, purification processes, such as repeated recrystallization of chromophores, can eliminate such effects. It is also reasonable to assume that simple precipitation of chromophores (aggregation) and the concentraton-dependent "effective dielectric constant" of the chromophore/polymer material would cause a nonlinear dependence of electro-optic coefficient on chromophore number density. However, it can be shown by mathematical modeling that such processes cannot account for the behavior shown in Figs. 7–9 and observed for other chromophores. As we shall now demonstrate, the experimentally observed behavior can be quantitatively simulated by considering intermolecular electrostatic interactions. We shall first focus on intermolecular interactions of a magnitude that do not lead to phase separation. Then we shall consider the case where intermolecular interactions are so strong that they lead to phase separation and light scattering.

5.1
Theoretical Treatment of Large Intermolecular Electrostatic Interactions

5.1.1
The Single Phase Region

In the following sections, we shall demonstrate that the observed behavior of electro-optic activity with chromophore number density can be quantitatively explained in terms of intermolecular electrostatic interactions treated within a self-consistent framework. We shall consider such interactions at various levels to provide detailed insight into the role of both electronic and nuclear (molecular shape) interactions. Treatments at several levels of mathematical sophistication will be discussed and both analytical and numerical results will be presented. The theoretical approaches presented here also provide a bridge to the fast-developing area of ferro- and antiferroelectric liquid crystals [219–222]. Let us start with the simplest description of our system possible, namely, that of the Ising model [223, 224]. This model is a simple two-state representation of the total orientational distribution. Before we begin to consider an Ising model treatment, a few comments about the importance of self-consistent treatment [225–229] of electrostatic interactions are appropriate. The chromophore will itself influence the effective electric field that it senses from application of the applied poling field. The electric field at a given chromophore will also be influenced by other chromophores and the host polymer lattice. Thus, effective fields must be calculated in an iterative and self-consistent manner. This, of course, means that experimental variables such as index of refraction and dielectric constant of the overall material will depend on the chromophore number density, N.

The average $<(\mu_{0z})^3> = <\cos^3\theta>$, or order parameter, is calculated over the equilibrium density matrix, $\rho = e^{-H/T}/Tr[e^{-H/T}]$, where Tr denotes the trace (sum over diagonal elements). The Ising Hamiltonian can be expressed as:

$$H = H_d - (F/\varepsilon)\Sigma(\mu_{iz}) \quad (11)$$

where $\mu_{iz} = \pm 1$ (Ising variable).

$$H_d = (\mu^2/2\varepsilon)\Sigma\{1 - 3(r_{ijz})^2/(r_{ij})^2\}(\mu_{iz}\mu_{jz}/(r_{ij})^3 \quad (12)$$

where μ is the chromophore dipole moment and r is the inter-chromophore separation. The Hamiltonian in this simple example has the form $H = -(\mu_{0z}\mu/\varepsilon)F_{eff}$ where $F_{eff} = E + F_d + F_L + F_b$. E is the applied poling field, F_d is the self-consistent field experienced by a selected dipole due to all other dipoles located inside a sphere around that dipole, $F_d = \mu\sigma\Sigma[3(r_{iz})^2/(r_i)^2 - 1](1/(r_i)^3)$ where $\sigma = <\mu_{0z}>$ is statisticall mechanicall average with the Hamiltonian. F_L is the Lorentz field [230] given by $F_L = (4\pi/3)N\mu\sigma = (4\pi/3)N_0 x \mu\sigma$. F_b is the depolarization field that depends on the shape of the sample [230].

To simplify our calculations, let us define a reference lattice by assuming that the shape of a chromophore can be approximated as an ellipsoid (see Fig. 10). Taking the long axis of the ellipse as l_0, we can define a closely packed cubic lattice of spheres of diameter l_0. The number density associated with this lattice is $N_0 = (l_0)^{-3}$. For chromophore concentrations (loading in a host matrix) below this reference number density (i.e., $N = N_0 x < N_0$), the chromophores will on average be arranged in a cubic lattice with lattice constant, $a = b = (N_0)^{-1/3}(x)^{-1/3}$. For $N > N_0$, only the intermolecular distances in the plane perpendicular to the principle ellipsoid axis can decrease with increasing concentration, x. We have a simple tetrahedral lattice with lattice constants, $a = (N_0)^{-1/3}(x)^{-1/3}$; $b = (N_0)^{-1/3}$.

Simple models such as the Ising model have the advantage of permitting analytical results to be obtained for limiting cases. Let us consider these.

Case 1. $N < N_0$ and $\mu F_{eff} < \xi T$

$$r_{33}(\text{electro-optic}) = N_0(x/\varepsilon)[1 + (2\pi/3)\zeta x/\varepsilon]^{-1}\xi \quad (13)$$

where $\xi = \mu E/T$ and $\zeta = 4\mu^2/T(l_0)^3$

Case 2. $N > N_0$ and $\mu F_{eff} < \xi T$

$$r_{33} = N_0(x/\varepsilon)\xi[1 + (\zeta/\varepsilon)\{x^{3/2} - 1 + (2\pi/3)x\}]^{-1} \quad (14)$$

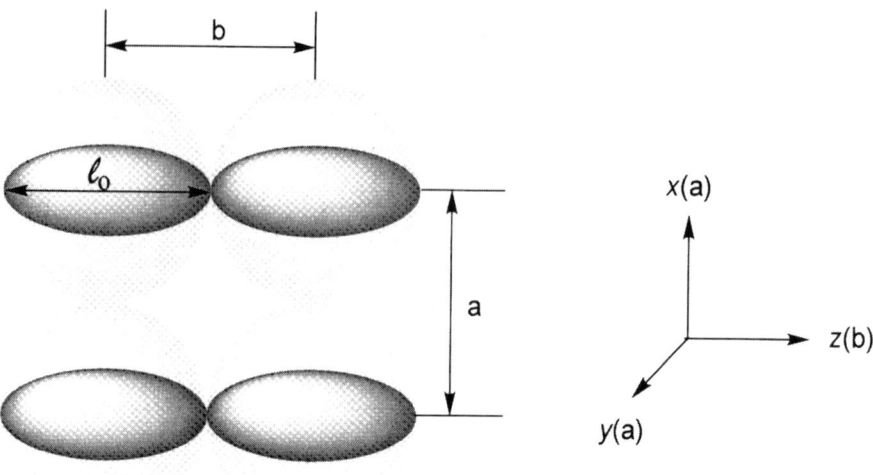

Fig. 10. Definition of unit cell structures for closely packed chromophores. This figure illustrates the problem of packing chromophores at high loading. Both shape (nuclear repulsive interactions) and electronic electrostatic interactions come into play and can be treated using the insights from this simple diagram

where for $x \gg 1$, r_{33} (EO coefficient) is proportional to $x^{-1/2}$ and $x_{max}=[(2\xi/\zeta)-2]^{2/3}$.

The position of the maximum in the plot of electro-optic activity versus chromophore number density is predicted by simple theory to be $N_{max}=N_0 x_{max}=N_0[(2\xi/\zeta)-2]^{2/3}$. Let us consider the case of the ISX chromophore where $l_0=2.3$ nm and N_0 is calculated by the above equation to be 8.2×10^{19} molecules/cm^3. For this chromophore, the dipole moment is 8 Debye and our simple theory yields $N_{max}=4.3\times10^{20}$ molecules/cm^3, which is in agreement with experiment. Thus, simple theory correctly predicts the optimum chromophore concentration from the knowledge of shape and electrostatic parameters. It is less good at predicting the detailed functional dependence of r_{33} upon N. In particular, the experimentally observed fall-off of electro-optic coefficient with number density at high number densities is sharper than that predicted by the Ising model. Although the Ising model is not very useful for quantitative analysis of data, it does illustrate important symmetry relationships valid for different concentration regions. In particular our earlier discussion indicates that there are two regions of consideration; namely, low and high concentration regions. At low concentrations, deviation of electro-optic activity from a linear dependence on chromophore number density arises entirely from electronic effects. The chromophores can be reasonably treated as spheres at the lowest concentrations. In the high concentration region, chromophores cannot access the full volume of the sample. Chromophore shape (nuclear repulsion effects) will become very important. The more spherical the chromophore, the better.

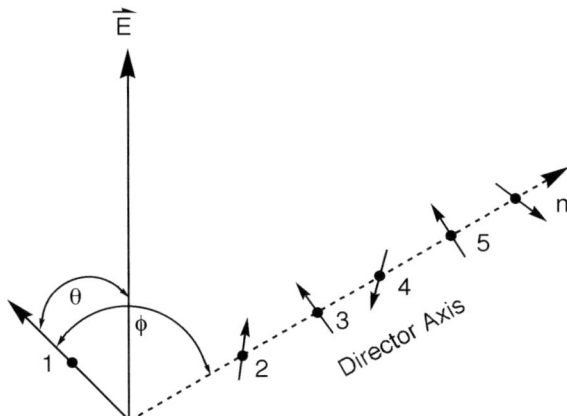

Fig. 11. Coordinate system appropriate for treating long-range, many-body chromophore intermolecular electrostatic interactions

Let us now turn our attention to more rigorous consideration of intermolecular electrostatic interactions [2, 3, 5, 64–66, 214–218, 231–239] permitting chromophores to access all orientations. Let us also focus on approximate methods of treating the many-body particle nature of intermolecular interactions. In particular, we must consider interactions beyond nearest neighbor interactions. The problem faced by treating long-range, many-body interactions can be realized by considering the coordinate system of Fig. 11 that defines the electrostatic interactions felt by a reference chromophore "1" in the presence of a poling field and intermolecular electrostatic interactions from other chromophores denoted "2" to "n" positioned along a "director axis". Self-consistent interactions between chromophore "1" and chromophores "2" through "n" must be computed for all orientations of the director axis. We can follow Piekara [235, 236] and approximate the potential energy of interaction for our reference chromophore as $U_{F+d} = -\mu F\cos\theta - W\cos\phi$. With this potential, the order parameter becomes [2, 3, 5, 64, 65]:

$$<\cos^3\theta> = L_3(\mu F/kT)[1 - L_1(W/kT)^2] \tag{15}$$

where L denotes the Langevin function and W is the intermolecular electrostatic (including chromophore dipole-dipole) interaction energy calculated according to the description of London [231, 232]. The functional dependence (recall that W is proportional to N^2) predicted by Eq. (15) is in much better agreement with experiment than the simple Ising model.

However, we must ask how reasonable are the approximations employed in its derivation. This can be answered by comparing the above results with those of Monte Carlo (molecular dynamics) methods [240]. With Monte Carlo calculations, no approximations are made other than that of restricting considerations to a finite system (e.g., 1000 chromophores in a polymer lattice of finite dielectric constant). The full Hamiltonian can be used and calculations can be carried be-

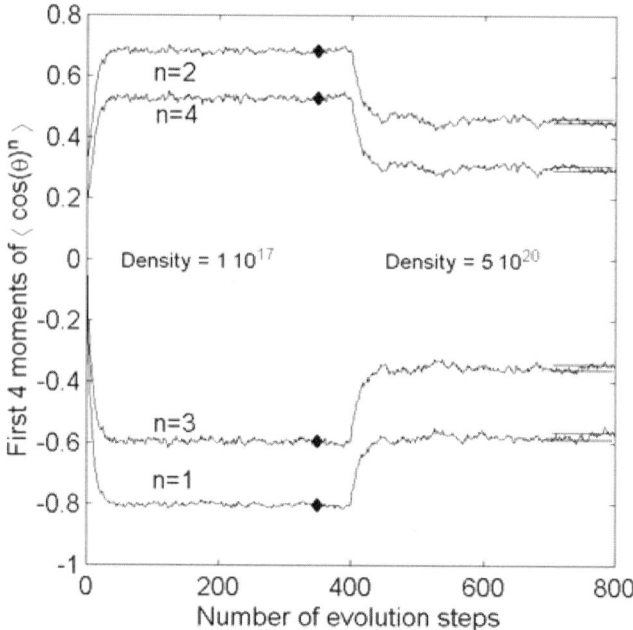

Fig. 12. Computation by Monte Carlo methods of the first four order parameters of an ensemble of 1000 chromophores (of dipole moment 13 Debye) existing in a medium of uniform dielectric constant. At the beginning of the calculation, the chromophores are randomly ordered; thus, $<\cos\theta>=<\cos^3\theta>=0$. During the first 400 Monte Carlo steps, an electric poling field (600 V/micron) is on but the chromophore number density ($=10^{17}$ molecules/cc) is so small that intermolecular electrostatic interactions are unimportant. The order parameters quickly evolve to well-known equilibrium values obtained analytically from statistical mechanics (*black dots* in figure; also see text). During steps 400–800 the chromophore number density is increased to 5×10^{20} and intermolecular electrostatic interactions act to decrease order parameters consistent with the results of equilibrium statistical mechanical calculations discussed in the text. Although Monte Carlo and equilibrium statistical mechanical approaches described in the text are based on different approximations and mathematical methods, they lead to the same result (i.e., are in quantitative agreement)

yond nearest neighbor interactions. Monte Carlo calculations have the added advantage of permitting examination of the full distribution of chromophores and not just a single order parameter. In the following, we start with a random distribution of chromophores (Fig. 12). Then, we turn on the electric poling field while keeping intermolecular electrostatic interactions small (by keeping chromophores concentrations small) and allow the system to evolve to equilibrium. After 200 steps, we turn on intermolecular electrostatic interactions (by increasing chromophore concentration) and allow the system to evolve to equilibrium. The equilibrium value of $<\cos^3\theta>$ is in good agreement with that obtained using the Piekara approximation [217, 218].

Let us consider the detailed chromophore distributions under various conditions (Fig. 13). In this figure, equilibrium chormophore distributions are shown

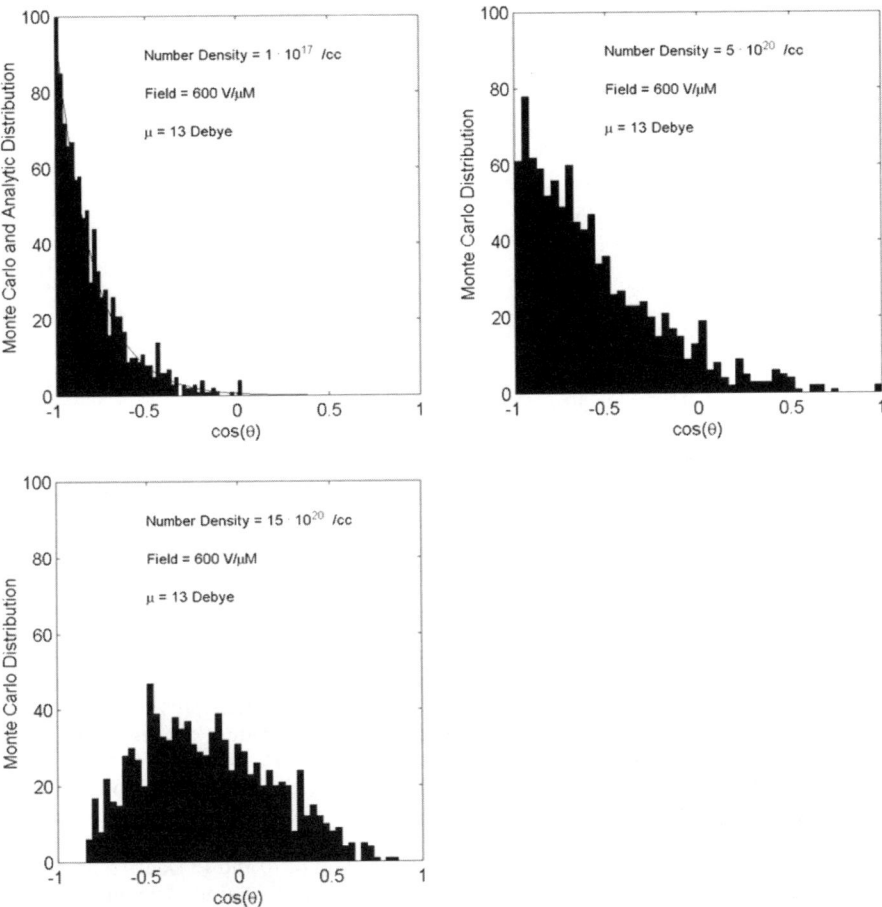

Fig. 13. Orientational distributions of chromophores for different magnitudes of intermolecular electrostatic interactions (achieved by varying chromophore number density). The distribution shown on the left corresponds to insignificant intermolecular electrostatic interactions (number density=10^{17} dipolar molecules per cubic centimeter) for chromophores interacting with a 600 V/μm poling field. As expected from analytical theory, the most probable chromophore orientation is along the applied field direction ($\theta=180°$ or opposing the applied electric field when signs of interactions are taken into account). When intermolecular electrostatic interactions are increased by increasing chromophore number density (*middle graph*, number density=5×10^{20} dipolar molecules per cubic centimeter), the most probable orientation shifts away from the applied field direction. The trend is shown to continue in the lower graph (number density=15×10^{20} dipolar molecules per cubic centimeter). The dipole moment for the interacting molecules in these calculations was taken as 13 Debye. For the conditions considered in these calculations an "effective field" model applies and aggregation effects are unimportant

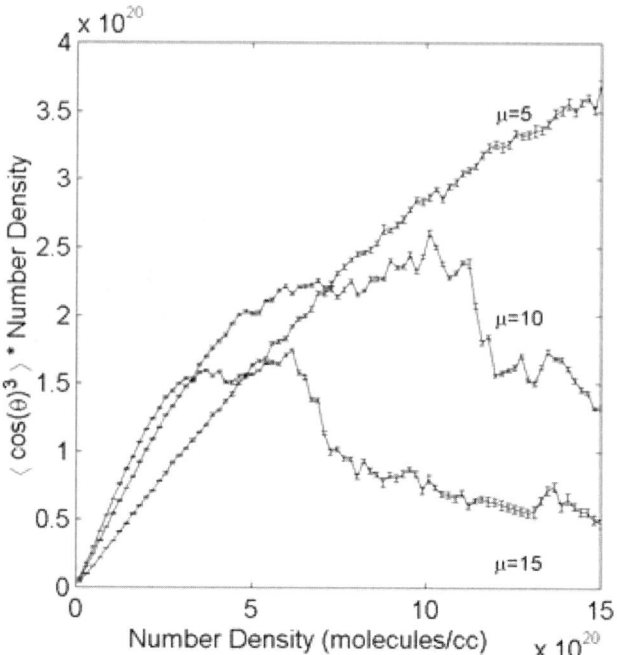

Fig. 14. Graphs of $N\langle\cos^3\theta\rangle$ vs. N obtained by Monte Carlo methods for various values of chromophore dipole moment. These graphs quantitatively reproduce the results obtained from equilibrium statistical mechanical treatments described in the text. Note that the position of the maximum shits to lower number density with increasing dipole moment (intermolecular electrostatic interactions). These graphs also illustrate a feature of Monte Carlo trajectories; namely, that they can reflect kinetic as well as equilibrium effects. The irregularities shown in the curves for dipole moment values of 10 and 15 Debye reflect transient false energy minima associated with chromophore aggregation. Averaging a large number of trajectories eliminates such effects. One can also let a calculation evolve from such a false minimum to the equilibrium value. However, such transient Monte Carlo effects are useful in alerting one to effects that actually happen in the sample. In many cases, Monte Carlo methods do an excellent job of identifying aggregation effects and the dependence of such effects on the various parameters of the experiment

for chromophore concentrations of 10^{17} molecules/cm^3 (Fig. 13, left), 5×10^{20} molecules/cm^3 (Fig. 13, middle), and 1.5×10^{21} molecules/cm^3 (Fig. 13, lower). With no intermolecular electrostatic interactions, the most probable chromophore orientation is predicted to be along the poling field direction. The effect of finite intermolecular electrostatic interactions is to tilt the most probable chromophore orientation away from the poling field direction; namely, the stronger the interactions, the greater that angle of tilt. The most probable chromophore orientation is defined in a vectorial manner by the competition of dipole-poling field and dipole-dipole interactions.

Note that no evidence of phase separation (second-order phase transition to a centric or antiferroelectric chromophore lattice) is found for conditions considered in these calculations. This is consistent with experimental results for comparable conditions. On the left in Fig. 14 we illustrate the variation of graphs of N<cos³θ> (proportional to electro-optic coefficient) versus N with chromophore dipole moment. In agreement with experiment, the position of the maximum shifts to lower chromophore loading with increasing dipole moment. The role of intermolecular electrostatic interactions in attenuating the linear dependence of electro-optic activity is clear from this figure. We have already alluded to the fact that good agreement is obtained between Monte Carlo results and equilibrium statistical mechanical calculations employing the Piekara approximation. This point is further illustrated by the comparison of results shown on the right in Fig. 14.

To this point in our discussion we have considered the full spatial anisotropy of intermolecular electrostatic interactions but have ignored chromophore shape. This corresponds to treating chromophores as spheres. Let us now turn our attention to consideration of the detailed shape of chromophores. This corresponds to taking into account nuclear repulsive forces (electrostatic interactions). In this treatment, two chromophores will not be permitted to occupy the same region of space at the same time. In our calculations, this means that the potential functions of a classical statistical mechanical treatment or Hamiltonian of a density matrix treatment will need to be modified to incorporate nuclear repulsive interactions. Moreover, the limits of integration for certain orientational integrals will now need to be changed to reflect the fact that chromophores cannot assess these angles for certain concentrations. Note that the limits of integration will now be concentration dependent with the lower limit defined as $\cos(\theta)=-c$ and the upper limit defined as $\cos(\theta)=+c$, where c is defined in the following manner. Consider a prolate ellipse with major and minor semi-axes of "a" and "b", respectively. When the inter-chromophore separation, r, is less than 2b, it is impossible to pack any more molecules into the available space. In this regime, there is a constraint on the angles that an ellipsoid can have before it collides with its neighbor (at the same orientation). The orientation of the molecule is found from the x and y distances from the center to the surface: $x^2+y^2=(r/2)$. The equation for the surface of an ellipse is $(x/b)^2+(y/a)^2=1$. From these two equations and two unknowns, the cut-off or minimum angle that can be obtained is $c=\cos(\theta_{min})=2y/r$. These two equations allow one to solve for c, which is a function of r (and thus N): $c=c(r)=\{1-(2b/r)^2/[1-(b/a)^2]\}$. To create a value for c that is a continuous function of r, take $c(r)=1$ for $r>2a$ and $c(r)=0$ for $2b>r$. When evaluating the average quantities, the integrals must be restricted so that $-c\leq\cos(\theta)\leq+c$. With these restrictions, our expression for order parameter becomes:

$$<\cos^3\theta>=c^3 L_3(c\mu F/kT)[1-L_1(W/kT)^2] \qquad (16)$$

In several of our publications, we have demonstrated quantitative simulation of experimental data for chromophores such as the FTC series [63–65]. In Fig. 15, we show the simulation of CLD-OMet data treating the chromophore as a sphere and treating the chromophore as a prolate ellipsoid. The prolate ellipsoidal shape used was obtained by quantum mechanical calculations.

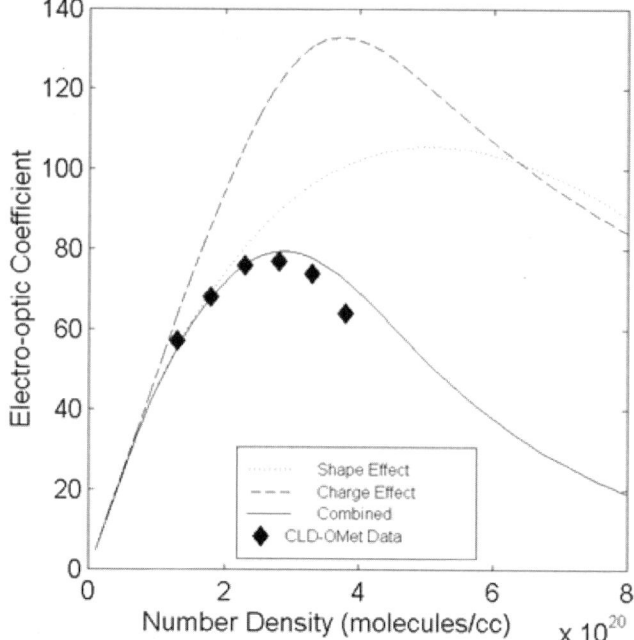

Fig. 15. Comparison of experimental (for the CLD-OMet chromophore in PMMA) and theoretical (equilibrium statistical mechanical calculations described in the text) data. Experimental data are denoted by *solid diamonds*. The *solid line* theoretical curve was computed without adjustable parameters. Quantitative agreement can be obtained by adjusting parameters (chromophore dipole moment, molecular polarizability, shape, and host dielectric constant) within reasonable limits. The theoretical curve can be broken down into two parts. The purely electronic part of the electrostatic interaction is shown by the *dashed line*. The steric effect of nuclear repulsive interactions is shown by the *dotted line*

The theoretical data was computed without adjustable parameters and provides a nearly quantitative simulation of the experimental data. By reasonable variation of electrostatic interaction potentials (within the limits of experimental error in determining electrostatic interaction and shape parameters), completely quantitative simulation can be obtained.

A more detailed and realistic treatment is to use the van der Waals surface for the chromophore of interest and use a $1/r^{12}$ repulsive potential analogous to Leonard-Jones treatment. The r^{12} is based on the point of closest approach and weighted by the surface-surface or volume overlap between chromophores. For the case at hand, such treatment gives results indistinguishable from those of the hard shell repulsion treatment discussed above.

Clearly, theory provides very good guidance for optimizing macroscopic electro-optic activity by control of chromophore shape and the magnitude of electrostatic interactions. One wants to keep chromophores as short and as spherical as possible. An extreme prolate ellipsoidal shape is the worst possible shape for optimizing electro-optic activity. Making the shape even longer by the addition of nonlinear optically inactive groups (such as protecting groups) is not a good idea. In contrast, adding bulky substituents to the middle (bridge region) of the chromophore will increase maximum achievable electro-optic activity.

In modifying chromophores by the addition of bulky substituents, it is important to keep in mind that bulky substituents can influence the conformation which the molecule adopts and can in turn influence π-orbital overlap and molecular hyperpolarizability and dipole moment. We have found this to be particularly true for chromophores utilizing a polyene bridge structure. Thus, it is dangerous to reason improvements simply from theoretical calculations that do not consider the possibility of a change in chromophore conformation with addition of substituents.

A detailed consideration of more advanced theoretical treatments clarifies the role played by the polymer dielectric constant. In the absence of intermolecular electrostatic interactions, one would desire the lowest possible dielectric constant, e.g., PMMA would be a better host matrix than polycarbonate. This is because the dielectric constant of the polymer host would act to attenuate the poling field felt by the chromophores. On the other hand, in the presence of intermolecular electrostatic interactions, optimum electro-optic activity will be achieved for polymer hosts of intermediate dielectric constant. The dielectric constant of the host acts not only to attenuate the externally applied poling field, but also fields associated with intermolecular electrostatic interactions.

Our detailed theoretical treatment tells us that the extremely optimistic early (pre-1985) predictions for electro-optic activity for polymeric materials will not be realized. Certainly, electro-optic activity will not increase in a linear manner with N and with μ. The quantity $\mu\beta$ divided chromophore molecular weight is not a good chromophore figure of merit as was assumed until recently. However, theoretical guidance provided by theories that explicitly take into account intermolecular interactions has permitted macroscopic electro-optic coefficients to be routinely achieved that significantly exceed those of lithium niobate.

Of course, for chromophores with dipole moments less than 6 Debye and corresponding small polarizability, chromophore-chromophore intermolecular electrostatic interactions are unimportant and independent particle analysis is completely appropriate. However, it is unlikely that such chromophores will ever have commercial relevance.

5.1.2
Phase Separation and Phase Diagrams

In the preceding section, we focused on conditions where only a single homogeneous phase exists. The reduction of electro-optic activity with increasing chromophore concentration can be associated with an effective field effect. The most probable orientation for a chromophore is tilted further from the poling field direction as chromophore intermolecular electrostatic interactions increase. For the application of the highest poling fields (>150 V/micron), for example as found with corona poling, this is the condition that applies. However,

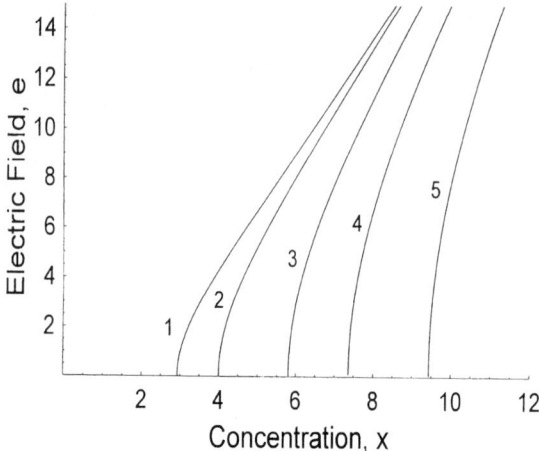

Fig. 16. Electric field/concentration phase diagrams separating the homogeneous (acentric or ferroelectric) phase region from phase-separated regions for various values of host dielectric constant (curves 1 through 5 correspond respectively to dielectric constant values of 2, 3, 5, 7, and 10). Field (e or ξ) and concentration (x) are in dimensionless units as described in the text

as the strength of the poling field is decreased and as intermolecular electrostatic interactions are increased, the regime of phase separation can be entered. Experimentally, this region results in time-dependent (non-equilibrium) variation of electro-optic coefficients and in the onset of light scattering from aggregates. Note that optical loss (in dB/cm) will be either independent or vary in a linear manner with chromophore number density throughout the homogeneous, single-phase region. A nearly exponential increase in optical loss (and loss values in excess of 2 dB/cm) with chromophore number density is evidence for chromophore aggregation (phase separation).

Detailed theoretical calculations establish that phase diagrams separating homogeneous from phase-separated regions will depend on intermolecular electrostatic interactions, the strength of the applied poling field, and on the dielectric constant of the host polymer. In Fig. 16, we show phase diagrams calculated as a function of N, ε, and E.

Again, we see that the lowest possible host polymer dielectric constant is not desirable. Figure 16 also illustrates the importance of keeping the potential of centrosymmetric ordering and aggregation in mind during processing. When spin casting, chromophores are dilute and thus randomly oriented in a homogenous phase. In the spin-casting process, thin films are produced in a rapid fashion trapping chromophores in a homogeneous state. We have verified by X-ray crystallography that if these same chromophores are permitted to slowly grow crystals from solution that centrosymmetric chromophore order is invariably observed in the crystal structures [183]. During poling, an electric field is applied and the thin film is heated permitting the chromophores to reorient under the influence of the applied field. Slow heating without application of a strong electric field can result in centrosymmetric aggregation. Indeed, we have observed that turning off the electric poling field when the sample is near the glass transition temperature of the chromophore-polymer matrix can result in centrosymmetric aggregation. In some cases, this is marked by the disappearance of electro-optic activity but retention of birefringence.

6
Processing Steps

6.1
Spin Casting, Electric Field Poling, and Lattice Hardening

The subjects of spin casting, electric field poling, lattice hardening, and relaxation of poling-induced order (electro-optic activity) were extensively investigated in the late 1980s through the mid-1990s. A number of reviews discuss extensively and in depth various aspects of these topics and these reviews cite the many hundreds of papers dealing with these aspects of the fabrication of electro-optic materials [2, 3, 5, 23, 34–44, 50, 63, 64]. Because such a treatment of these subjects already exists in the literature, we shall provide only an overview here with emphasis on lessons learned from past studies.

We have a number of material choices to make as we attempt to prepare device-quality electro-optic polymeric thin films. We can prepare composite films where the chromophore is simply physically incorporated into a polymer host such as PMMA, APC, PU, poly(amic acid), or polyimide. Such materials are referred to as guest/host materials. The polymer of such a composite material can be functionalized so that the polymer matrix can be crosslinked subsequent to electric field poling. We can covalently attach one end of the chromophore to the backbone of a polymer so that we are dealing with a homopolymer material. The chromophore can also be of a double-end-crosslinkable (DEC) type. With such a chromophore, a processable precursor polymer can be prepared and used with spin casting to prepare optical quality thin films. The other end of the chromophore can be attached to adjacent polymer chains subsequent to electric field poling (with the poling field still on) to achieve a three-dimensional, crosslinked lattice. Frequently, since the amine donor end has the functional form R_2N-, one is dealing with the situation where both R groups are functionalized with crosslinking reagents such as hydroxyl or isocyanate groups. Thus, two linking sites exist at the amine end of the chromophore. If a third crosslinking site exists at the acceptor end or on the chromophore bridge, the chromophore is referred to as a trilinking chromophore.

We can prepare functionalized chromophore and crosslinking reagents in a thermosetting composition; that is, the chromophores and crosslinking reagents can react to form a three-dimensional crosslinked polymer matrix. Thermosetting polyurethane, sol-gel, or epoxy chemistries are frequently used. Many variations on the above themes can be and have been executed; for example, chromophores can be prepared as components of interpenetrating polymer networks. The point of having so many material choices reflects attempts to find a compromise to the conflicting requirements of spin casting, electric field poling, and retention of poling-induced electro-optic activity.

With spin casting the following issues are of concern: (1) Volatility of the spin-casting solvent, (2) solubility of all components in the spin-casting solvent (a particular concern with guest/host systems where the solubilities of guest and host must be comparable), and (3) spinning speed. Rigid-rod, higher molecular weight polymers, and crosslinked polymers typically exhibit reduced solubility in the most common spin-casting solvents (dioxane, methylene bromide, 1,2-dichloroethane, 1,1,2-trichloroethane, 1,1,1-trichloroethane, cyclopentanone, etc.). Usually, spin-casting solvents have boiling temperatures around 100 °C. A problem with lower volatility solvents is that of trapping of solvent in the spin-cast films. Such solvent inclusion can lead to light scattering and unacceptably (optically) lossy films; moreover, solvent inclusion can be a source of thermal instability of materials during processing and device operation. On the other hand, the glass transition temperature of the final polymer matrix is important for the retention of poling-induced electro-optic activity after the poling field is removed. The conditions of electric field poling also influence the choice of materials and processing conditions. For example, to realize chromophore reorientation under the influence of an applied electric field, the chromophore-containing material must be poled near the glass transition temperature of the material. If the material is crosslinked, contains mostly rigid segments, or is of very high molecular weight, it may be impossible to achieve sufficiently high poling temperatures without decomposition of the chromophores or sublimation of the chromophores in the case of guest/host materials. On the other hand, too soft a material will result in dielectric breakdown under high, applied electric fields. Also, too soft a material will lead to orientational relaxation of poling-induced chromophore acentric order and will be more easily pitted by solvents used in the subsequent deposition of cladding layers. It is clear that we would prefer to deal with low molecular weight and flexible materials during the spin-casting and early poling stages, but would like to end up with a hardened polymer matrix on the completion of poling. The final polymeric material should have a glass transition temperature above 100 °C and ideally in the order of 150 °C. For applications with requirements for high thermal stability and survival in other harsh environments, a glass transition temperature on the range 150–250 °C is desirable. Higher thermal stability is typically achieved only with some sacrifice in electro-optic activity, due to the competing influences of lattice hardening and poling field interactions on chromophore reorientation.

Electric field poling is a complex subject and has been the subject of numerous articles [2, 3, 5, 63, 64, 69, 241–249]. Poling-related topics include methods of electric field poling, uniformity of fields across samples, charge dissipation, dielectric breakdown and poling-induced damage, and the correlation of poling and lattice hardening processes. Electric field poling can be accomplished employing a number of different poling configurations [2, 3, 5, 63, 64, 69, 241–250].

One of the most commonly employed configurations involves corona poling. Corona poling is attractive in that very high poling fields can be realized and large areas can be covered. For example, uniform chromophore orientation (electro-optic activity) can be effected over a 6-in. wafer; this can be an important consideration for integrating polymeric electro-optic circuitry with semiconductor VLSI circuitry. Corona poling also has the advantage of permitting very high poling fields to be achieved. Realization of such high fields is also an advantage of the liquid contact poling methods [247–249]. Corona poling has the disadvantage of not permitting simple measurement of the strength of the applied electric field.

Contact or electrode poling is typically carried out employing one of two configurations depending on whether the electrodes are on the same side (in-plane poling) of the EO polymer film or on opposite sides. Occasionally, the electrodes used for contact poling are also used to provide the radio frequency or microwave driving voltage. As such, the choice of contact poling configuration can be influenced by the device architecture. The spatial distribution of fields can be an issue in contact poling. Field distribution and current flow in particular are issues for in-plane poling [244].

A uniform electric field distribution across the sample is extremely important for achieving device quality materials. Unfortunately, real chromophore materials do not always behave as uniform insulator materials. We have already demonstrated that ionic impurities can dramatically reduce the effective electric field felt by chromophores. The presence of spatially and temporally varying nonuniform space charge distributions leads to nonuniform poling fields. The resulting nonuniform chromophore order can lead to light scattering.

A novel architecture, used recently by the TACAN Corporation (Carlsbad, CA) to achieve drive voltages of less than 1 V with the CLD-1 chromophore, is that of push-pull poling [6, 250, 252]. Push-pull poling is relevant to the fabrication of Mach-Zehnder interferometers and was previously demonstrated by Paul Ashley (Redstone Arsenal) and co-workers [251]. The basic configuration is shown in Fig. 17. The TACAN approach has the advantage of minimizing processing steps and preventing air dielectric breakdown.

During electric field poling (Fig. 17a, top), the chromophores in the two interferometric arms are aligned in opposite directions by the poling field. For device operation, a modulation drive voltage is applied to the top electrode while the two bottom electrodes are grounded. The driving voltage is along the poling direction in one arm and against the poling direction in the other arm. The index of refraction modulations in the two arms are always of opposite sign and therefore the total phase difference is twice as large as that in the single arm modulation case, Eq. (6). Since no other device parameters are changed, the value of V_π is thus reduced by a factor of 2. The three-section poling and driving electrodes can be easily adapted to a high-speed microstripline/slot ground electrode for broad band operation [252]. Recently, researchers at Lockheed-Martin have obtained results analogous to those of TACAN employing a push-pull modulator device fabricated with the FTC-1/APC (25 wt/wt%) materials [253]. They have

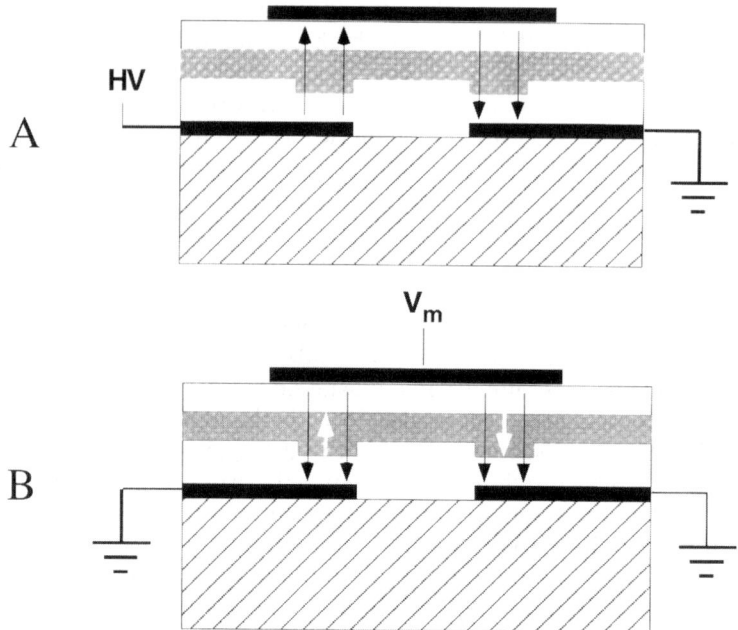

Fig. 17. A Electrode configuration and B operating configuration used for push-pull poling

achieved a V_π value of 1.35 V that has been observed to be stable for 3 months. Given the lower inherent hyperpolarizability of the FTC chromophore, this is an impressive result.

Although considerable effort has been expended investigating the relaxation of electric field induced acentric chromophore order, very little effort has been invested in the detailed characterization of chromophore order and dynamics during poling. With the theoretical effort focused on understanding intermolecular electrostatic interactions discussed above, the situation appears to be changing. A noteworthy theoretical effort aimed at understanding poling dynamics is that of Hayden and co-workers [254]. They have employed fully atomistic molecular dynamical modeling of the electric field poling of a guest/host [N,N-dimethyl-p-nitroaniline (DPNA)/PMMA] polymer system. They have analyzed the behavior of systems both above and below the glass transition temperature, T_g, of the system by calculating $<\cos\theta>$ and $<\cos^3\theta>$. Above T_g, the applied poling field was found to constrain the side group of the host PMMA. The shape of the volume swept by the dopant chromophores was found to be less cylindrical as the field was increased. The partial radial distribution functions indicated that the chromophores were likely to be located near the host polymer side group. No significant differences in the static conformational properties of the host were observed between poled and unpoled structures. Nearly equal polymer side group and backbone contributions were involved in chromophore orientation. Below the T_g, the $COCH_3$ side group motion was found to be more correlated with dopant contributions than with the atoms in the backbone units. These results are consistent with unpublished results of Robinson and Dalton (employing analogous modeling), although the chromophores studied by Hayden and co-workers have sufficiently low dipole moments and polarizabilities that intermolecular (chromophore-chromophore) electrostatic interactions are unimportant in contrast to the situation considered by Robinson and co-workers.

Kaino and co-workers [255] have investigated the effect of polymer polydispersity on electro-optic materials properties. No dependence on polydispersity was observed for guest host materials but for Disperse Red chromophores covalently attached to monodisperse polystyrene weaker absorption tails were observed. This result suggests that chromophore-chromophore interactions are modified by the polymer host.

A major topic that comes up when poling is considered is that of poling-induced optical loss. This is one of the most misunderstood topics involving polymeric electro-optic materials. What should be meant by poling-induced optical loss is optical loss associated with poling-induced material damage or phase separation (the electrophoretic effect: chromophore migration under the influence of an electric field). Optical loss due to processing (inclusion of dust particles) are often confused with poling-induced losses; in particular, a report by Teng and co-workers [256] is often misinterpreted. Teng and co-workers [256] correctly state that the index of refraction of a material will change with electric field poling and TE and TM modes are affected differently as is logical from consideration of their symmetries with respect to the symmetry of the poling field. They then consider a material consisting of regions that will be poling insensitive (e.g., dust particles) as well as regions sensitive to the poling field (the problem of nonuniform space charge fields mentioned earlier). They note that electric field poling will increase the difference between the index of refraction of these two regions and hence will increase light scattering and optical loss. They conclude that this results in an unfortunate relationship between electro-optic activity and optical loss. Their results have been interpreted by some scientists to mean that one must always have increased optical loss for materials with improved electro-optic activity. Of course, this is not the case in reality. Without dust particles or other factors leading to material heterogeneity, the Teng mechanism disappears. The Teng result does emphasize the need for care in purification of materials and control of material heterogeneity during processing (spin casting, poling, lattice hardening). The Teng mechanism should be fundamentally associated with processing-induced optical loss and not poling-induced optical loss.

Damage induced during poling is the most serious optical loss mechanism associated with electric field poling. Although the details of damage will vary with type of poling (corona versus contact), damage will depend on the strength of the electric field applied and on the hardness of the material being poled [63, 64, 137, 257]. Because it is desirable to use the largest possible poling field consistent with avoiding material damage, it is common to employ a stepped poling protocol (where both temperature and electric field strength are increased at periodic time intervals) coupled with lattice hardening by use of thermosetting reactions. When the lattice is being hardened by a temperature-dependent chemical reaction, realization of optimum poling efficiency depends on achieving the appropriate balance of kinetics associated with chromophore reorientation under the influence of the poling field and of lattice hardening to lock-in poling-induced order and to minimize dielectric breakdown. A variety of techniques have been used to assess poling-induced damage, including scanning electron microscopy (SEM) and atomic resolution imaging techniques [5].

The subject of lattice hardening and retention of poling-induced electro-optic activity is also a complex and frequently misunderstood topic. Most experimental and theoretical studies of the retention of poling-induced electro-optic activity have focused on guest/host and homopolymer materials where the chromophore is attached at one end to the linear chain backbone of an uncrosslinked polymer host. For such materials, it is well established that the retention of electro-optic activity is related to the glass transition temperature of the system [2, 3, 5, 50, 63, 64, 108]. Thermal and photochemical stability also relates to the hardness (glass transition temperature) of the final polymer material. Because of the dependence of acentric chromophore order and chemical stability on the glass transition temperature, high glass transition temperature polymers, such as polyimides, polyquinones, and polycarbonates, have been extensively investigated as polymer hosts [2, 3, 5, 50, 63, 64, 108, 121–126, 258–264]. Unfortunately, stiff polymers such as the polyimides exhibit poor solubility in traditional spin-casting solvents. Chromophores with high values of $\mu\beta$ typically cannot survive the harsh solvents required for polyimides. Moreover, the high poling temperatures required by the high glass transition temperatures of polyimides can result in problems with chromophore sublimation when guest/host materials are used.

On the other hand, low glass transition polymers, such as PMMA, which yield good optical quality films and good poling efficiencies, do not yield adequate stability of poling-induced order. Current research, which deals with polymeric systems where lattice hardening is not carried out during poling, focuses on finding a compromise between PMMA and the polyimides. The currently most investigated materials are various polyquinones [259–264] and amorphous polycarbonate (APC) from Aldrich. APC [poly(bisphenol A carbonate-co-4,4'-(3,3,5-trimethylcyclohexylidene)diphenol, see Fig. 18] is quite compatible with virtually all high $\mu\beta$ chromophores and has a glass transition temperature (T_g = 205 °C) in the appropriate processing range. Materials prepared using APC typically exhibit thermal stability of acentric chromophore order to 120 °C or greater. This is sufficient for some applications. APC also has the advantage of having a very favorable dielectric constant and is compatible with a variety of substrates and cladding materials.

The alternative to guest/host and homopolymer materials where the glass transition temperature of the system does not change during processing (spin casting, poling, and device fabrication) is induction of lattice hardening during

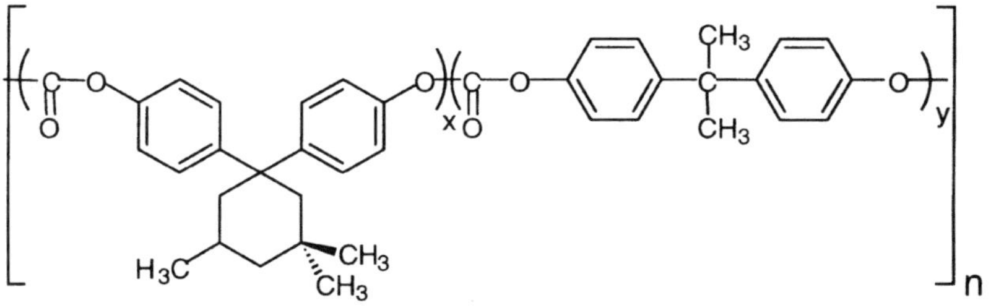

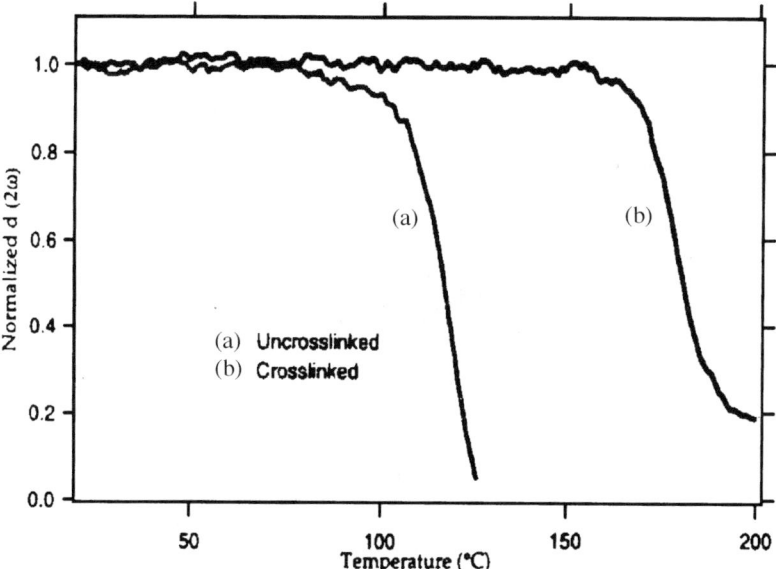

Fig. 18. Chemical structure of the APC (see text) polymer host

Fig. 19. Dynamical thermal stability of electric field poling-induced electro-optic activity for two samples. The data were obtained as described in [121] by slowly increasing temperature while monitoring second harmonic generation. The chromophore is a DEC chromophore described in [138]. "Uncrosslinked" refers to the precursor polymer where only one end of the DEC chromophore is attached to the polymer lattice. "Crosslinked" refers to the situation where both ends of the DEC chromophore have been reacted to achieve covalent coupling to the polymer lattice. The ends of this DEC chromophore are asymmetrically functionalized so that attachment reactions can be carried out independently

processing [2, 3, 5, 50, 63, 64, 108]. The idea is to have the solubility and processability where required in the early processing steps but also to harden the material up subsequent to induction of acentric order. As shown in Figs. 19–21, a dramatic improvement in the stability of electro-optic activity and long-term device stability can be achieved with such an approach.

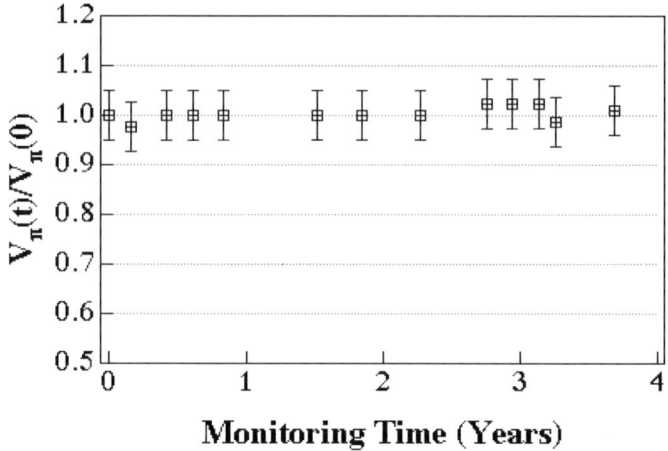

Fig. 20. The long-term stability (measured by TACAN Corporation) of a (PU-DR19) material. This material consists of a Disperse Red chromophore with one end attached to a polyurethane matrix. This long-term stability test was started in 1995 and no change in device halfwave voltage has been observed to the present time

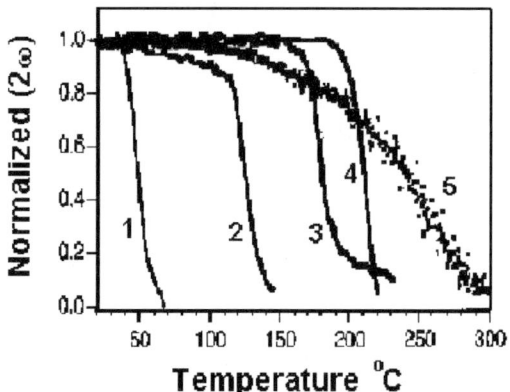

Fig. 21. Dynamic thermal stability [121] of a variety of polymeric electro-optic materials. All materials involve the Disperse Red chromophore. *Trace 1*: PMMA composite material; *trace 2*: chromophore covalently attached by one end to a "soft" (PMMA-like) matrix [138]; *trace 3*: DEC chromophore with both ends attached to a "soft" polymer matrix [138]; *trace 4*: Disperse Red chromophore covalently attached at one end to a polyimide polymer matrix [121]; *trace 5*: DEC-type chromophore with both ends attached to a sol-gel type matrix [139]

In Fig. 19, "uncrosslinked" refers to the situation where only one end of the chromophore is attached to the polymer backbone (prescursor polymer) while "crosslinked" refers to the situation where both ends of the chromophore are attached to the polymer backbone (final hardened polymer) [138]. The double-end crosslinking results in improvement in the dynamic thermal stability by more than 50 °C. A dramatic improvement in the stability of electro-optic activity is also observed in aging experiments. For example, when the double-end-crosslinked chromophore/polymer hardened material of Fig. 19 is held at 125 °C, no change in electro-optic activity is observed over a period of 3000 h. Even without coupling both ends of the chromophore to the polymer lattice, quite good thermal stability can be observed, as shown in Fig. 20. This figure shows data from the TACAN Corporation study of the electro-optic activity of a chromophore coupled at one end to a crosslinked polyurethane polymer matrix. No change in electro-optic activity was observed over a period of 4 years for a polymeric electro-optic modulator maintained at ambient temperatures. More accurately, it should be stated that no special precautions were taken to protect this material from temperature fluctuations during operation or storage, i.e., the modulator was always in contact with ambient air and no precautions were taken to maintain a fixed temperature. Obviously, considerable heating occurs during operation.

The thermal stability achieved by various hardening methods is straightforwardly understood in terms of the number of contact points between the electro-optic chromophore and the host polymer lattice and in terms of the stiffness of various segments of the host polymer. This point is demonstrated in Fig. 21. Guest/soft polymer host materials exhibit the poorest stability, typically losing electro-optic activity on heating to 50 °C (trace 1, Fig. 21). Of course, when the same chromophores are incorporated into a high T_g polymer matrix, such as polyimide, the thermal stability can be increased to 200 °C (e.g., to the glass transition temperature of the material). Covalent attachment of one end of the chromophore to a polymer matrix can significantly increase the thermal stability (traces 2 and 4). Double-end crosslinking the chromophore to the polymer lattice leads to further improvement in stability (traces 3 and 5). Trace 5 corresponds to incorporating a DEC chromophore into a sol-gel glass material. This particular sol-gel material is not fully hardened until temperatures considerably above 300 °C are reached. This accounts for the slow degradation of electro-optic activity with heating. Heating without the poling field on causes the crosslinking reaction of the sol-gel to continue, producing stresses that drive orientational relaxation of the poled chromophores. Improvement in thermal and photochemical stability is typically achieved at the cost of some reduction in poling efficiency and electro-optic activity because the lattice hardening can also inhibit chromophore reorientation under the influence of the poling field.

Finding the right trade-off between thermal stability and optimum electro-optic activity has frequently required the development of sophisticated poling and lattice hardening protocols. Typically, the best results are obtained when a stepped temperature, stepped electric field protocol is utilized. The concept is to keep the field low in the early stages of lattice hardening to prevent damaging the soft lattice but to gradually increase the poling field strength as the lattice hardens. With sol-gel materials, it is difficult to achieve temperatures sufficiently high to realize a fully hardened lattice; thus, the material may continue to evolve under high-temperature operation yielding complex thermal stability data such as those shown in Fig. 21.

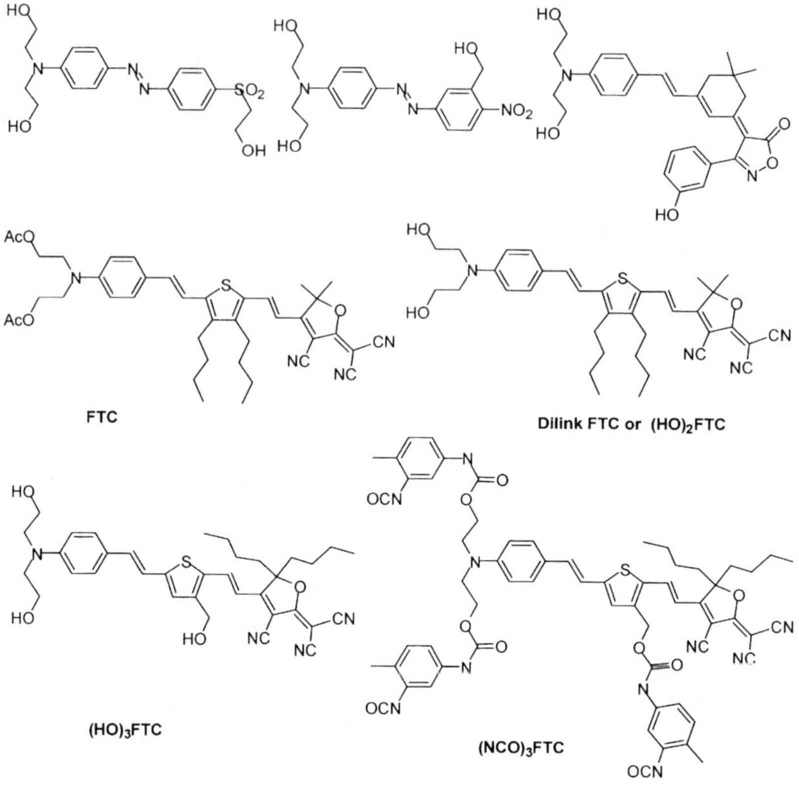

Fig. 22. Various di- and trilinking chromophores

With trilinking chromophore thermosetting materials (see Fig. 22), the situation is even more complex than with dilinking chromophore materials. For example, if the thiophene group of the FTC chromophore (see Fig. 22) is functionalized with a hydroxyl group to generate a trilinking chromophore, the hydroxyl group can react with the cyanofuran acceptor. Thus, hydroxyl coupling groups must be replaced by toluenediisocyanate groups to realize a useful trilinking chromophore. Moreover, poling efficiency can be improved (relative to a simple thermosetting trilinking protocol) by over 150% by developing a protocol that minimizes attachment to stiff polymer segments during electric field poling [265]. Clearly, the issue of lattice hardening remains one of the most challenging aspects of producing device quality electro-optic materials.

6.2
Fabrication of Buried Channel Waveguides

Electro-optic modulation typically requires concentration of co-propagating radio frequency and optical fields in a narrowly defined region of space called a waveguide [2, 3, 5, 63, 64]. Electro-optic modulation requires single mode (rather than multi-mode) optical propagation. The requirement of single-mode propagation defines the permitted dimensions of waveguides. The requirement of large electrical fields (e.g., a volt per micron) means small electrode spacing that, in turn, means rectangular waveguides and elliptical optical modes.

6.2.1
Reactive Ion Etching Techniques

The production of buried channel waveguides by reactive ion etching has been extensively discussed elsewhere [2, 3, 5, 63, 64, 267, 268] and we will not repeat that discussion here. Waveguides can also be fabricated by electron beam irradiation [266] and by photochemical/photophysical processing methods (see below). Buried channel waveguides can be fabricated by either etching a ridge or trough structure. The critical issue is to produce surfaces of sufficient smoothness to avoid unwanted optical loss. In other reviews [2, 3, 5, 63, 64, 267], we have shown that optical loss due to wall roughness can be keep below 0.1 dB/cm by maintaining the conditions of a chemical etch. A chemical etch corresponds to the reactive ions of the etching plasma possessing inconsequential kinetic energy. High kinetic energy ions have been shown to causing pitting. Surface smoothness can be verified by a number of techniques including, in addition to optical loss measurements, electron microscopy and atomic level resolution imaging techniques [2, 3, 5, 63, 64, 267,268].

Later in this review we will return to the topic of reactive ion etching as it pertains to the topic of fabricating three-dimensional optical circuits that require the production of vertical transitions.

6.2.2
Photolithographic Techniques

By photolithographic techniques, we most specifically mean multi-color photolithography where spatially controlled index of refraction variations are created by photo-initiation of physical and chemical changes in the material. Again, this topic has been reviewed in considerable depth elsewhere so we will only briefly review the central concept here [2, 3, 5, 63, 64].

The critical concept of multi-color photolithography is that the penetration depth of light into a sample depends on optical absorption by the sample. Thus, light of different colors will exhibit different penetration depths in a sample that has a strong absorption. If the photochemical absorption in question leads to a photophysical or photochemical change in the material that alters the index of refraction, then the color of light and the exposure time can be used to control the extent of photo-initiated reaction that occurs in a given spatial region of the sample. The processing of a buried channel waveguide by multi-color photolithography can be envisioned as follows: (1) a mask is used to protect the high index electroactive polymer region that is to become the buried channel waveguide. The region on either side of the protected electroactive region is bleached by "long-time" exposure to deeply penetrating radiation, i.e., light of a wavelength removed from the λ_{max} of the photoactive absorption. Such light produces a uniform reduction in the index of refraction of the material on either side of the protected region of electroactive material. (2) In a second processing step, the protecting mask is removed and the sample is exposed for a short period of time to light that shallowly penetrates the sample, i.e., light near the λ_{max} of the photoactive absorption. In this step, only the upper-most region of the sample is affected. The net result of the above two processing steps is that a buried channel region of high index of refraction electro-optic material is produced completely surrounded by lower index bleached regions.

Production of buried channel waveguides by multi-color photolithographic techniques has several advantages relative to reactive ion etching processes. It is a completely dry process and involves a smaller number of steps. It can be carried out with automated scanning and hence well-defined circuits can be rapidly fabricated. The process can be modified to incorporate two-photon processing techniques. Unfortunately, it does depend on the existence of a photoactive absorption, although photochemical decomposition can almost always be used to effect an index change. However, except for initial feasibility studies, it has largely been utilized as a technique for trimming of reactive ion-etched circuits such as splitters and Mach-Zehnder modulators [269–273].

6.3
The Unique Problem of Fabricating Tapered Transitions

While propagation loss in electro-optically active waveguides remains a great concern, typically the greatest contribution to device insertion loss comes from mode mismatch between silica fibers and electro-optic waveguides. At 1.3 microns, the core of a silica optical fiber is spherical with a dimension of about 10 microns. On the other hand, the mode in an electro-optic waveguide is elliptical with a vertical dimension of 1–2 microns. The small vertical dimension is necessary for minimum electrode separation demanded by the need for small drive voltages. If active (electro-optic) and passive (silica fiber) waveguides are simply butted together per-connection-loss is in the order of 5 dB or more. A number of solutions have been proposed to the mode mismatch problem. Pacific Wave Corporation (Los Angeles, CA) has achieved low coupling loss by positioning a spherical lens between the silica fiber and the active waveguide. A more common approach is to use tapered transition waveguides where the optical mode is allowed to expand or compress in transition regions. The fabrication, experimental characterization, and theoretical analysis of such tapered transitions have been discussed at length by Steier and co-workers [268, 274].

Another approach to low loss coupling makes use of the production of vertical waveguide transitions (to be discussed later with respect to packaging [65, 272–278]). Essentially, a section of high index electro-optic waveguide is fabricated on top of a very low loss passive polymer waveguide [65, 273, 275]. The optical mode is coupled up into the active layer where signal processing occurs and then the mode is coupled back to the passive mode at the end of the upper active layer section. Such a configuration permits realization of very low insertion loss. Other researchers have pursued similar approaches to reducing insertion loss [279].

6.4
Electrodes and Claddings

Electrodes and claddings are topics that are essentially ignored in polymeric electro-optic materials research even though these components of active circuits have a dramatic effect on performance parameters such as bandwidth, drive voltage, and optical loss. At 100 GHz, resistive loss in metal electrode structures is a very serious problem and the operational bandwidth of an actual device is typically determined by electrode design and not by the electro-optic material. Fetterman and co-workers have achieved device performance to 113 GHz (with unpublished performance to even higher frequencies) and have discussed a number of electrode and electrode coupling designs [280–282]. We have already noted how electrode configuration affects performance for amplitude modulators. Of course, a propagating optical mode must not be permitted to see metal electrodes or very high optical loss will occur. Cladding layers are used to confine optical modes and prevent optical modes from sensing lossy metal electrodes. Cladding layers must exhibit an appropriate index of refraction to yield the desired optical confinement. Cladding materials must be highly transparent to the propagating optical mode as loss due to the optical mode penetrating into the cladding contributes to the total propagation optical loss. Just as in the case with the active layer, the cladding layer must exhibit excellent thermal and photochemical stability. The author of this review has had the unfortunate experience of observing the output of an electro-optic modulator go from single mode to multi-mode operation as mode confinement was degraded by photochemical decomposition of a cladding layer. Ideally, one would like cladding layers to exhibit greater electrical conductivity than the active electro-optic layer. Greater conductivity permits the poling field to be dropped across the cladding layer resulting in a greater effective field at the chromophores of the active layer and greater poling efficiency. We have already seen in our theoretical discussion how important it is to achieve a large effective poling field. If electrical conductivity at radio, microwave, and millimeter wave frequencies could be obtained, the V_π voltage required would also be lowered. To date, no materials that exhibit low optical loss and high frequency conductivity have been reported. The issue of conducting cladding has only recently been addressed. The results reported by Dr. James Grote at the Air Force Research Laboratory are very encouraging [152, 283, 284], resulting in significant improvements in electric field poling efficiencies and material electro-optic activity.

Deposition of cladding layers must be compatible with the active electro-optic layer. In particular, the solvents used in spin casting of the cladding layer must not dissolve the active layer. Since the cladding layer must exhibit the same robustness as the active layer, it must be hard. UV-curable epoxies have frequently been used as cladding materials. However, the UV radiation used must not cause photochemical damage of the active layer. To date, off-the-shelf cladding materials such as epoxylite 9653 [285] and NOA73 [286] have frequently been used as cladding materials. In the future, custom-designed cladding materials are likely to be required for optimized device performance.

7
Sophisticated Packaging

Certain applications of electro-optic modulator technology require dense packaging of a large number of modulators and the integration of individual modulators with very large-scale integration (VLSI) semiconductor electronics and a variety of passive and active optical circuit elements. For example, one might well envision optical circuits composed of modulators, splitters, optical amplifiers, etc. [2, 3, 5, 6–65, 287–289]. Indeed, one of the attractive features of polymeric electro-optic materials is their adaptability to sophisticated integration and packaging. For example, circuits can be fabricated on virtually any substrate including flexible substrates such as MYLAR. Over the past several years, Professor William Steier and co-workers have executed an ever-increasingly sophisticated integration and packaging of polymeric electro-optic modulators [2, 3, 5, 63–65, 271–278, 290–297]. In the following sections, we briefly review this effort.

7.1
Vertical and Horizontal Integration with VLSI Semiconductor Chips

For electro-optic modulators to be widely used in telecommunication and data processing industries, they must be seamlessly integrated with high-speed electronic circuitry. As discussed elsewhere [2, 3, 5, 63–65, 271–278, 290–297], this integration can either be accomplished as side-by-side (horizontal) integration or in a stacked (vertical) integration architecture. Vertical integration is more space efficient but requires dealing with the irregular surface topology of semiconductor chips. Deposition of active polymer layers directly onto semiconductor chips would result in unacceptably high optical loss. Moreover, a scheme for poling electro-optic chromophore-containing polymer films without damaging neighboring semiconductor circuitry must be employed. Finally, the interconnection of electro-optic and semiconductor circuitry must be effected.

Fortunately, over the past several years, the problems associated with each of the above requirements have been overcome and now integrated "opto-chips" are now relatively routinely fabricated [2, 3, 5, 63, 64, 271–278, 290–297]. The problem of irregular VLSI surface topology has been overcome by use of planarizing polymers such as Futerrex PC3-6000. The reflow properties of this polymer reduce the 1–6 micron semiconductor circuit features to surface variations of 0.2 microns after planarization. The optical quality of planarized surfaces is comparable to that of polished silicon and optical loss due to the roughness of the planarized surface is immeasurably low. Steier and co-workers [3, 63, 64] have shown that appropriate grounding can be used to protect underlying VLSI circuitry when corona poling the electro-optic material. Indeed, poling can be achieved with no detectable change in semiconductor I/V curves [3, 63, 64]. Interconnection between semiconductor circuitry and modulators requires fabrication of deep vias (channels) through the planarizing polymer layer. Steier and co-workers have shown that this can be accomplished by use of a combination of CF_4 reactive ion etching and spin-on-glass technology. Sophisticated vertically integrated opto-chips have been fabricated and the outputs of various electronic circuits have been used to independently drive multiple modulators [3, 63, 64]. The cross-talk between adjacent modulators is immeasurably low, i.e., less than 45 dB [298].

7.2
Integration with Passive Optical Circuitry and Diode Lasers

If sophisticated systems architectures containing a large number of modulators are to be fabricated, a stacked and vertically integrated optical circuit architecture will be required. The key to vertical integration of optical circuitry is the ability to achieve a vertical transition, e.g., to fabricate vertical polarization and power splitters. Recently, Steier and co-workers [273, 275] have shown that this can be achieved by employing shadow, gray scale, and offset lithographic mask techniques. Representative examples are given in Figs. 23–25. In Fig. 26, we show how these masking techniques can be combined with spin casting to produce a vertically integrated waveguide circuit.

- Shadow Masking of Ions
  - Angle ∝ RF Power, Gas Pressure, Time, Mask Dimensions
  - Angles: 0.1-3°
  - Heights: 1-9µm
  - Lengths: 200-2,000µm
- Fast Prototyping
  - Various Angles From Single Mask
  - No Extensive Fabrication Steps
- Repeatable Quality

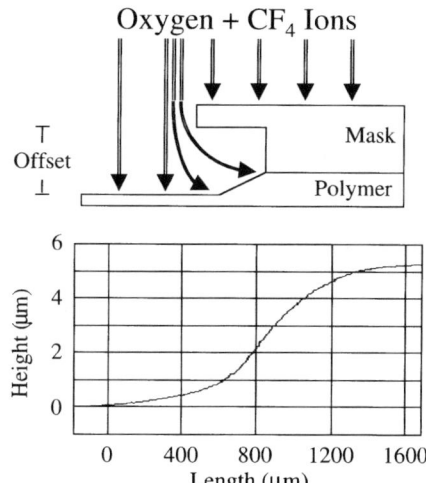

Fig. 23. Production of a vertical slope by shadow mask processing

- Computer Generated Layout
- Variable Transmission Exposure
  - Height ∝ Exposure Level
  - Angles: 0.1-3°
  - Heights: 1-15µm
  - Lengths: 100-2,000µm
- Entire Device Contoured
  - Complex Patterns Possible
  - 10µm Resolution
- Precision of Mask Aligner
- Repeatable Quality

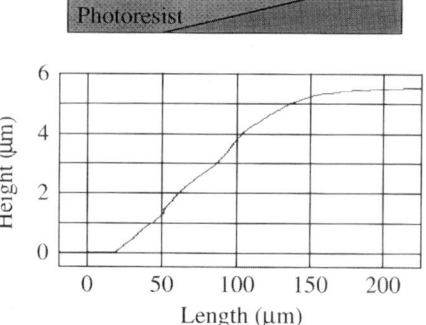

Fig. 24. Production of a vertical slope by gray scale mask processing

Steier and co-workers have used such processing techniques to fabricate a variety of devices such as the 1×4 power splitter and polarization splitter shown in Figs. 27 and 28, respectively.

Figures 23–28 show the process of producing basic circuit elements of vertically stacked circuits. The use of the three lithographic masking techniques mentioned results in very low loss transitions between optical circuits in different vertical stacks. These techniques can also be used to fabricate the low loss electro-optic modulator configuration shown in Figs. 29 and 30.

In this configuration, light is coupled into the active electro-optic material only when signal processing is desired. Otherwise, light propagates in very low loss polymeric passive waveguides. Such low loss passive waveguides have been the subject of a number of research efforts including those of Koike and co-workers [299, 300].

7.3
Special Electrode Structures for High (100 GHz and above) Frequencies

As already noted, resistive loss in metal electrode structures and in transition from millimeter wave waveguides to the electrode structure is the greatest problem in achieving 100 GHz and higher electro-optic modulation. Fetterman and co-workers [301] have shown by pulse techniques that the 3-dB bandwidth of polymeric electro-optic materials is typically in the order of 360 GHz for 1 cm of material. Stripline electrode structures have been used to achieve operation to somewhat above 100 GHz. Fetterman and co-workers [282] have recently described a novel finline transition between a millimeter waveguide and such elec-

- Straight-Edge Diffraction
  - Angle ∝ Vertical Offset, Photoresist Thickness, Exposure Energy Density
  - Angles: 1-13°
  - Heights: 1-5μm
  - Lengths: 30-2,000μm
- Fast Prototyping
  - Various Angles from Single Mask
- Precision of Mask Aligner
- Repeatable Quality

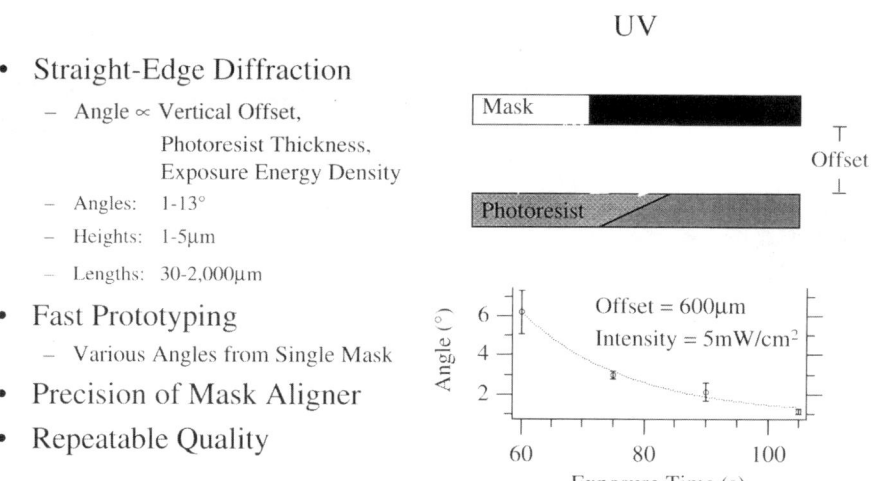

Fig. 25. Production of a vertical slope by offset lithography

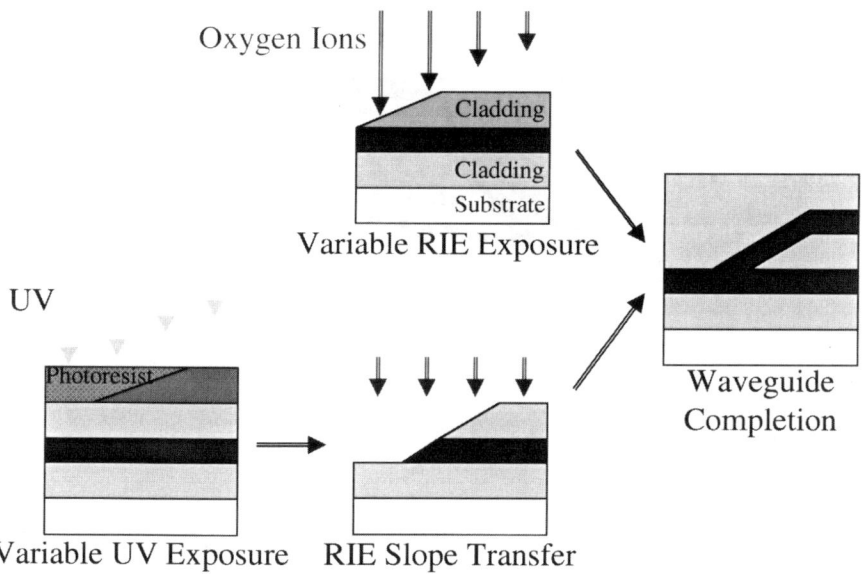

Fig. 26. Processing steps associated with vertically integrated optical circuit fabrication

trode structures. Very recently, researchers at IPITEK (TACAN) have proposed device structures that may permit operation for telecommunication applications to 220 GHz [302].

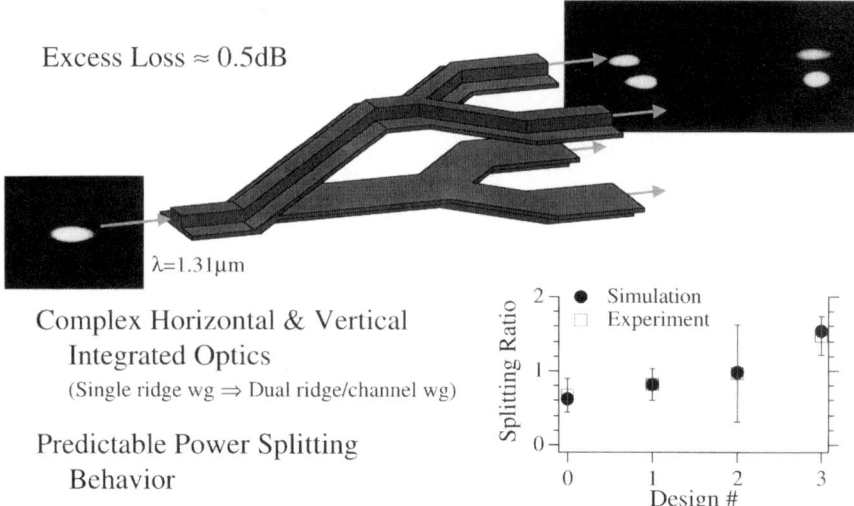

Fig. 27. Vertically integrated 1×4 power splitter

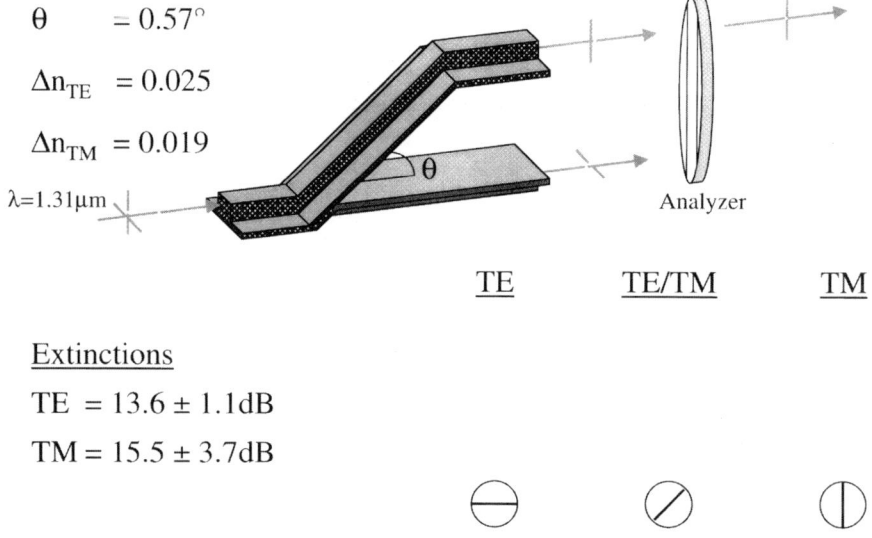

Fig. 28. Vertically integrated polarization splitter

Total length = 6.5cm
- Active core = 2cm
- Electrode = 1cm

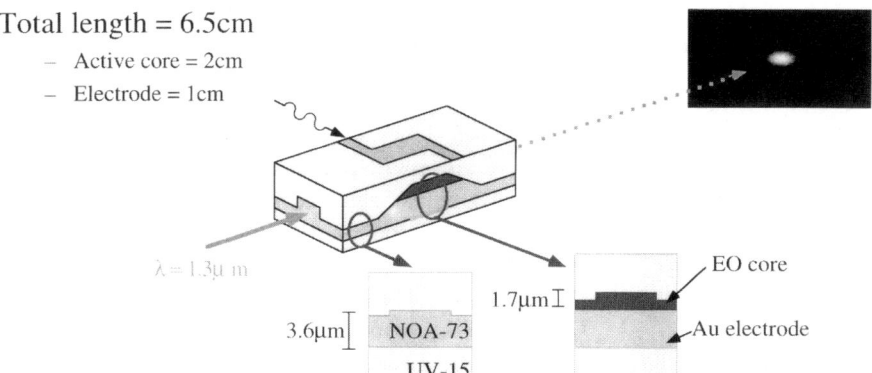

- TM insertion loss ⇒ 8dB improvement
- TE insertion loss ⇒ 9dB improvement

Fig. 29. Low insertion loss electro-optic modulator structure

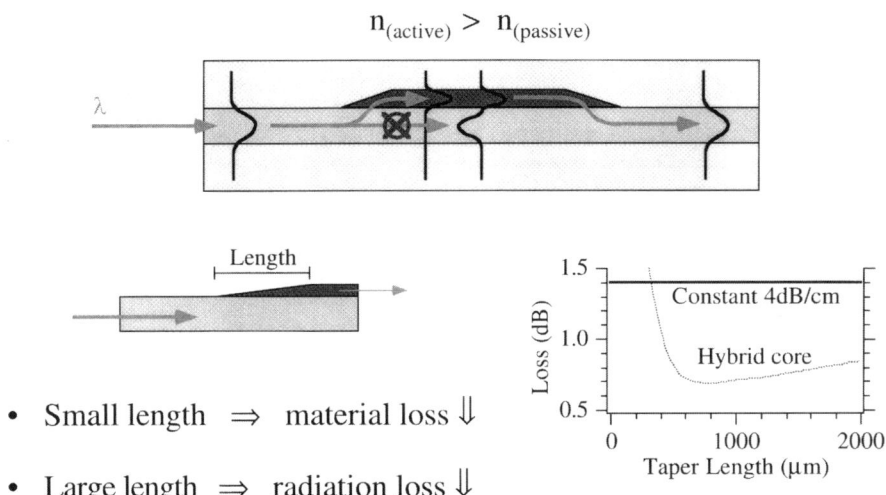

- Small length ⇒ material loss ⇓
- Large length ⇒ radiation loss ⇓

Fig. 30. Blow-up of the active modulator region of Fig. 29

8
Evaluation of Device and System Performance

Of course the details of device and system performance will depend on the particular device or system under consideration. Our discussion will, however, focus primarily on limitations to system performance associated with material limitations rather than that of a particular device configuration. Parameters of particular interest include drive (V_π) voltage, bandwidth, waveguide propagation loss, total device insertion loss, drive voltage stability, bias voltage stability, and optical power handling capability.

8.1
Bandwidth

Measurement of device bandwidths in the order of 100 GHz typically requires heterodyne detection and a stripline electrode configuration such as that illustrated in Fig. 31. The response of polymeric electro-optic modulators is typically flat to 100 GHz. Fall off above that frequency (Fig. 32) can be traced to resistive losses in millimeter wave transmission structures and in the metal electrodes.

The bandwidth performance of polymer modulators clearly surpasses those of other materials although by clever device engineering lithium niobate modulators have been demonstrated to somewhat above 70 GHz.

8.2
Drive Voltage

IPITEK (formerly TACAN) Corporation (Carlsbad, CA) has achieved 0.8 V V_π values using 3-cm interaction length push-pull Mach-Zehnder modulators, as illustrated in Fig. 33 [6, 252, 302]. With 2-cm interaction length modulators, an average V_π value of 1.1 V was obtained [6, 252, 302].

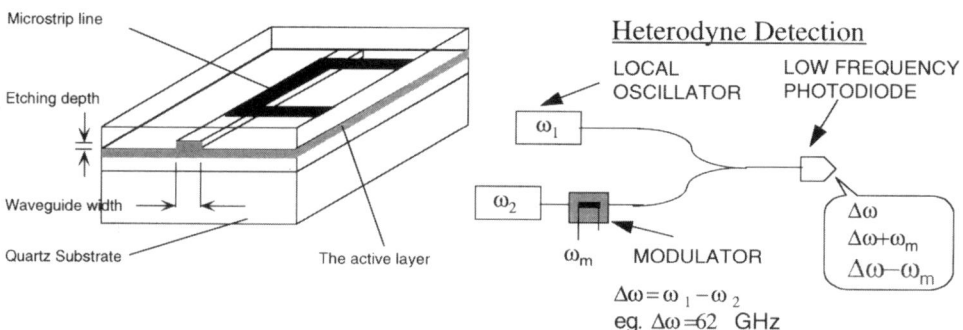

Fig. 31. Heterodyne apparatus for measuring high frequency (e.g., 100 GHz) electro-optic modulation

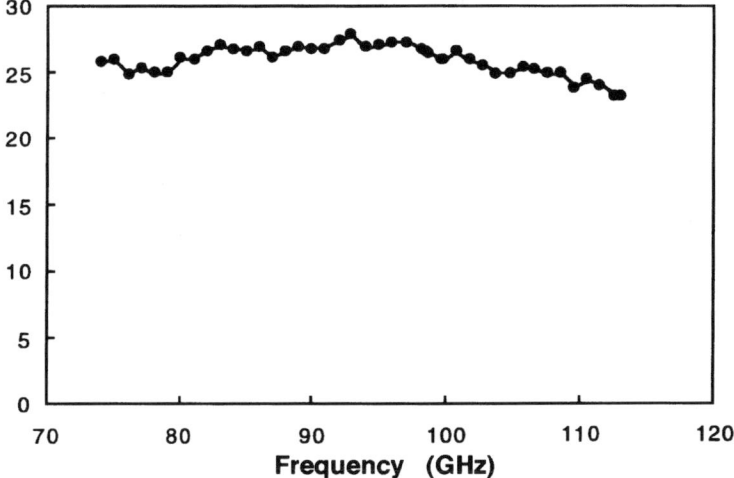

Fig. 32. Operation of a polymeric electro-optic modulator to 113 GHz

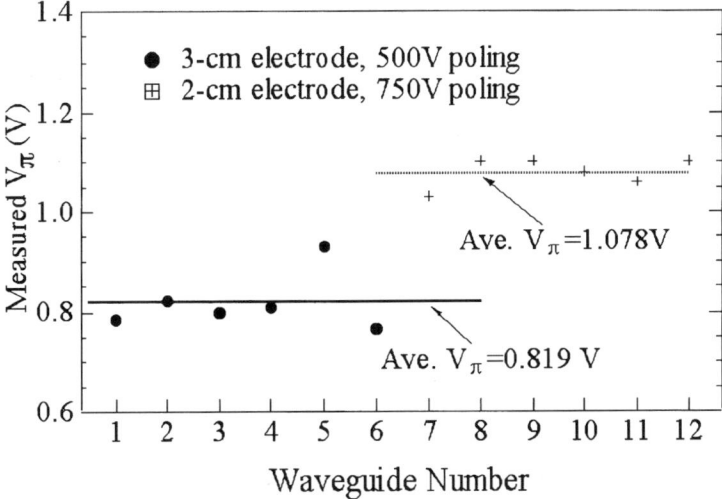

Fig. 33. Halfwave voltages (V_π) measured by the TACAN Corporation for various push-pull electro-optic modulators. Twelve modulators (six 3-cm and six 2-cm modulators) were fabricated. The poling voltages and waveguide modulator performance are given in the figure

The CLD-1 chromophore in a PMMA polymer host was used to obtain the TACAN results. The V_π results are consistent with the electro-optic coefficients determined by ATR methods and discussed earlier in this review. Pacific Wave Industries (Los Angeles, CA) have constructed single modulation arm Mach-Zehnder interferometers using CLD-1 in either PMMA or APC polymer matrices and have measured V_π voltages of 2 V at 1.3 microns wavelength and 3 V at 1.55 microns wavelength. These results are also consistent with those already mentioned in this article. The results for the CLD-1 chromophore represent an improvement of 2–3 over values obtained with use of lithium niobate modulators.

Researchers at Lockheed-Martin have fabricated push-pull Mach-Zehnder modulators using FTC in APC. They observed drive voltages near 1 V and these modulators exhibited good stability for periods of 3 months (with stability testing continuing) [253].

Comparable electro-optic coefficient values appear to have been obtained at US Government Research Laboratories by Geoffrey Lindsay (US Navy, NAWCWPNS, China Lake), Paul Ashley (US Army, Redstone Arsenal), and James Grote (AFRL, Wright Patterson AFB) with CLD-type chromophores in various matrices [303].

To this point in time, the FTC, CLD, and GLD chromophore series appear to be useful for fabricating a variety of prototype modulator devices operating at telecommunication wavelengths with drive voltages significantly less than those of lithium niobate modulators explored for the same applications. These results represent an approximate order of magnitude improvement over more conventional organic chromophores such as the Disperse Red and diaminonitrostilbene (DANS) series. Results being obtained at the time of writing for the best CLD, GLD, and CWC variants represent another 25–50% improvement. Nearly a factor of two improvement relative to the above results has been obtained for chromophore-containing dendrimers, as will be discussed at the very end of this article.

8.3
Optical Loss

Optical propagation loss for polymeric electro-optic materials is typically in the order of 1 dB/cm when care is taken to avoid scattering losses associated with processing and poling-induced damage [2, 3, 5, 63, 64, 257]. Lower loss values can be obtained by isotopic replacement of protons with deuterium and with halogens [211, 304, 305]. With effort, electro-optic material losses can be reduced to approximately 0.2 dB/cm for the telecommunication wavelengths of 1.3 and 1.55 microns.

Propagation losses through active materials are a serious concern; however, these typically contribute only a small fraction to the total insertion loss. The most serious problem relating to minimization of optical loss with use of electro-optic modulators is that of loss associated with mode mismatch between passive and active optical circuitry. When tapered transitions and other device structures discussed in this review are used to reduce optical loss associated with mode mismatch, total device insertion losses in the order of 4–6 dB are obtained. Without such adequate attention to coupling losses, insertion loss can be 10 dB or greater.

8.4
Stability under Harsh Conditions

The stability of drive voltage, optical loss, and bias voltage under operating conditions that can include high temperatures and optical powers is a subject of great concern. Definitive data are not available at this time for polymer materials containing the latest high μβ chromophores. Although beta site testing has not been conducted, there is some hope that polymer modulators will exhibit satisfactory performance with respect to stability for operation under harsh conditions including exposure to high-energy radiation.

The TACAN Corporation has evaluated the stability of drive voltage and bias voltage under conditions of extended operation and for irradiation with light of different wavelengths and optical power levels [306, 307]. The performance of polymer modulators was reasonably good and systematically improved as lattice hardness increased (see Figs. 6, 20 and 34). We have observed comparable stability.

TACAN has prepared a summary table of the performance of polymeric electro-optic materials and comparison with inorganic materials (Table 5). Note that, although this compilation by TACAN is only a year old, the "future" performance of polymeric electro-optic materials has already been realized.

Stability to gamma irradiation has been investigated by researchers at Lockheed-Martin Corporation, the Air Force Research Laboratory (Dayton, OH), and by ourselves. The results (no detectable change in performance) suggest that polymer modulators are suitable for space applications. We have, by electron

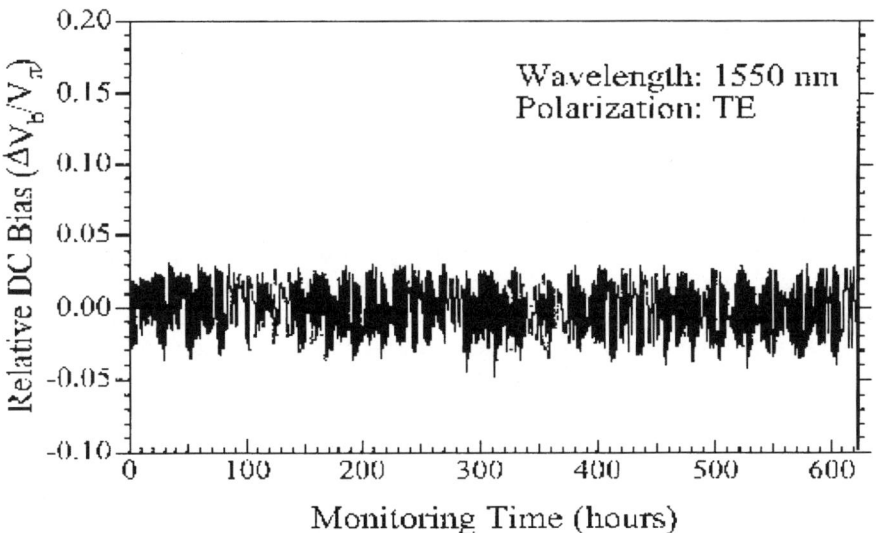

Fig. 34. DC bias voltage stability (measured by the TACAN Corporation) of a polymeric electro-optic modulator operating at 1550 nm

Table 5. Comparison of materials performance by the TACAN Corp.

Parameters	EO polymer Current	EO polymer Future	LiNbO$_3$ Shelf	LiNbO$_3$ Lab	GaAs Shelf
Vπ (V)	3.5–5	1–3	5–7	2–5	5–7
Pπ (dBm)	15–18	4–13	>32	12–18	>24
VπL (Vcm)	8–10	5	10	10	>10
Bandwidth (GHz)	40	65–110	18–20	30–70	50
Insertion loss (dB)	12–14	6–9	3–5	3–5	9–12
Max. storage temp. (°C)	60–90	70–100	90 °C pkg limit	NA	80 °C pkg limit
Max. optical power (mW)	NA	NA	250	NA	30
Stability	NA	NA	small drift	low drift	low drift
Cost (1 unit)	NA	NA	$6500	NA	$17,000

paramagnetic resonance, observed damage to certain host polymers such as PMMA but not to the electro-optic chromophores.

We have carried out pulsed (including femtosecond) laser studies of two-photon absorption and other potential photochemical decay mechanisms for a variety of polymeric electro-optic materials including FTC, CLD, and GLD guest/host systems [185, 308]. Such studies do not reveal problems associated with operation with light of telecommunication wavelengths, i.e., two-photon-initiated photochemistry. The greatest single concern for polymeric electro-optic materials involves photochemical damage arising from radiation into the charge-transfer band. Clearly, if enough energy is deposited into this band, photochemical decomposition will be observed for virtually any organic chromophore, although the harder the lattice the higher the energy required. Packaging polymeric electro-optic modulators to protect them from visible light seems to be a useful precaution. Researchers at Pacific Wave Industries and our research collaborator Prof. W. H. Steier have both observed a dramatic improvement in the photochemical stability of polymer electro-optic modulators with hermetic sealing. Protected from air (and singlet oxygen decomposition mechanisms), modulators have been operated with 10–50 mW incident power for periods of months with no detectable degradation in performance. Of course, the same problem is faced with organic (molecular and polymeric) light emitting diode materials and photochemical decomposition does not appear to be a debilitating problem for OLED display technology when appropriate packaging is used.

A number of researchers have recently commenced study of the thermal and photochemical stability of organic EO chromophores [87, 92, 309–312]. It is not clear that an agreed upon mode of assessment has been arrived at that will be applicable to a wide range of device applications. Many studies are being performed on materials that are not likely relevant to the fabrication of devices. The final assessment of thermal and photochemical stability may well have to be performed on actual devices under specific application operating conditions. Nevertheless, identification of photodecomposition mechanisms and quantification of environmental effects will be of utility in guiding the further development of materials.

9
Applications and Commercialization

At the present time, no significant commercialization of polymeric electro-optic modulators exists. However, that situation appears be changing rapidly. Pacific Wave Industries now offer a variety of broad bandwidth modulators for purchase and firms such as Radiant Research, IPITEK and Lumera Corporations are dramatically expanding their activities. Figure 35 shows the PWI 40 GHz modulator fabricated from CLD-1/APC polymer material.

A number of firms [including Lockheed-Martin, Allied Signal, Pacific Wave Industries, IPITEK (TACAN) Corporation, Radiant Research, Lumera Corporation, Microvision, and Lucent] and government laboratories (including US Army Redstone Arsenal; US Navy NAWCWPNS, China Lake, CA; Air Force Research Laboratory/Materials & Manufacturing Directorate, WPAFB, Dayton, OH) are developing prototype devices and systems that utilize polymeric electro-optic modulator technology. Prototype devices and demonstrations include high voltage sensing relevant to the electric power industry [313], electrical-to-optical signal transduction of multiple television signals relevant to cable (CATV) and satellite television transmission [43, 44, 63, 64, 314], broad bandwidth acoustic spectrum analyzers [315, 316], optical gyroscopes [317], phased array radar [43, 44, 63, 64, 318–322], photonically detected radar [5], time stretching and ultrafast analog-to-digital conversion [4], components for fiber optical and satellite telecommunications [3, 43, 44, 63, 64, 253, 302, 323–327], generation and detection of ultrafast electrical fields [5, 63, 63, 328–330], simple electric field sensors and land mine detection [3, 63, 64, 331], backplane inter-

Fig. 35. Pacific Wave Industries 40-GHz polymeric modulator. PWI also offers a version that is designed for 100-GHz operation

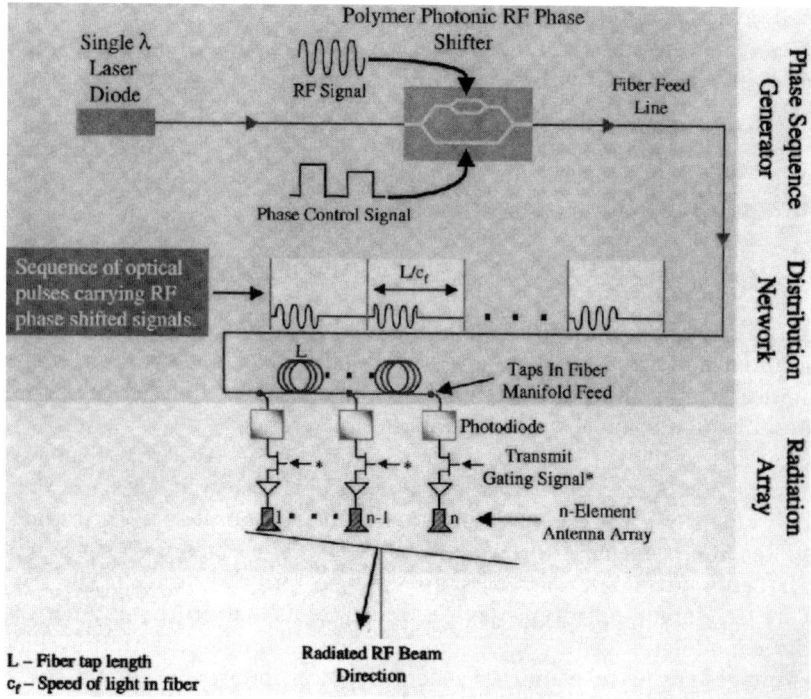

Fig. 36. Phased array system based on a polymeric photonic rf phase shifter

connections for high-speed computation [3, 63, 64], wavelength division multiplexing applications [3, 63, 64], optical switching [2, 3, 5, 43, 44, 63, 64], and spatial light modulation (beam steering) applications [332, 333]. Relevant to telecommunications applications, it can be noted that both polarization-sensitive and polarization-insensitive [323] polymeric electro-optic modulators have been demonstrated. Polarization-insensitivity is desired for some applications, as it is difficult to maintain polarization during transport over long distances. Polarization-insensitivity has been considered to be one of the arguments for the use of electro-absorptive modulators.

Unfortunately, describing all the above prototype device applications would make this review overly long. Thus, we will limit our discussion to just two applications that involve steering of radio-frequency (phased array radar) and optical (spatial light modulation) beams. In phased array radar, a radar beam is steered by controlling the interference of beams radiated from different elements of an antenna. This, in turn, requires controlling the phase of the radio-frequency signal arriving at these different elements. Electro-optic modulators provide a convenient method of controlling the radio-frequency phase during the distribution processes. The radio-frequency signal is converted to an optical signal by an electro-optic modulator and the phase is controlled photonically. In Figs. 36–39 we show two configurations that have been used both by us and our

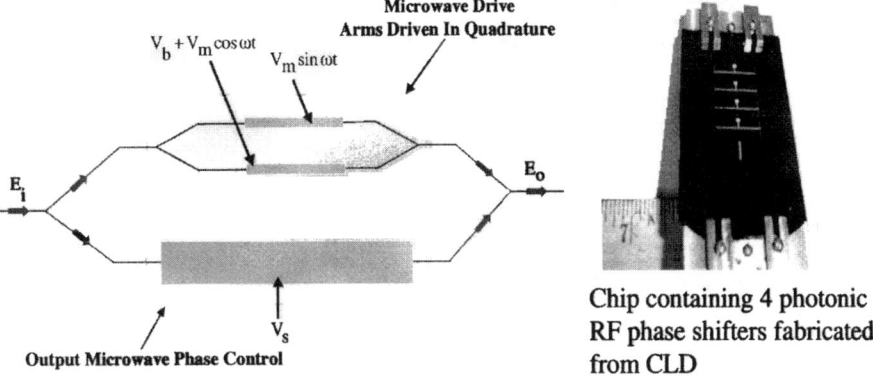

- V_s DC or low frequency bias controlling output microwave phase
- V_b Bias generating an optical phase of $\pi/2$
- Microwave arms driven in quadrature

Fig. 37. Photonic rf phase shifter

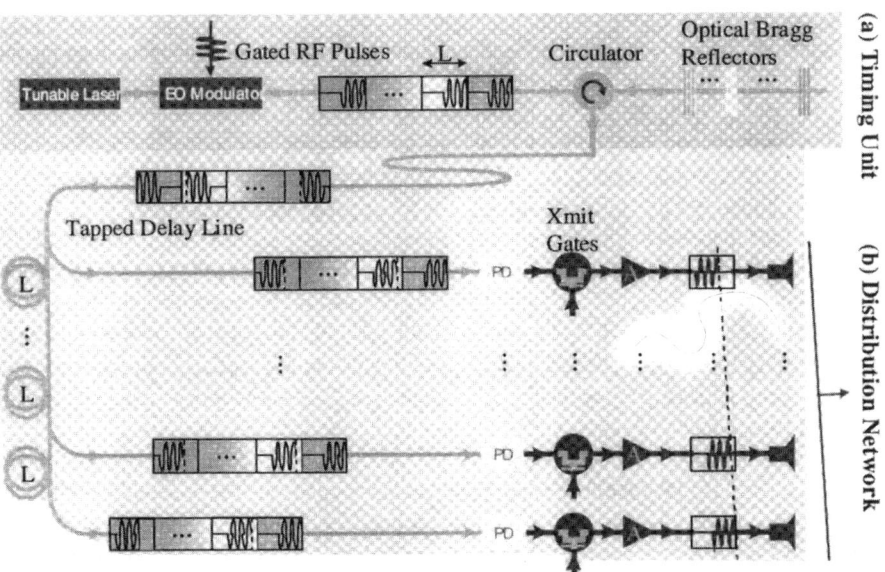

Fig. 38. Serial feed phased array transmitter

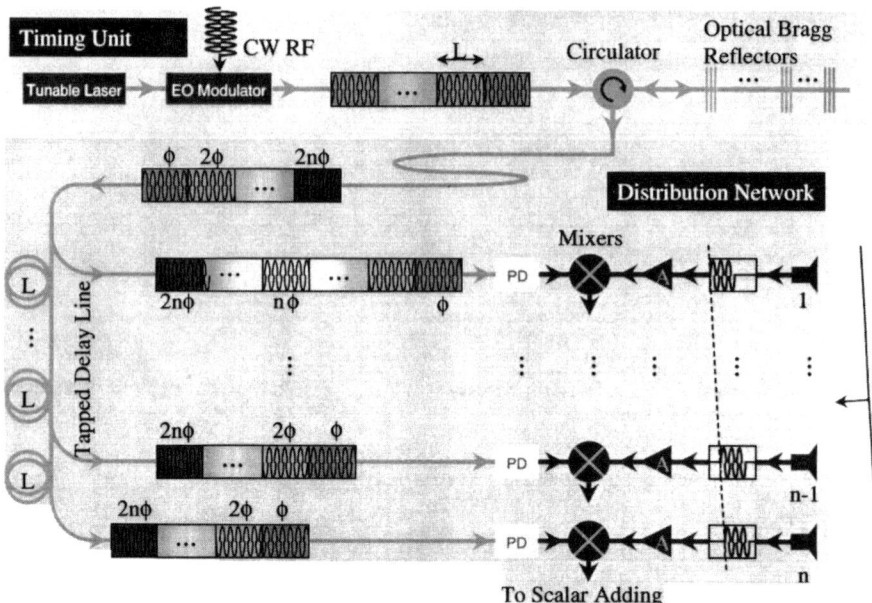

Fig. 39. Serial feed phased array receiver

collaborators to demonstrate phased array radar utilizing polymeric EO modulators.

With both serial feed and photonic radio-frequency phase shifter configurations, polymeric EO modulators are used to produce the desired phase shifts in signals ultimately routed to the different antenna elements. It is interesting to note that the photonic radio-frequency phase shifter looks like a Mach-Zehnder inside of a Mach-Zehnder [64, 318]. Recently, we have shown that the linearity of output phase versus control voltage can be improved by going to an even higher order Mach-Zehnder inside a Mach-Zehnder design. Photonically controlled radar offers many advantages relative to conventional radar and electrically controlled radar including increased beam agility (and no problems with overshoot due to momentum effects with mechanical movement), immunity from radio-frequency interference, and reduced weight. For a photonically controlled phased array radar to be commercially successful, expensive low noise amplifiers must be avoided. Lossless radio-frequency to optical to radio-frequency conversion relies heavily on devices operating with V_π values of 1 V or less. If even lower drive voltage modulators can be produced gain can be realized for the radio-frequency signal. While 1–3 V values have recently been obtained with systems based on polymeric modulators, a low noise amplifier is still required after conversion back to radio frequency to boost the power to required levels for radiation from the antennae elements.

As shown in Fig. 40, we have used a cascaded prism arrangement to achieve spatial light modulation [333]. Further amplification has recently been achieved

Large Angle Laser Beam Scanner

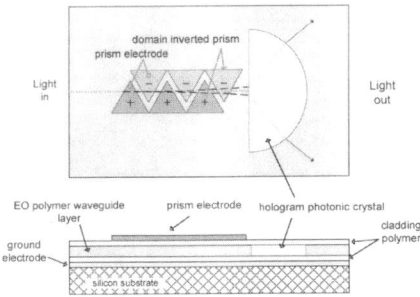

EO waveguide prism introduces a small deflection angle to initialize the beam scanning. The half-circle 2-D photonic crystal region is imbedded into the waveguide, so that the deflection angle is "amplified" as the light pass through the crystal region. 3D scanning can also be provided if a 3-D structure is built

The experimental observation of the angle sensitivity

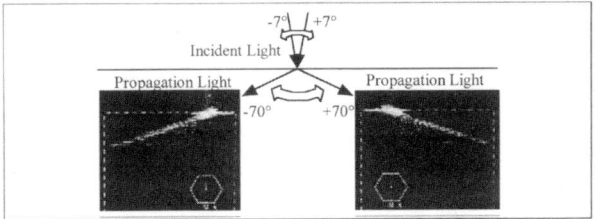

Fig. 40. Schematic representation of our electro-optic polymer beam steering device with a demonstration of its performance

by addition of a polymeric photonic bandgap material to the output of the cascaded prism array. Although beam steering has been accomplished with significantly smaller drive voltages than required by lithium niobate devices, drive voltages still remain larger than those required for liquid crystal spatial light modulators. From early experiments, it is clear that the fabrication of the cascaded prism array has not been optimized. It is likely that detailed analysis and characterization of the actual fields produced by this configuration will permit further reduction in drive voltage requirements.

While the performance characteristics of prototype phased array radar, spatial light modulators, high bandwidth oscillators, high bandwidth analog-to-digital converters, etc. require optimization, the ease with which preliminary results have been obtained using polymers is impressive and speaks to the processability of such materials. Moreover, the performance of prototype devices strongly argues for the ultimate commercial viability of devices based on polymeric materials.

10
Future Prognosis

It should by now be clear to the reader that chromophore design has not yet been optimized. Chromophores to be produced in the near future might look like the

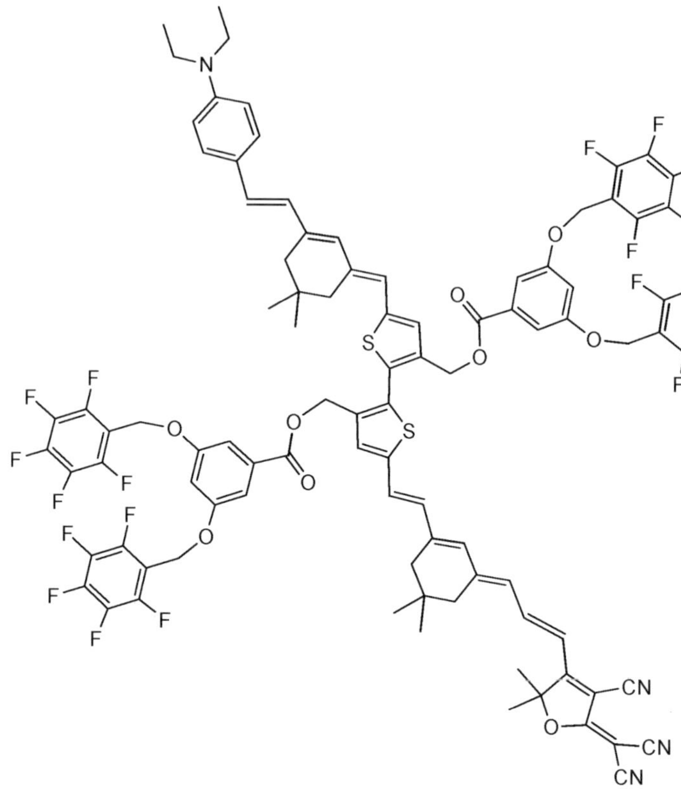

Fig. 41. Hypothetical improved CWC chromophore

one shown in Fig. 41. With such a design, considerable serendipity is required to achieve improved electro-optic activity. The added substituents must not perturb the π-electron structure of the chromophore thus reducing the values of μ and β. For example, we have used a dendritic synthetic approach [64, 334, 335] to successfully inhibit chromophore-chromophore intermolecular electrostatic interactions. Chromophore loading of 50% (wt/wt) has been achieved without dramatic attenuation of the linear dependence of electro-optic activity on loading. Unfortunately, perturbation of the chromophore structure has, in some instances, resulted in sufficient reduction in μ and β, so that no net gain in material electro-optic activity was realized. Also, the chromophore must assume a structure that actually inhibits intermolecular electrostatic interactions. Again, some dendrimer structures can rearrange so that the chromophore becomes exposed. Quantum mechanical and kinetic Monte Carlo modeling of structures is an excellent idea before attempting synthesis. Nevertheless, most attempts at derivatization do lead to significant improvement in achievable electro-optic activity.

Indeed, knowledge of the role of chromophore shape in defining maximum achievable electro-optic activity has permitted dramatic improvement in electro-optic activity of polymeric materials to the point of significantly exceeding the performance of lithium niobate. With chromophores dissolved in polymers such as APC, it is likely that devices can be fabricated, in the near future, which will require halfwave voltages on the order of 0.5 V. Moreover, over the next several years, it is very reasonable to assume that chromophores with even greater $\mu\beta$ values (and yet adequate stability) will be prepared. When these chromophores are modified to near optimum shape and incorporated into appropriate polymer matrices, V_π voltages of 0.1 V or less may be realized. It is likely that the performance of polymeric electro-optic circuitry will also be dramatically improved by the development of custom hardened hosts and cladding materials.

An extremely interesting new direction in the preparation of organic electro-optic materials involves development of multi-chromophore-containing dendritic materials such as that shown in Fig. 42 [335]. Note that for such materials a polymer host is not required. The dendrimer is of sufficient molecular weight that appropriate viscosity can be achieved in traditional spin-casting solvents. The crosslinking functionalities on the periphery of the dendrimer permit lattice hardening subsequent to electric field poling. By use of fluorinated and cyanurate dendrons, material optical loss can be kept to below 0.5 dB/cm at telecommunications wavelengths. The dendrimer shown in Fig. 42 has yielded electro-optic materials characterized by an electro-optic coefficient of 60 pm/V at 1.55 microns wavelength. This large electro-optic coefficient reflects the fact that high chromophore loading (33% wt/wt) is achieved without attention to electro-optic activity by intermolecular chromophore interactions. Under thermal testing, this electro-optic activity was observed to be unchanged after 1000 hours at 85 °C. The material optical loss for the dendritic material shown in Fig. 42 was found to be comparable to that of CLD in APC. Although this material has not been extensively evaluated in prototype devices, as is the case for CLD in APC, it does appear to equal or exceed the performance of CLD/APC in all respects. The sequential synthetic scheme represented by multi-chromophore-containing dendrimers appears to hold the possibility of another material breakthrough. However, further work is required to ascertain if this is indeed true.

Sequential synthetic techniques proceeding from substrates, such as those being developed by Professor Tobin Marks and others [336–340], may also lead to sub-1 V drive voltage devices. Sequential synthesis from functionalized surfaces has the advantage of utilizing ionic interactions and covalent bond formation to overcome intermolecular interactions that oppose acentric order. As such, this approach is not limited by dielectric breakdown, as is the case for electric field poling. Of course, intermolecular electrostatic interactions still need to be taken into account in computing expected electro-optic activity. Such interactions act to cause the chromophore axis to tilt relative to the normal to the deposition surface and to increase disorder. Such a tilt will, of course, reduce macroscopic electro-optic activity in a manner analogous to increased tilt (with increasing electrostatic interactions) of the average chromophore orientation with respect to the poling field axis in electric field poling.

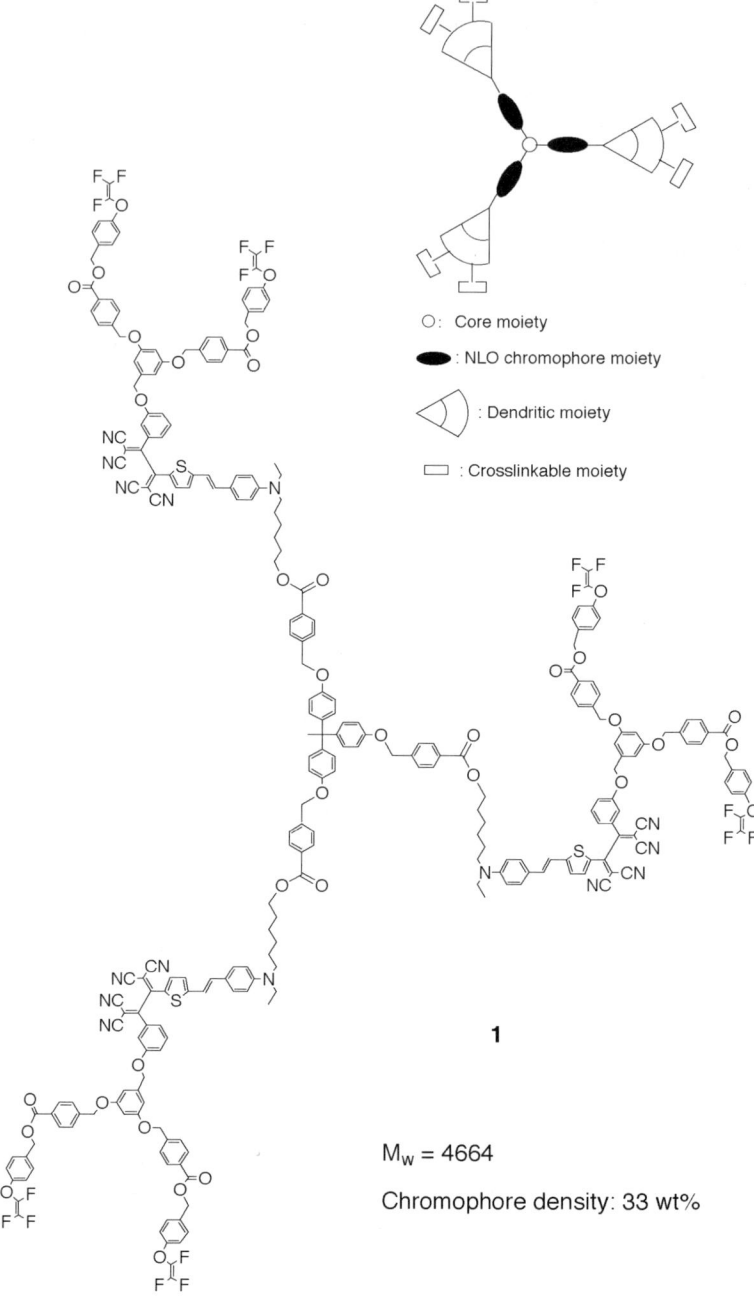

Fig. 42. Multi-chromophore-containing electro-optic dendrimer. After electric field poling and lattice hardening this material yields an electro-optic coefficient of 60 pm/V at 1.55 microns

Self-assembly methods such as crystal growth may also hold promise if new principles of chromophore design are exploited. Obviously, it is important to consider more spherical chromophores in an effort to attenuate unwanted intermolecular electrostatic interactions. It will also be important to strategically position ionic charges to drive acentric ordering. Unfortunately, although development of chromophores designed for acentric self-assembly appears a worthwhile research goal, it is unlikely to be systematically pursued in the next few years. Thus, prototype device fabrication is likely to rely on materials prepared by electric field poling and by sequential synthesis methods.

The bandwidth (>100 GHz) and drive voltage specifications (<3 V at 1.3 and 1.55 microns) of organic electro-optic materials will stimulate development of some commercial products for use at these wavelengths [341, 342]. The broader application of modulators will require addressing the two potential limitations of organic materials; namely, optical loss and stability. It is clear that insertion losses in the range 5–6 dB can now be achieved with current materials when appropriate attention is paid to processing and when novel coupling architectures are utilized. Very low loss applications will require development of low loss electro-optic and cladding materials. Such materials will almost certainly involve replacement of hydrogen atoms or by exploitation of chemical structures with low hydrogen content. Moreover, attention will likely have to be paid to integration of active and cladding materials with low loss passive waveguides including polymeric waveguides. Dendrimeric materials appear to be a particularly attractive route to the development of low optical loss materials. For operation at telecommunications wavelengths, organic materials should ultimately be very competitive with lithium niobate and quite superior to gallium arsenide in terms of insertion loss.

The most difficult issue facing organic electro-optic materials is that of thermal and photochemical stability. In principal and in practice, very stable materials can be fabricated. The issue is the price that is typically paid in electro-optic activity to achieve a high degree of stability. Because both poling and thermosetting lattice-hardening reactions depend on temperature, it is no simple matter to define compatible poling and lattice-hardening protocols. In like manner, it is difficult to exploit photo-initiated lattice-hardening chemistry due to absorption by the chromophores. Perhaps in the future chromophores and crosslinking reagents can be developed that do not have overlapping optical absorption. It may also be possible to develop two-photon activated crosslinking reagents [343–345]. Until such advances in lattice-hardening chemistry are developed, it is likely that lattice hardening will be achieved at the price of a 20–40% reduction in electro-optic activity. However, current poling and lattice-hardening protocols can be fine-tuned to permit the production of electro-optic materials that exhibit thermal stability of greater than 170 °C and are capable of handling telecommunications level optical powers for periods of several years, while at the same time affording halfwave voltages of 2–4 V. It may well be possible to achieve such stability with drive voltages in the order of 1 V. It is clear from preliminary studies of packaged and unpackaged organic electro-optic materials that hermetic sealing to exclude oxygen is important for enhancing photochemical stability. Evidence also exists that photochemical stability can be improved by covalent incorporation of free radical and singlet oxygen scavengers [92]; however, we have found that the positioning of such scavengers is important and effective exploitation of this route to improved stability materials is likely to require further study. Photochemical stability is also difficult to define in the sense that requirements vary from application to application and power levels used in telecommunications applications may increase in the future. It is doubtful that organic materials can standup for long periods of time to incident power levels of a watt or greater. High power levels can also give rise to instabilities that do not relate to photochemical degradation. For example, ultra-high bandwidth oscillators demand high power levels and problems with noise and bias voltage drift have been identified for oscillators made from certain polymeric active core and cladding materials. In some cases, this drift has been found to arise from photochemically induced charge injection from the active material into the cladding material. The photocurrent flowing in the cladding varies with optical power in the active core waveguide leading to inherent instability. This problem can be rectified by modification of materials; however, it illustrates the complexities that must be faced as a broader range of operational conditions is explored.

A strong point for the use of polymeric electro-optic materials is the adaptability of these materials to sophisticated processing techniques and the fabrication of integrated devices. Moreover, it is clear that polymeric electro-optic circuitry can be fabricated on a variety of substrates including flexible substrates. In the next several years, I anticipate dramatic exploitation of this aspect of polymeric electro-optic materials. It is very likely that demonstration of an ever-increasing array of novel multi-modulator devices and systems will be effected. Development of integrated opto-chip packages will likely make polymeric electro-optic materials increasingly attractive for fiber and satellite telecommunications applications, electronic counter measure applications, sensing, and applications in the fields of computation and data processing.

At the present time, polymeric electro-optic materials are currently not suited for applications that require operation below 1-micron wavelength. Many applications in the field of displays, including flat panel and virtual reality headset displays, could benefit from use of low-drive-voltage, low-optical-loss modulators operating at wavelengths in the range 400–1000 nm. Such operation would require that new materials be developed. To accomplish this one would have to identify a family of chromophores with a particularly attractive electro-optic coefficient versus λ_{max} (optical nonlinearity/optical loss trade-off) curve [3, 341]. To be useful at visible wavelengths, the λ_{max} of the electro-optic material would have to be <250 nm. While it is possible that appropriate materials could be developed, a substantial research effort would be required and is unlikely in the present funding climate. Thus, it is likely that lithium niobate will be the material of choice for applications below 1000 nm for some time to come.

When all factors are considered, polymeric and lithium niobate electro-optic and gallium arsenide electro-absorptive materials will very likely remain competitive over the next 2–10 years. Each technology is likely to find niche markets. The large bandwidths, low-drive voltages, and ease of integration of polymeric electro-optic materials provide a strong argument for increased utilization. Of course, it is never easy to displace an established technology such as that based on lithium niobate. Adoption of polymeric modulators will require that they meet thermal and optical power handling standards.

Displacement of lithium niobate will be made even more difficult due to business decisions made over the past 5 years. These include: (1) the financial investment by companies such as Lucent and Boeing in construction of lithium niobate modulator production facilities and (2) the perceived failure of companies such as ROI Tech, Hoechst-Celanese, IBM, 3M, etc. to commercialize polymeric electro-optic modulator technology in the early 1990s. These two factors are very likely to cause large corporations to leave the development of polymeric electro-optic materials to smaller start-up companies. This is consistent with an increasing trend of large American corporations to act only as packagers and marketers of technology leaving the actual production of materials and discrete devices to smaller vendors. A more serious consequence of this philosophy is that large corporations typically define systems design strategies. As a young scientist (3 decades ago) trying to execute replacement of vacuum tube technology with semiconductor component technology in the analytical instrumentation industry, I witnessed the difficulty of implementing a new technology in the face of specifications written for an older technology. In the present case of attempting to adapt polymeric electro-optic materials, it is clear that these materials must satisfy demands determined by other materials. Just as the famous IBM "drop test" had to be abandoned for personal computers and "high voltage" circuit concepts had to be modified for semiconductor devices, it is likely that some modification of design concepts and material specifications will be required if polymeric electro-optic circuitry is to be most effectively exploited. For such modification to occur, polymeric electro-optic devices will have to offer dramatic advances in information and signal processing, displays, sensing, etc. As is usually the case with high technology, the demand for the advanced capabilities of polymeric modulator devices will likely initially come from military applications.

Acknowledgements. The author would like to thank the National Science Foundation and the Air Force Office of Scientific Research for support of portions of the research described in this review. He would also like to express his sincere appreciation to his students and collaborators for making many of the advances described here possible by their exceptional efforts. The helpful comments of these individuals are also gratefully acknowledged.

11
Appendix. Conversion Factors and Useful Relationships

Conversion Factors

Units of dipole moment, μ: 1 Debye (10^{-18} esu/cm)=3.336×10^{-30} C/m
Units of hyperpolarizability, β: 1 (cm^5/esu or esu)=4.19×10^{-10} m^4/V
Units of second-order nonlinear optical susceptibility, $\chi^{(2)}$= 1 cm^2/esu=4.19×10^{-4} m/V

Useful Relationships

The Sellmeier equation for index dispersion: $n^2(\lambda)=a+b/[\lambda^2-(\lambda_{max})^2]$ where the constants a and b are defined by experimental measurement

Dispersion equation for hyperpolarizability: $\beta_{EO}=\lambda^2 [3\lambda^2-(\lambda_{max})^2]\beta_0/3[\lambda^2-(\lambda_{max})^2]$ where β_0 is the value of β at zero frequency and β_{EO} is the value of β at the wavelength of interest

Dispersion equation for electro-optic activity: $r_{EO}= \lambda^2 [3\lambda^2-(\lambda_{max})^2]r_0/3[\lambda^2-(\lambda_{max})^2]$

Zero frequency local field factor for poling field: $f_0=\varepsilon_r(n^2+2)/[2\varepsilon_r+n^2]$ where ε_r is the relative dielectric constant

Local field factor for light wave field: $f_\lambda=(n^2+2)/3$

Relation of number density to weight fraction: $N=(w\rho N_A/M)$ where ρ is the density of the polymer and N_A is Avogadro's number

Equation for electro-optic coefficient: $|r_{33}|=2Nf_0(f_\lambda)^2\beta_{EO}<\cos^3\theta>/n^4$

Equation for EO coefficient: $|r_{33}|=2N(f_0)^2(f_\lambda)^2\mu\beta_{EO}E/5kTn^4$

Equation for EO coefficient: $|r_{33}|= 2N(f_0)^2(f_\lambda)^2\mu\beta_0 E[3-(\lambda_{max})^2/\lambda^2]/15kTn^4\{1-(\lambda_{max})^2/\lambda^2\}^2$

Bandwidth·length product equation: $\Delta f \cdot L=c/4|(\varepsilon_r)^{1/2} - n|$

Figure of merit, $FOM=n^3r/\varepsilon$

References

1. Zyss J, Ledoux I (1994) Chem Rev 94:77; Andraud C, Zabulon T, Collet A, Zyss J (1999) Chem Phys 245:243
2. Dalton LR, Harper AW, Wu B, Ghosn R, Laquindanum J, Liang Z, Hubbel A, Xu C (1995) Adv Mater 7:519
3. Dalton LR (1998) Polymers for electro-optic modulator waveguides. In: Wise DL, Wnek G, Trantolo DJ, Cooper TM, Gresser JD (eds) Electrical and optical polymer systems. Marcel Dekker, New York, chap 18
4. Chang DH, Erlig H, Oh MC, Zhang C, Steier WH, Dalton LR, Fetterman HR (2000) IEEE Photon Tech Lett 12:537
5. Dalton LR, Harper AW, Ghosn R, Steier WH, Ziari M, Fetterman HR, Shi Y, Mustacich RV, Jen AKY, Shea KJ (1995) Chem Mater 7:1060
6. Shi Y, Zhang C, Zhang H, Bechtel JH, Dalton LR, Steier WH (2000) Science 287:119
7. Perry JW, Marder SR, Perry KJ, Sleva ET (1991) Proc SPIE1560:302
8. Thakur M, Xu J, Bhowmik A, Zhou L (1999) Appl Phys Lett 74:635

9. Pan F, McCallion K, Chiapetta M (1999) Appl Phys Lett 74:492
10. Pan F, Wong MS, Bosshard C, Gunter P (1996) Adv Mater 8:592; Meier U, Bosch M, Bosshard Ch, Pan F, Gunter P (1998) J Appl Phys 83:3486; Liakatas I, Wong MS, Bosshard Ch, Ehrensperger M, Gunter P (1997) Ferroelectrics 202:299; also ICRS Symposium and Symposium on Future Aspects of Photonics Technology, Tohoku University, November 24–26, 1999; Sohma S, Takahashi H, Taniuchi T, Ito H (1999) Chem Phys 245:539
11. Zyss J, Oudar JL (1982) Phys Rev A 26:1977
12. Kleinman DA (1962) Phys Rev 126:1977
13. Healy D, Thomas PR, Szablewski M, Gross GH (1995) Proc SPIE 2527:32
14. Meyrueix R, Tapolsky G, Chan YP, Lecomte JP (1995) Mol Cryst Liq Cryst S&T B Nonlin Opt 9:293
15. Cao XF, Yu LP, Dalton LR (1990) Opt Soc Am Ann Meet Tech Dig 165
16. Sekkat Z, Wood J, Aust EF, Knoll W, Volksen W, Miller RD (1996) J Opt Soc Am B 13:1713
17. Yitzchaik S, Marks TJ (1996) Acc Chem Res 29:197
18. Roscoe SB, Kakkar AK, Marks TJ, Malik A, Durbin MK, Lin W, Wong GK, Dutta P (1996) Langmuir 12:4218
19. Lin W, Lin W, Wong GK, Marks TJ (1996) J Am Chem Soc 118:8034
20. Roscoe SB, Yitzchaik S, Kakkar AK, Marks TJ, Xu Z, Zhang T, Lin W, Wong GK (1996) Langmuir 12:5338
21. Lin W, Yitzchaik S, Lin W, Malik A, Durbin MK, Richter AG, Wong GK, Dutta P, Marks TJ (1995) Angew Chem Int Ed Engl 34:1497
22. Zhou H, Hahn D, Wong GK, Roxcoe SB, Marks TJ (1997) J Opt Soc Am B 14:391
23. Lundquist PM, Lin W, Zhou LH, Hahn DN, Yitzchaik S, Marks TJ, Wong GK (1997) Appl Phys Lett 70:1941
24. Marks TJ, Ratner MA (1995) Angew Chem Int Ed Engl 34:155
25. Prasad PN, Williams DJ (1991) Introduction to nonlinear optical effects in molecules and polymers. Wiley, New York
26. Williams DJ (ed) (1993) Nonlinear optical properties of organic polymeric materials. American Chemical Society, Washington DC
27. Zyss J (ed) (1994) Molecular nonlinear optics. Academic, New York
28. Shen YR (1984) The principles of nonlinear optics. Wiley, New York
29. Bosshard C, Sutter K, Pretre P, Hulliger J, Florsheimer M, Kaatz P, Gunter P (1995) Organic nonlinear optical materials, vol 1. Gordon and Breach, New York
30. Chemla DS, Zyss J (eds) (1987) Nonlinear optical properties of organic molecules and crystals. Academic, Orlando
31. Boyd RW (1992) Nonlinear optics. Academic, New York
32. Hornak LA (ed) (1992) Polymers for lightwave and integrated optics. Marcel Dekker, New York
33. Williams DJ (1984) Angew Chem Int Ed Engl 23:690
34. Lindsay GA, Singer KD (eds) (1995) Polymers for second-order nonlinear optics. American Chemical Society, Washington DC
35. Sienicki K (ed) (1993) Molecular electronics and molecular electronic devices, vol 2. CRC Press, Boca Raton FL
36. Wessels BW, Marder SR, Walba DM (eds) (1995) Thin films for integrated optics applications, vol 392. Materials Research Society, Pittsburgh
37. Emerson JA, Torkelson JM (eds) (1991) Optical and electrical properties of polymers, vol 214. Materials Research Society, Pittsburgh
38. Garito AF, Jen AKY, Lee CYC, Dalton LR (eds) (1994) Electrical, optical, and magnetic properties of organic solid state materials, vol 328. Materials Research Society, Pittsburgh
39. Jen AKY, Lee CYC, Dalton LR, Rubner MF, Wnek GE, Chiang LY (eds) (1996) Electrical, optical, and magnetic properties of organic solid state materials, vol 413. Materials Research Society, Pittsburgh

40. Reynolds JR, Jen AKY, Rubner MF, Chiang LY, Dalton LR (eds) (1998) Electrical, optical, and magnetic properties of organic solid state materials, vol 488. Materials Research Society, Pittsburgh
41. Messier J, Kajzar F, Prasad P (eds) (1991) Organic molecules for nonlinear optics and photonics. Kluwer, Dordrecht
42. Kajzar F, Agranovich VM, Lee CYC (eds) (1996) Photoactive organic materials: science and application. Kluwer, Dordrecht
43. Nalwa SW, Miyata S (eds) (1997) Nonlinear optics of organic molecules and polymers. CRC Press, Boca Raton FL
44. Wise DL, Wnek GE, Trantolo DJ, Cooper TM, Gresser JD (eds) (1998) Electrical and optical polymer systems. Marcel Dekker, New York
45. Miyata S, Sasabe H (eds) (1997) Poled polymers and their applications to SHG and EO devices. Gordon and Breach, The Netherlands
46. Grote JR (1998) Design and fabrication of nonlinear optic polymer integrated optic devices. In: Wise DL, Wnek GE, Trantolo DJ, Cooper TM, Gresser JD (eds) Electrical and optical polymer systems. Marcel Dekker, New York, chap 16
47. Lee MH, Lee HJ, Han SG, Kim HY, Won YH (1998) Fabrication and characterization of electro-optic polymer waveguide modulator for photonic application. In: Wise DL, Wnek GE, Trantolo DJ, Cooper TM, Gresser JD (eds) Electrical and optical polymer systems. Marcel Dekker, New York, chap 17
48. Dalton LR (1996) Nonlinear optical materials. In: Kroschwite JI, Howe-Grant M (eds) Kirk-Othmer encyclopedia of chemical technology. Wiley, New York
49. Dalton LR, Sapochak LS, Chen M, Yu LP (1993) Ultrastructure concepts of optical integrated microcircuits and polymeric materials. In: Sienicki K (ed) Molecular electronics and molecular electronic devices. CRC Press, Boca Raton FL
50. Burland DM, Miller RD, Walsh CA (1994) Chem Rev 94:31
51. Kanis DR, Ratner MA, Marks TJ (1994) Chem Rev 94:195
52. Nie W (1993) Adv Mater 5:520
53. Meredith GR, Van Dusen JG, Williams DJ (1982) Macromolecules 15:1385
54. Singer KD, Sohn JE, King LA, Gordon HM, Katz HE, Dirk CW (1989) J Opt Soc Am B 6:1339
55. Morichere D, Chollet PA, Fleming W, Jurich M, Smith BA, Swalen JD (1993) J Opt Soc Am B 10:1894
56. Aust E, Knoll W, Hickel W, Knobloch H, Orendi H (1993) Proc SPIE 2025:255
57. Singer KD, Kuzyk MG, Sohn JE (1987) J Opt Soc Am B 4:236
58. Herminghaus S, Smith BA, Swalen JD (1991) J Opt Soc Am B 8:2311
59. Chollet PA, Gadret G, Kajzar F, Raimond P (1991) Thin Solid Films 1:7
60. Marder S, Kippelen B, Jen A, Peyghambarian N (1997) Nature 388:845
61. Dalton LR, Harper AW (1988) Polym News 23:114
62. Ren A, Dalton LR (1999) Curr Opin Coll Interface Sci 4:165
63. Dalton LR, Harper AW, Ren A, Wang F, Todorova G, Chen J, Zhang C, Lee M (1999) Ind Eng Chem Res 38:8
64. Dalton LR, Steier WH, Robinson BH, Zhang C, Ren A, Garner S, Chen A, Londergan T, Irwin L, Carlson B, Fifield L, Phenlan G, Kincaid C, Amend J, Jen A (1999) J Mater Chem 9:1905
65. Steier WH, Chen A, Lee SS, Garner S, Zhang H, Chuyanov V, Dalton LR, Wang F, Ren AS, Zhang C, Todorova G, Harper A, Fetteman HR, Chen D, Udupa A, Bhattacharya D, Tsap B (1999) Chem Phys 245:487
66. Robinson BH, Dalton LR, Harper AW, Ren A, Wang F, Zhang C, Todorova G, Lee M, Aniszfeld R, Garner SM, Chen A, Steier WH, Houbrecht S, Perssons A, Ledoux I, Zyss J, Jen AKY (1999) Chem Phys 245:35
67. Ticknor AJ, Lipscomb GF, Lytel R (1994) Proc SPIE 2285:386
68. Drost KJ, Jen AKY, Rao VP (1995) CHEMTECH 25:16

69. Teng CC (1997) High-speed electro-optic modulators from nonlinear optical polymers. In: Nalwa HS, Miyata S (eds) Nonlinear optics of organic molecules and polymers. CRC Press, Boca Raton FL, p 441
70. Rao VP, Cai YM, Jen AKY (1995) Proc SPIE 2527:84
71. Lytel R (1995) Mater Res Soc Symp Proc 392:65
72. Ashley PR (1994) Proc SPIE 2290:114
73. Ermer S, Lipscomb GF, Lytel R (1994) Mol Cryst Liq Cryst Sci Technol B 7:283
74. Burland DM, Miller RD, Twieg RJ, Volksen W, Walsh CA (1994) Mol Cryst Liq Cryst Sci Technol B 7:189
75. Cai YM, Jen AKY, Liu YJ, Chen TA (1995) Proc SPIE 2528:128
76. Khararian G, Sounik J, Allen D, Shu SF, Walton C, Goldberg H, Stamatoff JB (1996) J Opt Soc Am B 13:1927
77. Yardley JT (1996) Adv Nonlinear Opt 3:607
78. Roesky HW (ed) (1996) Polymers as electrooptical and photooptical active media. VCH, Weinheim
79. Van der Vorst CPJM, Picken SJ (1996) Electric field poling of nonlinear optical side chain polymers. In: Shibaev VP (ed) Polym Electroopt photoopt Act Media. Springer, Berlin Heidelberg New York, p 173
80. Saadeh H, Yu L, Wang M, Yu LP (1999) J Mater Chem 9:1865
81. Painelli A (1999) Chem Phys 245:185
82. Stegeman GI, Hagan DJ, Torner L (1996) Opt Quantum Electron 28:1691
83. Oh MC, Lee HJ, Lee MH, Oh MC, Ahn JH, Han SG (1998) IEEE Photon Technol Lett 10:813
84. Alain V, Redoglia S, Blanchard-Desce M, Lebus S, Lukaszuk K, Wortmann R, Gubler U, Bosshard C, Gunter P (1999) Chem Phys 245:51
85. Nalwa HS (1993) Adv Mater 53:341
86. Rojo G, de la Torre G, Garcia-Ruiz J, Ledoux I, Torres T, Zyss J, Agullo-Lopez F (1999) Chem Phys 245:27
87. Filpse MC, Van der Borst CPJM, Hofstratt JW, Woudenberg RH, Van Gassel RAP, Lamers JC, Van der Linden GM, Veenis WJ, Diemeer MBJ, Donckers MCJM (1996) NATO ASI Ser 3 9:227
88. Skindoej J, Perry JW, Marder SR (1994) Proc SPIE 2285:116
89. Zhang Q, Canva M, Stegeman GI (1998) Appl Phys Lett 73:912
90. Dubois A, Canva M, Brun A, Chaput F, Boilot JP (1996) Appl Opt 35:1996
91. Cai C, Liakatas I, Wong MS, Bosshard Ch, Gunter P (1998) Polym Prepr 39:1111
92. Galvan-Gonzalez A, Belfield KD, Stegeman GI, Canva Chan KP, Paek K, Sukhomlinova L, Twieg RJ (2000) Appl Phys Lett 77:2083
93. Pretre P, Kaatz P, Bohren A, Gunter P, Zysset B, Ahlheim M, Stahelin M, Fehr F (1994) Macromolecules 27:5476
94. Jungbauer D, Teraoka I, Yoon DY, Reck B, Swalen JD, Twieg RJ, Wilson CG (1991) J Appl Phys 69:8011
95. Dionisio MS, Moura-Ramos JJ, Williams G (1994) Polymer 35:1705
96. Dhinojwala A, Wong GK, Torkelson JM (1994) J Chem Phys 100:6046
97. Dhinojwala A, Hooker JC, Torkelson JM (1994) J Non-Cryst Solids 172:286
98. Kohler W, Robello DR, Dao PT, Willand CS, Williams DJ (1990) J Chem Phys 93:9157
99. Dhinojwala A, Wong GK, Torkelson JM (1993) Macromolecules 26:5943
100. Ghebremichael F, Kuryk MG, Dirk CW (1993) Nonlin Opt 6:123
101. Stutz SJ, Brower SC, Hayden LM (1998) J Polym Sci B Polym Phys 36:901
102. Ghebremichael F, Lackritz HS (1997) Appl Opt 36:4081
103. Goodson T III, Wang CH (1996) J Phys Chem 100:13920
104. Heldmann C, Neher D, Winkelhahn HJ, Wegner G (1996) Macromolecules 29:4697
105. Heldmann C, Schulze M, Wegner G (1996) Macromolecules 29:4686
106. Burland DM, Verbiest T (1996) Mol Cryst Liq Cryst S&T B Nonlin Opt 15:1058

107. Pinsard-Levenson R, Liang J, Toussaere E, Bouadma N, Carenco A, Zyss J, Froyer G, Guilbert M, Pelous Y, Bosc D (1993) Mol Cryst Liq Cryst S&T B Nonlin Opt 4:233
108. Kaatz P, Pretre Ph, Meier U, Stalder U, Bosshard Ch, Gunter P (1997) Adv Nonlin Opt 4:165
109. Dureiko RD (1998) PhD thesis, Case Western Reserve University, Cleveland OH
110. Lee KS, Choi SW, Woo HY, Moon KJ, Shim HK, Jeong M, Lim TK (1998) J Opt Soc Am B 15:393
111. Ahumada O, Weder C, Neuenschwander P, Suter UW, Herminghaus S (1997) Macromolecules 20:3256
112. Geng L, Wang JF, Mark TJ, Lin W, Lundquist PM, Wong GK (1995) Proc SPIE 2528:96
113. Wang JF, Geng L, Lin W, Marks TJ, Wong GK (1995) Mater Res Soc Symp Proc 392:85
114. Firestone MA, Ratner MA, Marks TJ (1995) Macromolecules 28:6296
115. Firestone MA, Ratner MA, Marks TJ, Lin W, Wong GK (1995) Macromolecules 28:2260
116. Firestone MA, Park J, Minami N, Ratner MA, Marks TJ, Lin W, Wong GK (1995) Macromolecules 28:2247
117. Geng L, Wang J, Marks TJ, Lin W, Zhou H, Lundquist PM, Wong GK (1996) Mater Res Soc Symp Proc 413:135
118. Ma HM, Jen AKY, Wu J, Wu X, Liu S, Shu CF, Dalton LR, Marder SR, Thayumanavan S (1999) Chem Mater 8:2218
119. Wu JW, Valley JF, Ermer S, Binkley ES, Kenney JT, Lipscomb GF, Lytel R (1991) Appl Phys Lett 58:225
120. Lindsay G, Chafin A, Gratz R, Hollins R, Nadler M, Nickel E, Stenger-Smith J, Yee R, Herman W, Ashley P (1997) Proc SPIE 3006:390
121. Becker M, Sapchak L, Ghosn R, Dalton LR, Shi Y, Steier WH, Jen AKY (1994) Chem Mater 6:104
122. Verbiest T, Burland DM, Jurich MC, Lee VY, Miller RD, Volksen W (1995) Science 268:1604
123. Verbiest T, Burland DM, Jurich DM, Lee MC, Lee VY, Miller RD, Volksen W (1995) Macromolecules 28:3005
124. Yu D, Charavi A, Yu L (1995) J Am Chem Soc 117:11680
125. Yu D, Gharvai A, Yu L (1995) Macromolecules 28:784
126. Yu D, Yu L (1994) Macromolecules 27:6718
127. Zysset B, Ahlhaim M, Staheilin M, Lehr F, Pretre P, Gunter P, (1993) Proc SPIE 2025:70
128. Chen TA, Jen AKY, Cai YM (1996) Macromolecules 29:535
129. Chen TA, Jen AKY, Cai YM (1995) J Am Chem Soc 117:7295
130. Jen AKY, Liu YJ, Cai Y, Rao VP, Dalton LR (1994) J Chem Soc Chem Commun 2711
131. Woo YW, Jin JI, Jin MY, Lee KS (1999) Chem Mater 11:218
132. Crumpler ET, Li D, Marks TJ, Lin W, Lundquist PM, Wong GK (1995) Polym Mater Sci Eng 72:289
133. Crumpler ET, Reznichenko JL, Li D, Marks TJ, Lin W, Lundquist PM, Wong GK (1995) Pure Appl Chem 67:213
134. Kowalcyk TC, Kosc TZ, Singer KD, Beuhler AJ, Wargowski DA, Cahill PA, Seager CH, Meinhardt MB, Ermer S (1995) J Appl Phys 78:5876
135. Liang Z, Dalton LR, Garner SM, Kalluri S, Chen A, Steier WH (1995) Chem Mater 7:941
136. Liang Z, Dalton LR, Garner SM, Kalluri S, Chen A, Steier WH (1995) Chem Mater 7:1756
137. Mao SSH, Ra Y, Guo L, Zhang C, Dalton LR, Chen A, Garner S, Steier WH (1998) Chem Mater 10:146
138. Xu C, Wu B, Todorowa O, Dalton LR, Shi Y, Ranon PM, Steier WH (1993) Macromolecules 26:5303
139. Oviatt HW Jr, Shea KJ, Kalluri S, Shi Y, Steier WH, Dalton LR (1995) Chem Mater 7:493
140. Woo HY, Shim HK, Lee KS (2000) Polymer J 32:8
141. Lee SB, Lee KS (1999) Nonlin Opt 22:43
142. Min YH, Mun JH, Yoon CS, Kim HK, Lee KS (1999) Elect Lett 35:1770
143. Woo HY, Shim HK, Lee KS (1998) Macromol Chem Phys 199:1427

144. Lee KS (1997) Macromol Symp 118:518
145. Moon KJ, Lee KS, Shim HK (1996) Mol Cryst Liq Cryst 280:39
146. Moon KJ, Shim HK, Lee KS, Zieba J, Prasad PN (1996) Macromolecules 29:861
147. Wang X, Yang K, Kumar J, Tripathy S (1996) Chem Mater 10:146
148. Wang X, Kumar J, Tripathy S, Li L, Chen J, Marturunkakul S (1997) Macromolecules 30:219
149. Jiang HW, Kakkar AK (1998) Macromolecules 31:4170
150. Jiang HW, Kakkar AK (1998) Macromolecules 31:2501
151. Drumond JP, Clarson SJ, Zetts JS (1999) Proc SPIE 3623:130
152. Grote JG, Drummond JP, Zetts JS, Nelson RL, Hopkins FK, Zhang C, Dalton LR, Steier WH (2000) Mater Res Soc Symp Proc 597:109
153. Ermer S private communication
154. Levine BF, Bethea CG (1975) J Chem Phys 63:2666
155. Kajzar F, Messier J (1987) Rev Sci Instrum 58:2081
156. Jen AKY, Rao VP, Drost KJ, Cai YM, Mininni RM, Kenney JT, Binkley ES, Dalton LR, Marder SR (1994) Proc SPIE 2143:321
157. Serbutoviez C, Bosshard Ch, Knopfie G, Wyss P, Pretre P, Gunter P, Schenik K, Chapuis G (1995) Chem Mater 7:1198
158. Blanchard-Desce M, Alain V, Midrier L, Wortmann R, Lebus S, Glania C, Kramer P, Fort A, Muller J, Barzoukas M (1997) J Photochem Photobiol A 105:115
159. Clays K, Persoons A (1991) Phys Rev Lett 66:2980
160. Clays K, Persoons A (1994) Rev Sci Instrum 65:2190
161. Verbiest T, Clays K, Samyn C, Wolff J, Reinhoudt D, Persoons A (1994) J Am Chem Soc 116:9320
162. Stadler S, Bourhill G, Brauchle C (1996) J Phys Chem 100:6927
163. Flipse MC, de Jonge R, Woodenberg RH, Marsman AW, van Walree CA, Jenneskens LW (1966) Chem Phys Lett 245:297
164. Noordman OF, van Hulst NF (1996) Chem Phys Lett 253:145
165. Olbrechts G, Strobbe R, Clays K, Persoons A (1998) Rev Sci Instrum 69:2233
166. Olbrechts G, Wostn K, Clays K, Persoons A (1998) Opt Lett 24:403
167. Stadler S, Bourhill G, Brauchle C (1996) Proc SPIE 2852:142
168. Schmalzlin E, Bitterer U, Langhals H, Brauchle C, Meerholz K (1999) Chem Phys 245:73
169. Zhang C, Ren AS, Wang F, Zhu J, Dalton LR, Woodford JN, Wang CH (1999) Chem Mater 11:1977
170. Levine BF, Bethea CG, Wasserman E, Leenders L (1978) J Chem Phys 68:5042
171. Teng CC, Man HT (1990) Appl Phys Lett 56:1734
172. Michelotti F, Nicolao G, Tesi F, Bertolotti M (1999) Chem Phys 245:311
173. Levy Y, Dumont M, Chastaing E, Robin P, Chollet PA, Gadret G, Kajzar F (1993) Mol Cryst Liq Cryst S&T B 4:1
174. Levy Y (1993) Mol Cryst Liq Cryst S&T B 5:1
175. Dentan V, Levy Y, Dumont M, Robin P, Chastaing E (1989) Opt Commun 69:379
176. Chen A (1998) PhD thesis, University of Southern California
177. Chen A, Chuyanov V, Garner S, Steier WH, Dalton LR (1997) Modified attenuated total reflection for the fast and routine electro-optic measurements of nonlinear optical polymer films. In: Organic thin films for photonic applications, vol 14. Optical Society of America, Washington DC, p 158
178. Ziari M, Kalluri S, Garner S, Steier WH, Liang Z, Dalton LR, Shi Y (1995) Proc SPIE 2527:218
179. Kalluri S, Garner S, Ziari M, Steier WH, Shi Y, Dalton LR (1996) Appl Phys Lett 69:275
180. Shin MJ, Cho HR, Kim JH, Han SH, Wu JW (1997) J Korean Phys Soc 31:99
181. Chen A, Chuyanov V, Zhang H, Garner S, Lee SS, Steier WH, Chen J, Wang F, Zhu J, He M, Ra Y, Mao SSH, Harper AW, Dalton LR, Fetterman HR (1998) Proc SPIE 3281:94
182. Chen A, Chuyanov V, Zhang H, Garner S, Steier WH, Chen J, Zhu J, He M, Mao SSH, Dalton LR (1998) Opt Lett 23:478

183. Wang F (1998) PhD thesis, University of Southern California
184. Teng CC (1993) Appl Phys Lett 32:1051
185. Drenser KA, Larsen RJ, Strohkendl FP, Dalton LR (1999) J Phys Chem 103:2301
186. Ward J (1965) Rev Mod Phys 37:1
187. Orr JB, Ward JF (1971) Mol Phys 20:513
188. Agrawal GP, Flytzanis C (1976) Chem Phys Lett 44:366
189. Oudar JL, Chemla DS (1977) J Chem Phys 66:2664
190. Oudar JL (1977) J Chem Phys 67:446
191. Lalema SJ, Garito AF (1979) Phys Rev A 20:1179
192. Heflin JR, Wong KY, Zamani-Kharmiri O, Garito AF (1988) Phys Rev B 38:1573
193. Garito AF, Wong KY, Cai YM, Man HT, Zamani-Khamiri O (1986) Proc SPIE 682:2
194. Morley JO, Pugh D (1989) Spec Publ-R Soc Chem 69:28
195. Bredas JL, Adant C, Tackx P, Persoons A, Pierce BM (1994) Chem Rev 94:243
196. Champagne B, Kirtman B (1999) Chem Phys 245:211
197. Lipinski J, Bartkowiak W (1999) Chem Phys 245:263
198. Painelli A (1999) Chem Phys 245:185
199. Tretiak S, Chernyak V, Mukamel S (1999) Chem Phys 245:145
200. Di Bello S, Fragala I, Ratner MA, Marks TJ (1995) Chem Mater 7:400
201. Albert IDL, Marks TJ, Ratner MA (1996) J Phys Chem 100:9714
202. Albert IDL, di Bella S, Kanis DR, Marks TJ, Ratner MA (1995) ACS Symp Ser 601:57
203. Marder SR, Beratan DN, Cheng LT (1991) Science 252:103
204. Gorman CB, Marder SR (1993) Proc Natl Acad Sci USA 90:11297
205. Marder SR, Perry JW (1994) Science 263:1706
206. Bourhill G, Cheng LT, Lee G, Marder SR, Perry JW, Perry MJ, Tiemann BG (1994) Mater Res Soc Symp Proc 328:625
207. Marder SR, Groman CB, Meyers F, Perry JW, Bourhill G, Bredas JL, Pierce BM (1994) Science 265:632
208. Jen AKY, Cai Y, Bedworth PV, Marder SR (1997) Adv Mater 9:132
209. Harper AW (1997) PhD thesis, University of Southern California, Los Angeles CA
210. Chen J (1998) PhD thesis, University of Southern California, Los Angeles CA
211. Ren AS (1999) PhD thesis, University of Southern California, Los Angeles CA
212. Zhang C (1999) PhD thesis, University of Southern California, Los Angeles CA
213. Melikian G, Rouessac FP, Alexandre C (1995) Syn Commun 25:3045
214. Dalton LR, Harper AW, Robinson BH (1997) Proc Natl Acad Sci USA 94:4842
215. Dalton LR, Harper AW, Chen J, Sun S, Mao S, Garner S, Chen A, Steier WH (1997) Proc SPIE CR68:313
216. Harper AW, Sun S, Dalton LR, Garner SM, Chen A, Kalluri S, Steier WH, Robinson BH (1998) J Opt Soc Am B 15:329
217. Robinson BH, Dalton LR (2000) J Phys Chem 104:4785
218. Zhang C, Lee M, Winkleman A, Northcroft H, Lindsey C, Jen A KY, Londergan T, Steier WH, Dalton LR (2000) Mater Res Soc Symp Proc 598:BB4.2.1
219. Boulbitch A, Toledano P (1999) Phys Lett A 237:271
220. Lalanne PJ, Marcerou J (1995) Phys Rev E 52:1846
221. Toledano P, Amf N, Aa B (1999) Phys Rev E 59:6785
222. Mottram NJ, Elston SJ (1999) Liq Cryst 26:457
223. Chandler D (1987) Introduction to modern statistical mechanics. Oxford University Press, Oxford, New York
224. Loginov AA, Rereverzev YV (1998) Low Temp Phys 24:652
225. Uzunov DI (1993) Introduction to the theory of critical phenomena: mean fields, fluctuations and renomalization. World Scientific, Singapore
226. Ma SK (1976) Modern theory of critical phenomena. Benjamin, Reading MA
227. Smart JS (1966) Effective field theories of magnetism. Saunders, Philadelphia
228. Prezhdo OV (1999) J Chem Phys 111:8366
229. Prezhdo OV, Kisil VV (1997) Phys Rev A 56:162

230. Kittel C (1996) Introduction to solid state physics. Wiley, New York
231. London F (1937) Trans Faraday Soc 33:8
232. London F (1930) Z Phys 63:245
233. Debye P (1935) Phys Z 35:100
234. Fowler RH (1935) Proc Roy Soc London A149:1
235. Piekara A (1938) Z Phys 108:395
236. Piekara A (1939) Proc Roy Soc London A172:360
237. Isrealachvili JN (1985) Intermolecular and surface forces. Academic, London
238. Hansen JP, McDonald IR (1976) Theory of simple liquids. Academic, London
239. Ehrenson S (1989) J Comp Chem 10:77
240. Allen MP, Tildesley DJ (1987) Computer simulation of liquids. Clarendon Press, Oxford
241. Giacometti JA, DeReggi AS, Davis GT, Dickens B, Leal Ferreria GF (1996) J Appl Phys 80:6407
242. Sprave M, Blum R, Eich M (1996) Appl Phys Lett 69:2962
243. Sprave M, Blum R, Eich M (1997) Appl Phys Lett 70:2056
244. Cohen R, Berkovic G (1994) Mol Cryst Liq Cryst S&T A 252:87
245. Park KH, Shin DH, Lee SD, Lee CJ, Kim N (1999) Mol Cryst Liq Cryst S&T A 327:23
246. Bauer-Gogonea S, Bauer S, Wirges W (1999) Chem Phys 245:297
247. Tang H, Cao G, Maki JJ, Taboada JM, Tang S, Chen RT (1997) Proc SPIE 3006:472
248. Tang H, Taboada JM, Cao G, Li L, Chen RT (1997) Appl Phys Lett 70:538
249. Tang H, Maki JJ, Taboada JM, Cao G, Sun D, Chen RT (1997) Proc SPIE 3147:156
250. Wang W, Shi Y, Olson DJ, Lin W, Bechtel JH (1999) IEEE Photon Technol Lett 11:51
251. Tumolillo TA Jr, Ashley PR (1992) IEEE Photon Technol Lett 4:142
252. Shi Y, Lin W, Olson DJ, Bechtel JH, Wang W (1999) Microstrip line-slot ground electrode for high-speed optical push-pull polymer modulators. In: Organic thin films for photonic applications. Optical Society of America, Washington DC, p 20
253. Ermer S, Girton DG, Dries LS, Taylor RE, Eades W, Van Eck TE, Moss AS, Anderson WW (2000) Proc SPIE 3949:148
254. Kim WK, Hayden LM (1999) J Chem Phys 111:5212
255. Takam U, Kyoji K, Shuji O, Kaino T (1999) Mol Cryst Liq Cryst S&T A 327:13
256. Teng CC, Mortazavi MA, Boughoughian GK (1995) Appl Phys Lett 66:667
257. Zhang C, Wang C, Yang J, Dalton LR, Sun G, Zhang H, Steier WH (2001) Macromolecules 34:235
258. Twieg RJ, Burland DM, Hedrick J, Lee VY, Miller RD, Moyland CR, Seymour CM, Volksen W, Walsh CA (1994) Proc SPIE 2143:2
259. Chen TA, Jen AKY, Cai YM (1996) Chem Mater 8:607
260. Jen AKY, Wu XM, Ma H (1998) Chem Mater 10:471
261. Ma H, Wang XJ, Wu XM, Liu S, Jen AKY (1998) Macromolecules 31:4049
262. Ma H, Jen AKY, Wu J, Wu X, Liu S, Shu CF, Dalton LR, Marder SR, Thayumanavan S (1999) Chem Mater 11:2218
263. Wu JW, Valley JF, Ermer S, Binkley ES, Kenney JT, Lipscomb GF, Lytel R (1991) Appl Phys Lett 58:225
264. Jen AKY, Ma H, Wu J, Wu X, Liu S, Dalton LR, Marder SR (2000) High performance side-chain aromatic polyquinolines for E-O devices. In: Organic thin films for photonic applications. Optical Society of America, Washington DC, p 3
265. Zhang C, Wang C, Dalton LR, Zhang H, Steier WH (2001) Macromolecules 34:253
266. Hideki N, Hisashi F, Egami C, Okihiro S, Ryoka M, Naomichi O (1998) Appl Opt 37:1213
267. Steier WH, Kalluri S, Chen A, Garner S, Chuyanov V, Ziari M, Shi Y, Fetterman H, Jalali B, Wang W, Chen D, Dalton LR (1996) Mater Res Soc Symp Proc 413:147
268. Chen A, Kaviani K, Remple A, Kalluri S, Steier WH, Shi Y, Liang Z, Dalton LR (1996) J Electrochem Soc 143:3648
269. Chen A, Chuyanov V, Marti-Carrera FI, Garner S, Steier WH, Chen J, Sun S, Dalton S (1997) IEEE Photon Technol Lett 9:1499

270. Chen A, Chuyanov V, Marti-Carrera FI, Garner S, Steier WH, Dalton LR (1997) Proc SPIE 3147:268
271. Lee SS, Garner S, Chen A, Chuyanov V, Steier WH, Guo L, Dalton LR, Shin SY (1998) Appl Phys Lett 73:3052
272. Chen A (1998) PhD thesis, University of Southern California, Los Angeles CA
273. Garner SM (1998) PhD thesis, University of Southern California, Los Angeles CA
274. Chen A, Chuyanov V, Marti-Carrera FI, Garner SM, Steier WH, Chen J, Sun SS, Dalton LR (1999) Opt Eng 38:2000.
275. Garner SM, Lee SS, Chuyanov V, Chen A, Yacoubian A, Steier WH, Dalton LR (1999) IEEE J Quant Electron 35:1146
276. Garner SM, Chuyanov V, Lee SS, Chen A, Steier WH, Dalton LR (1999) IEEE Photon Tech Lett 11:842
277. Garner SM, Lee SS, Chuyanov V, Yacoubian A, Chen A, Steier WH, Zhu J, Chen J, Wang F, Ren AS, Dalton LR (1998) Proc SPIE 3491:421
278. Garner S, Chuyanov V, Chen A, Lee SS, Steier WH, Dalton LR (1998) Proc SPIE 3278:259
279. Morand A, Ho-Quoc A, Tedjini S, Benech P, Bosc D, Loisel B (1998) Proc SPIE 3278:63
280. Chen D, Fetterman HR, Chen A, Steier WH, Dalton LR, Wang W, Shi Y (1997) Proc SPIE 3006:314
281. Chen D, Fetterman HR, Chen A, Steier WH, Dalton LR, Wang W, Shi Y (1997) Appl Phys Lett 70:3335
282. Chen D, Bhattacharya D, Udupa A, Tsap B, Fetterman HR, Chen A, Lee SS, Chen J, Steier WH, Dalton LR (1999) IEEE Photon Tech Lett 11:54
283. Grote JG, Zetts JS, Drummond JP, Nelson RL, Hopkins FK, Zhang C, Dalton LR, Steier WH (2000) Proc SPIE 3950:108
284. Drummond JP, Clrson SJ, Zetts JS, Hopkins FK, Caracci SJ (1999) Proc SPIE 3623:130
285. Epoxylite, Irvin, CA 92713–9671
286. Norland Products, Brunswick, NJ 08902
287. Dalton LR (1997) Chem Ind 14:510
288. Dalton LR, Harper AW (1998) Polym News 23:114
289. An D, Yue Z, Chen RT (1998) Appl Phys Lett 72:2806
290. Kalluri S, Chen A, Chuyanov V, Ziari M, Steier WH, Dalton LR (1995) Proc SPIE 2527:375
291. Kalluri S, Chan A, Ziari M, Steier WH, Liang Z, Dalton LR, Chen D, Jalali B, Fetterman HR (1995) Opt Soc Am Tech Dig Ser 21:317
292. Kalluri S, Ziari M, Chen A, Chuyanov V, Steier WH, Chen D, Jalali B, Fetterman HR, Dalton LR (1996) IEEE Photon Tech Lett 8:644
293. Tumollilo TA Jr, Ashley PR (1993) Appl Phys Lett 62:3068
294. Hikita M, Shuto Y, Amano M, Yoshimura R, Tomaru S, Kozawaguchi H (1997) Appl Phys Lett 63:1161
295. Yoshimura R, Hikita M, Usui M, Tomaru S, Imamura S (1997) Electron Lett 63:1311
296. Wachter C, Hennig Th, Bauer Th, Brauer A, Karthe W (1998) Proc SPIE 3278:102
297. Kalluri S (1997) PhD thesis, University of Southern California, Los Angeles CA
298. Udupa AH, Erlig H, Tsap B, Chang Y, Chang D, Fetterman HR, Zhang H, Lee SS, Wang F, Steier WH, Dalton LR (1999) Electron Lett 35:1702
299. Tanio N, Kato H, Koike Y, Bair HE, Matuoka S, Blyler LL Jr (1998) Polym J 30:56
300. Sato M, Hirai M, Ishigure T, Koike Y (1999) Thermal stability of high bandwidth GI POF. In: Organic thin films for photonic applications. Optical Society of America, Washington DC, p 86
301. Fetterman HR (1998) Review meeting: MURI on RF Photonics, UCLA, October 22, 1998
302. Bechtel JH, Shi Y, Zhang H, Steier WH, Zhang CH, Dalton LR (2000) Proc SPIE 4114:58
303. Lindsay GA, Grote JG private communication of unpublished results
304. Park KH, Shin DH, Lee SD, Lee CJ, Kim N (1999) Mol Cryst Liq Cryst S&R A 327:23

305. Lee HJ, Lee MH, Oh MC, Ahn JH, Hwang WY, Han SG (1999) Low-loss optical polymer waveguide applications based on crosslined fluorinated poly(arylene ether)s. In: Organic thin films for photonic applications. Optical Society of America, Washington DC, p 197
306. Shi Y, Wang W, Bechtel JH, Chen A, Garner S, Kalluri S, Steier WH, Chen D, Fetterman HR, Dalton LR, Yu L (1996) IEEE J Sel Top Quant Electron 2:289
307. Bechtel JH, Shi Y (1998) Review meeting: MURI on RF Photonics, UCLA, October 22, 1998
308. Dalton LR, Strohkendl FP, Zhang C, Todorova G unpublished data
309. Bosch M, Fischer C, Cai C, Liakatas I, Jager M, Bosshard Ch, Gunter P (1999) Photochemical stability of highly nonlinear optical chromophores for electro-optic applications. In: Organic thin films for photonics applications. Optical Society of America, Washington DC, p 75
310. Kowalczyk TC, Zhang XQ, Lackritz HS, Galvan-Gonzalez A, Canva M, Stegeman GI, Twieg R, Marder S, Thayumanavan S (1999) Systematics of NLO chromophore photochemical stability. In: Organic thin films for photonics applications. Optical Society of America, Washington DC, p 66
311. Galvan-Gonzalez A, Canva M, Stegeman GI, Twieg R, Kowalczyk TC, Lackritz HS, Marder S, Thayumanavan S (1999) Effect of environmental factors on the photodegradation of azobenzene doped polymers. In: Organic thin films for photonics applications. Optical Society of America, Washington DC, p 67
312. Le Duff AC, Ricci V, Pliska T, Canva M, Stegeman G, Chan K P, Twieg R (1999) Effects of the host matrix on the near-infrared red tail of the absorption of chromophore doped polymeric waveguides. In: Organic thin films for photonics applications. Optical Society of America, Washington DC, p 70
313. Skinhoek J, Perry JW, Marder SR (1994) Proc SPIE 2285:116
314. Smith BA, Jurich M, Moerner WE, Volksen W, Best ME, Fleming W, Swalen JD, Bjorklund GC (1993) Proc SPIE 2025:450
315. Yacoubian A, Chuyanov V, Garner SM, Steier WH, Ren AS, Dalton LR (2000) IEEE J Sel Top Quant Electron 6:810
316. Yacoubian A, Chuyanov V, Garner SM, Zhang H, Steier WH, Ren AS, Todorova G, Dalton LR (1999) Acoustic spectrum analysis using polymer integrated optics. In: Organic thin films for photonics applications. Optical Society of America, Washington DC, p 101
317. Ashley PR, Cites JS, (1997) Opt Soc Am Tech Dig Ser 14:196
318. Lee SS, Udupa AH, Erlig H, Zhang H, Chang Y, Zhang C, Chang DH, Bhattacharya D, Tsap B, Steier WH, Dalton LR, Fetterman HR (1999) IEEE Microwave and Guided Wave Lett 9:357
319. Li RLQ, Tang H, Cao G, Chen RT (1997) Appl Opt 36:4269
320. Chen R, Li B, Chen Y, Fu Z, Steier L, Dalton LR, Fetterman HR, Lee C (2001) Proc PSAA in press
321. Steier WH, Oh MC, Zhang H, Szep A, Dalton LR, Zhang C, Fetterman HR, Chang DH, Erlig H, Tsap B, Shi Y, Bechtel JH, Lin W, Chen R, Lee CYC (2001) Proc PSAA in press
322. Fetterman H, Chang Y, Erlig H, Tsap B, Oh M, Zhang C, Steier W, Dalton L, Chen R, Lee C (2001) Proc PSAA in press
323. Donval A, Toussaere E, Hierle R, Zyss J (2000) J Appl Phys 87:3258
324. Oh MC, Zhang H, Szep A, Chuyanov V, Steier WH, Zhang C, Dalton LR, Erlig H, Tsap B, Fetterman HR (2000) Appl Phys Lett 76:3525
325. An D, Tang S, Shi Z, Sun L, Taboada JM, Zhou Q, Lu X, Chen RT, Zhang H, Steier WH, Ren A, Dalton LR (2000) Proc SPIE 3950:90
326. Wang W, Shi Y, Olson DJ, Lin W, Bechtel JW (1997) Proc SPIE 2997:114
327. Kondo M, Toba Y, Tokano Y, Hayeiwa K, Sugimoto T (1999) Radio wave receiving system with optical remote antenna. Symposium on future aspects for photonics technology, ICRS, Tohoku University, Sendai, Japan, November 24, p 14
328. Nahata A, Weling AS, Heinz TF (1997) Appl Phys Lett 69:2321

329. Nahata A, Heinz TF (1998) Opt Lett 23:867
330. Nahta A, Heinz TF (1999) Ultrafast optoelectronics using poled polymers. In: Organic thin films for photonics applications. Optical Society of America, Washington DC, p 30
331. Sivaraman R, Clarson SJ, Burcham K, Naghski D, Boyd JT (1999) Fabrication and characterization of electro-optic polymeric E-field sensors. In: Organic thin films for photonics applications. Optical Society of America, Washington DC, p 60
332. Yatagai T (1999) Spatial light modulator using high T_g poled polymers. In: Organic thin films for photonics applications. Optical Society of America, Washington DC, p 228
333. Sun L, Kim JH, Jang CH, Maki JJ, An D, Zhou Q, Lu X, Taboada JM, Chen RT, Tang S, Zhang H, Steier WH, Ren AS, Dalton LR (2000) Proc SPIE 3950:98
334. Ma H, Chen B, Dalton LR, Jen AKY (2000) Proc PMSE 83:165
335. Ma H, Chen B, Takafumi S, Dalton LR, Jen AKY (2001) J Am Chem Soc 123:986
336. Wang X, Balasubramanian S, Li L, Jiagn X, Sandman DJ, Rubner MF, Kumar J, Tripathy SK (1997) Macromol Rapid Commun 18:451
337. Delcorte A, Bertrand P, Wischerhoff E, Laschewsky A (1997) Langmuir 13:5125
338. Lenahan KM, Wang Y, Liu Y, Claus RO, Heflin JR, Marciu D, Figura C (1998) Adv Mater 10:853
339. Lindsay GA, Roberts MJ, Stenger-Smith JD, Zarras P, Hollins RA, Chafin AP, Yee RY, Wynne KJ (2001) J Mater Chem 11:924
340. Roberts MJ (1999) Nonlinear optical film formed layer-by-layer using alternating polyelectrolyte deposition. In: Organic thin films for photonics applications. Optical Society of America, Washington DC, p 38
341. Geppert L (2000) IEEE Spectrum 37:28
342. Ackerman RK (2000) Signal 54:21
343. Albota M, Beljonne D, Bredas JL, Ehrlich JE, Fu JY, Heikal AA, Hess SE, Kogej T, Levin MD, Marder SR, McCord-Maughon D, Jerry JW, Rockel H, Rumi M, Subramaniam G, Webb WW, Wu XL, Xu C (1998) Science 281:1653
344. Cumpston BH, Ananthavel S, Barlow S, Dyer DL, Ehrlich JE, Erskine LL, Heikal AA, Kuebler SM, Sandy IY, McCord-Maughon LD, Qin J, Rockel H, Rumi M, Wu XL, Marder SR, Perry JW (1999) Nature 398:51
345. Perry JW, Ananthavel S, Carmmack K, Kuebler SM, Marder SR, Rumi M, Cumpston BH, Heikal AA, Ehrlich JE, Erskine LL, Levin MD (1999) Materials for two-photon 3D lithography. In: Organic thin films for photonics applications. Optical Society of America, Washington DC, p 174

Received: January 2001

Quadratic Parametric Interactions in Organic Waveguides

Michael Canva[1], George I. Stegeman[2]

[1] Groupe d'Optique Non-Linéaire, Laboratoire Charles Fabry de l'Institut d'Optique – CNRS UMR 8501, Bâtiment 503, Université d'Orsay/Paris-Sud, B.P. 147, 91403, ORSAY Cedex,
e-mail: *michael.canva@iota.u-psud.fr*

[2] School of Optics/CREOL, University of Central Florida, 4000 Central Florida Boulevard, Orlando, FL 32816–2700, USA
e-mail: *george@creol.ucf.edu*

This chapter addresses issues related to quadratic parametric interactions in organic waveguides. It attempts to give an overview of the field, especially emphasizing the initial experimental milestones together with the most recent results.
 This field of research is very challenging in terms of the technology required to turn a good candidate material into a useful optical device demonstration. The keys to this problem are the phase-matching conditions that must be fulfilled by the different interacting waves. These points are illustrated by the bibliography that, still to this day, contains mainly demonstrations of the potential of various materials with very few subsequent demonstrations of efficient optical device operation. Nevertheless, this is still a relatively young field of research and there is no doubt that, given time, effort and investment, it will grow and mature along the lines of its successful electro-optic counterpart.

Keywords. Electro-optic chromophores, Polymers, Waveguides, Quadratic nonlinearities, Phase-matching

1	General Introduction – NLO Materials and Optical Applications .	88
2	Second Order Nonlinear Optics – Introduction	88
2.1	Wavevector-Matching Techniques	93
2.2	Interaction Cross-sections and Figures of Merit	95
2.2.1	All-Guided Waves	95
2.2.2	Cerenkov Geometries	97
3	Important Physical and Structural Quantities	98
3.1	Material Requirements	98
3.2	Spatial Overlap Integral	99
3.3	Phase-Matching Length (Temporal Overlap)	100
4	Experimental Realization Using Different Phase-Matching Techniques	100
4.1	Cerenkov Phase-Matching	100
4.1.1	Langmuir Blodgett Media	101

4.1.2	Poled Polymers	102
4.1.3	Organic Crystals	103
4.2	Counter-Propagating Phase-Matching	103
4.3	Anomalous Dispersion Phase-Matching	104
4.4	Waveguide (and bulk) Birefringence Phase-Matching	105
4.5	Modal-Dispersion Phase-Matching	105
4.6	Quasi-Phase-Matching	108
4.6.1	Nonlinearity Modulation	109
4.6.2	Index Modulation Only	111
4.7	Non-Collinear Wavevector Matching	111
4.8	Self Light Organization	111
5	**Summary and Perspectives**	112
References		118

1
General Introduction – NLO Materials and Optical Applications

One of the most important aspects in the evolution of nonlinear optics has been the search for materials in which practical implementations of the many diverse and fascinating nonlinear phenomena can be realized. It has been known from the early days of nonlinear optics that organic materials have some of the largest nonlinear coefficients, both second and third order [1]. Chemical engineering can further optimize them in any specified direction. In the case of second order effects, charge transfer states, enhanced by electron delocalization, have produced in non-centrosymmetric structures second order nonlinearities of many hundreds of pm V^{-1}, larger than any other known material to date, including semiconductors [2]. Electron delocalization, especially in diacetylenes has also led to record large third order nonlinearities.

However, as is well known in the nonlinear optics community, just large coefficients are not sufficient for making a material useful. For all nonlinear interactions to generate large effects, optical quality is always necessary and despite great progress, material processing in organic materials does not yet compare to semiconductors, for example. For coherent interactions, such as second harmonic generation, phase-matching must also be realized which places stringent constraints on uniformity of the refractive index, and so on.

In this chapter we will specifically examine the state-of-the art in the utilization of organic $\chi^{(2)}$ materials for nonlinear optics. (Electro-optics and $\chi^{(3)}$ are treated in other chapters in this book and will not be discussed here.)

2
Second Order Nonlinear Optics – Introduction

Second order nonlinear optics involving frequency conversion encompasses a spectrum of interactions including second harmonic generation (SHG), optical rectification, sum and difference frequency generation, optical amplification etc. [3]. Of these, second harmonic generation is the simplest to implement and has historically been the most frequently studied. This is also the case in the evolution of polymers and single crystal organics for parametric interactions where the focus has been, and still is, on frequency doubling.

Second order materials have been used either as bulk media or in waveguide formats. The earliest work dealt with single crystals, including organics. It was recognized relatively early that organic molecules consisting of a strong electron donor and acceptor group separated by a strong electron delocalized bridge had strong charge transfer states that can be the origin of very large second order nonlinearities on the molecular scale [1]. Some 20% of these can be crystallized into non-centrosymmetric single crystals [4].

All coherent nonlinear interactions such as SHG require that the nonlinear polarization field generated by the mixing of two or more optical fields travels at the same velocity as the field generated by that interaction. In SHG, to allow optimum energy transfer between the fundamental and harmonic waves all along the waveguide, the phase of the locally radiating dipoles must match that of the harmonic electric field. Although efficient conversion is usually discussed in terms of phase-matching, wavevector conservation is more convenient for dealing with the different waveguide geometries and we use these nomenclatures interchangeably. That is, the harmonic wavevector must equal the vector sum of the fundamental wavevectors. We will list the different ways of achieving this requirement later.

Charge transfer states not only create large nonlinearities, but their inherently one-dimensional nature also leads to large dipole moments and substantial birefringence. On the one hand, this makes the classical birefringent wavevector-matching scheme feasible in which multiple polarizations for the interacting waves are involved. However, for low symmetry crystals, the usual case in organics, wavevector-matching requires operation away from crystal axes and frequently results in large walk-off angles between the different polarizations, an undesirable feature. The equivalent of quasi-phase-matching that allows operation along crystal axes still has to be discovered for organic crystals. Nevertheless, a number of phase-matchable crystals have been developed and interesting applications demonstrated [5].

Using waveguides for SHG has been shown to offer new options for wavevector conservation because, at a given frequency, there are a discrete number of guided wave modes with different propagation wavevectors available, and these wavevectors can be tuned by changing the waveguide geometry [6]. The simplest case is for an isotropic slab waveguide formed from a substrate of index n_s, a film of index n_f and a cladding of index n_c as shown in Fig. 1.

An example of the corresponding variation in the effective index N_{eff} (the propagation wavevector is $\beta=N_{eff}k_0$ where k_0 is the vacuum wavevector) for a number of modes "m" is shown in Fig. 2 versus film thickness.

Note that the mode number "m" indicates the number of zeros in the guided wave field, and the thicker the film, the larger the number of guided modes

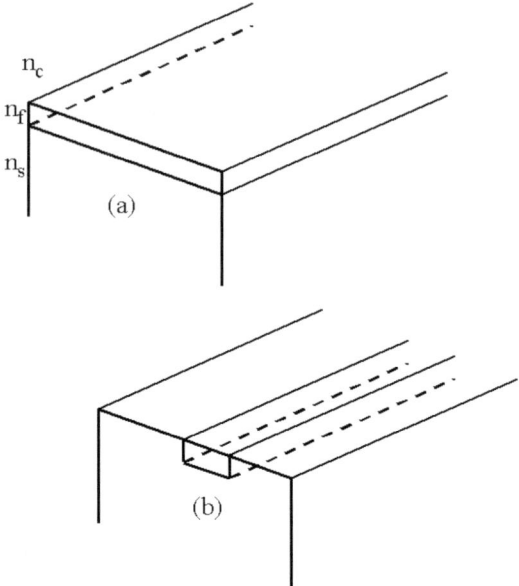

Fig. 1. Examples of the two common integrated optics waveguides used SHG in organic materials: (a) slab waveguide, (b) channel waveguide. The dimensions along the confinement directions are typically a few optical wavelengths. The core optical index made from the polymer film n_f is higher that the substrate n_s and cladding n_c indices

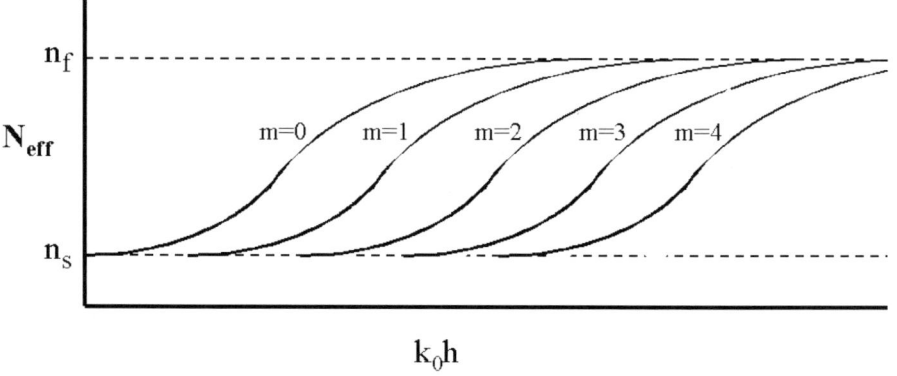

Fig. 2. Typical guided wave modal dispersion in the effective index N_{eff} versus film thickness for a slab waveguide with $n_c < n_s < n_f$

which exist. Examples of the corresponding field distributions are shown in Fig. 3, for m=0 and m=1.

There is a complete set of unique modes for each polarization, parallel (TE) and orthogonal (TM) to the film surface. For channel waveguides with optical confinement in two dimensions (like a fiber), shown in Fig. 1, the modes are de-

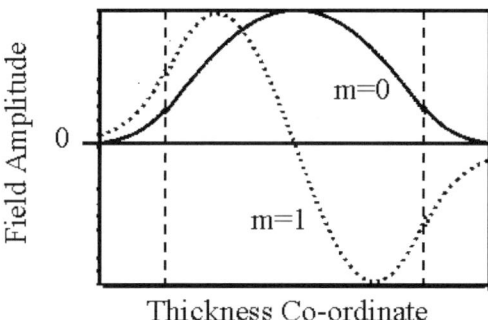

Fig. 3. Field distributions for the two lowest order TM modes. The mode number "m" identifies die number of field zeros that occur within the high index film

scribed by a pair of mode numbers (m, n) and the fields now have oscillatory distributions along both transverse axes. This variety of wavevectors can participate in the wavevector matching, an added degree of freedom. On the other hand, the product of different guided wave fields can result in interference effects during SHG between different cross-sectional regions of a waveguide and hence large reductions in the conversion efficiency: this will be discussed in detail later in terms of the overlap integral.

The applicable fundamental concepts of nonlinear integrated optics for SHG were outlined decades ago and can be found in a number of review papers [6–8]. The basic theory as applied to organic materials and polymers is of course unchanged from that for dielectric materials and these papers are still very useful. Some twenty plus years ago, nonlinear integrated optical experiments started to be conducted, but mostly on inorganics and crystals. The specific field of amorphous and semi-ordered organics came when the chemical engineering of nonlinear chromophores was developed.

Although single crystal films of a number of non-centrosymmetric organics have been grown, it has proven difficult to use them as the waveguide core for efficient SHG at specific wavelengths, in the telecommunications windows for example. The key problems are the control of the dimensional tolerances needed to achieve wavevector-matching over centimeter distances, and the optical loss due to the limited transparency windows available with organic materials. On the other hand, thin, high-index, isotropic films have been deposited on organic single crystals and the evanescent guided wave fields can be used to access the organic crystal's nonlinearity for doubling. However, because the evanescent fields only carry a fraction of the guided wave field, this reduces the conversion efficiency correspondingly.

Nonlinear polymers in which second harmonic active molecules are incorporated into a polymer matrix offer interesting possibilities for waveguide SHG by virtue of their fabrication technology. The multi-layer films needed for waveguiding can be fabricated by simple spinning techniques already in use for decades in the electronics industry, and systematic calibration of the process allows tolerances of ±10 nm to be achieved. Spin-coating of diverse films places restrictions on the solvents needed to maintain the integrity of adjacent layers, but this has not proven to be a severe problem. Furthermore, polymer mixtures can be used to fine-tune the film refractive index. The non-centrosymmetric molecules (chromophores) can be "dissolved" (called guest-host) in the host polymer matrix, or attached to it by various chemical means. When the polymer composite is heated to near the glass transition temperature, the chromophores become orientationally mobile and can be aligned. Electric field poling has proven successful for molecules with large permanent dipole moments, resulting in a $\chi^{(2)}$ oriented normal or parallel to the film surface, depending on the electrode geometry. Alignment normal to the surface is also achieved by corona poling obtained by "spraying" charge onto one surface. All-optical poling has also been implemented for molecules that undergo *trans-cis* conformational changes on illumination near their charge transfer state absorption maxima. Details will be discussed later.

Reviews addressing specifically poled polymers for frequency doubling of a laser diode have been published emphasizing the different material aspects and potential phase matching techniques [9]. The best materials available at that time were also reviewed. There has also been a more general, recent review specifically concerned with the work performed in Japan and addressing the question of materials and phase-matching techniques [10].

The analysis of organic polymer waveguides with quasi-phase matched structures has also been theoretically addressed, taken into account pump depletion and attenuation [11].

Another successful approach has been to use Langmuir-Blodgett (LB) techniques, or self-assembly methods to make films. The LB approach has been used to make oriented films with excellent registry, and even multiple oppositely ordered layers. Although SHG, and even low optical losses, have been obtained by a number of researchers, the LB approach still has to be demonstrated as a cost-effective fabrication technique.

The net result is that in the last decade there has been a great deal of activity in second harmonic generation in organic waveguides. Film fabrication techniques, especially spin-coating, have proven to be well-suited to implementing wavevector-matching techniques such as modal dispersion and quasi-phase-matching. The progress in these areas will be especially critically reviewed here.

In this chapter we deal primarily with experimental results that have been reported dealing with parametric interactions in nonlinear poled polymers, i.e. mostly on second harmonic generation in phase-matched configurations. Because the theoretical analysis associated with these processes has been known for some time and has been independently reviewed many times, we will only briefly overview these basics. Also, the polymeric materials developed for similar applications are reviewed in another section of this book and we refer the reader to that for details.

2.1
Wavevector-Matching Techniques

The key problem in all coherent interactions, such as SHG, is to achieve wavevector matching. This requires material engineering. The possible phase-matching scenarios depend critically on the relative values of the refractive indices of the materials making up the waveguide, their frequency dispersion, and the waveguide dimensions. Here we discuss the different phase-matching techniques implemented to date in organic SHG devices.

In bulk media and specifically for SHG, optimum conversion efficiency requires $\Delta k=\{\sum k_p(\omega)-k(2\omega)\pm q\kappa\}=0$ where the $k_p(\omega)$ are the individual fundamental beam wavevectors, $|\kappa|=2\pi/\Lambda$ if a periodic modulation in the index or nonlinearity exists (Quasi-Phase-Matching, QPM) and q is an integer. In polymer waveguides, many different techniques allow the wavevector-matching of the different waves interacting in a parametric process. Although the case of slab waveguides (Fig. 1a) is the easiest to implement, the most efficient interactions occur in channel waveguides (Fig. 1b) in which the input optical beams are confined in two dimensions and the output beam is also wave-guided. The (wavevector-matching) techniques include anomalous dispersion (ADM), waveguide birefringence (WBM), modal dispersion (MD, MDPM) and quasi-phase-matching (QPM), all shown schematically in Fig. 4.

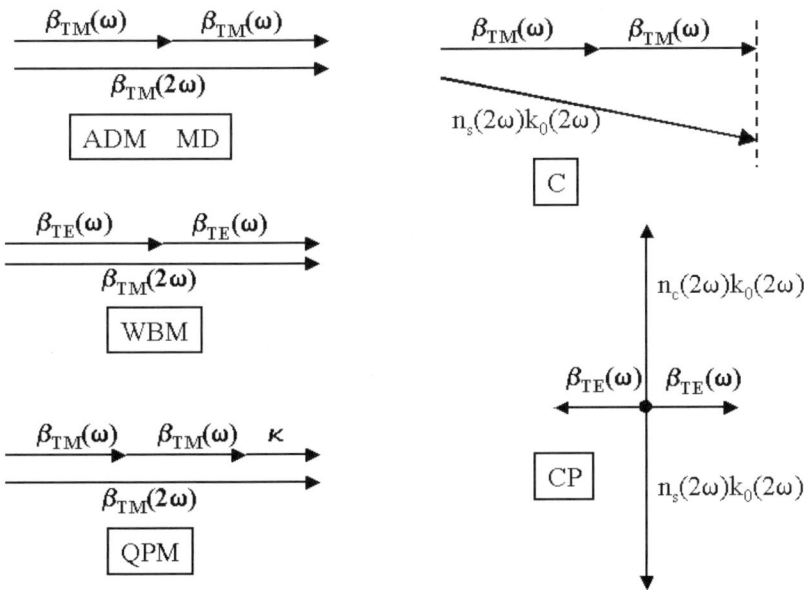

Fig. 4. Schematic representation of the different common phase-matching techniques in the k space representation. (*ADM*) anomalous dispersion; (*WBM*) waveguide birefringence; (*MD*) modal dispersion; (*QPM*) quasi-phase-matching; (*C*) Cerenkov; and (*CP*) counter propagating Cerenkov

In these cases, wavevectors must be matched along the axis of the channel, i.e. for propagation along the z-axis $\Delta\beta=\Sigma\,\{\beta_{z,p}(\omega)-\beta_z(2\omega)\pm q\kappa\}=0$ where the $\beta_p(\omega)$ are the fundamental guided wavevectors. In slab waveguides, non-collinear (NC) SHG is an option that requires two intersecting fundamental beams producing a harmonic beam in a direction intermediate to the fundamentals with wavevector conserved only in the plane of the surface, i.e. $\Delta\beta=\Sigma\,\{\beta_{2p}(\omega)-\beta_2(2\omega)\pm q\kappa\}=0$. Wavevector-matching in the plane of the surface, or along a channel axis is also possible when the signal beam leaves the waveguide region into the surrounding media, such as occurs for Cerenkov (C) and counter propagating (CP) SHG, also shown in Fig. 4. In this case $\Sigma\{\beta_p-k_{2p}\}=0$ where k_{2p} are the wavevector components of the radiation fields parallel to the waveguide surface. Only co-propagating beams (such as found in the AD, WB, MD and QPM cases) will insure quadratic growth of the harmonic with propagation length and make full use of the long interaction distances that can be provided by guided optics. Other techniques (C, NC and CP) inherently limit the possible interaction distance and are bound to be less efficient.

A number of the different wavevector matching conditions unique to waveguides can be illustrated in a plot of the effective refractive index N_{eff} versus film thickness for both the fundamental and harmonic waves, Fig. 5.

The example shows the typical dispersion in N_{eff} with both frequency and film thickness for a few pertinent TM_m modes. Note that $TM_0(\omega)$ and $TM_0(2\omega)$ never intersect, and can therefore not be phase-matched to each other. This is a consequence of the dispersion in the refractive index and due to waveguiding. Note however that $TM_0(\omega)$ and $TM_1(2\omega)$ do intersect indicating that phase-matching can take place between these two modes (which have different symmetries in the field distribution) and this is the MD case. However, for WBM it is possible to have an intersection between $TM_0(\omega)$ and $TE_0(2\omega)$ in a birefringent film. For QPM, there is a vertical effective index shift of $\pm q[\kappa/2k_0(\omega)]$ which

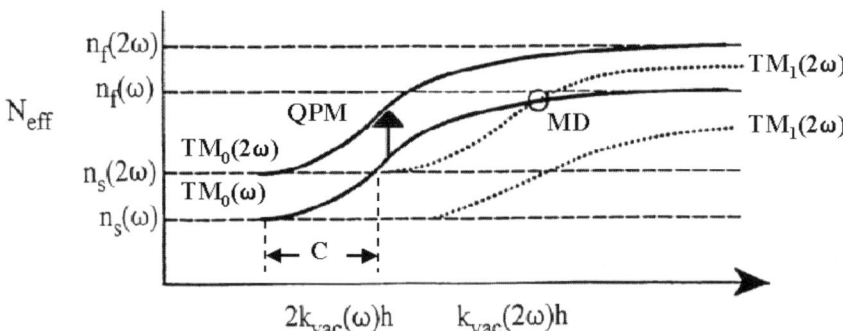

Fig. 5. Effective index of the TM0 and TM, modes at both the fundamental and harmonic frequencies versus film thickness for a slab waveguide. MD identifies modal dispersion wavevector matching for $TM_0(\omega)\rightarrow TM_1(2\omega)$. The solid vertical arrow identifies QPM for $TM_0(\omega)\rightarrow TM_0(2\omega)$. The region ←C→ identifies film thicknesses for which Cerenkov SHG into the substrate can occur

can now wavevector match $TM_0(\omega)$ and $TM_0(2\omega)$ (or $TE_0(\omega)$ and $TE_0(2\omega)$ with a different value of κ). Since the Cerenkov condition is defined by $n_s(2\omega) > N_{eff}(\omega)$, this occurs for the region marked C. Similar arguments can be made for TE polarizations. And in fact all of these wavevector matching conditions have been exploited for polymer waveguides.

2.2
Interaction Cross-Sections and Figures of Merit

2.2.1
All-Guided Waves

The SHG case, in which two photons of the same energy combine into one of twice that energy, is certainly the most experimentally investigated parametric interaction at this time. It is generally studied in the low conversion limit. In the case of a 2D channel waveguide, the second harmonic power P_2 may be expressed as a function of the material characteristics, device structure, propagation interaction length L and fundamental input power P_1 as:

$$P_2(L) = \frac{\varepsilon_0^2 \omega^2 P_1^2(0) d_0^2 L^2}{4} \left| \frac{1}{L} \int_0^L dz\, d(z) e^{i\Delta\beta z} \right|^2 \tag{1}$$

$$\left| \iint dx\,dy\, d_{ijk}(x,y) e_{1j}(x,y) e_{1k}(x,y) e_{2i}^*(x,y) \right|^2$$

In the first integral, the integrated effect of the wavevector mismatch is quantified for the lossless case and is given by $\mathrm{sinc}^2[\Delta\beta L/2]$. This is a characteristic common to all nonlinear coherent processes. As $\Delta\beta L/2 \to 0$, this integral gives one. In the presence of loss, this term is instead given by

$$e^{-(\alpha_1+\alpha_2/2)L} \frac{\cosh(\Delta\alpha L) - \cos(\Delta\beta L)}{(\Delta\alpha)^2 + (\Delta\beta)^2} \tag{2}$$

where α_1 and α_2 are the intensity attenuation coefficients at the fundamental and harmonic respectively, and $\Delta\alpha = (\alpha_1 - \alpha_2/2)$. It is also assumed that the second order nonlinearity $d_{ijk}(\mathbf{r})$ can be written as $d_0 d(z) d_{ijk}(x,y)$ where d_0 is the amplitude, $d(z)$ describes any modulation in the nonlinearity with propagation distance with $|d(z)| \leq 1$ and $d_{ijk}(x,y)$ is the transverse distribution of the nonlinearity across the cross-sectional area of the waveguide. In the second (overlap) integral calculated over the field distributions, $e_{1j}(x,y)$, $e_{1k}(x,y)$ and $e_{2i}(x,y)$ are respectively the unit vectors of the interacting electric field transverse profiles. This integral is especially important for waveguides since the guided wave fields can have both positive and negative regions. Maximizing the SHG conversion efficiency is all about maximizing the product of these two integrals.

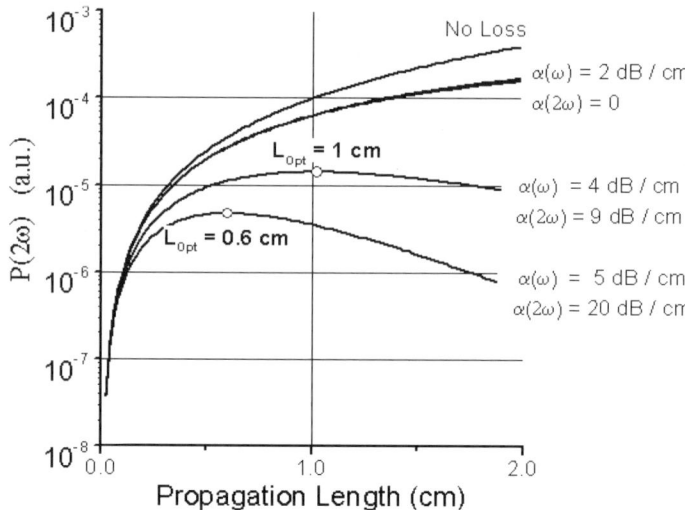

Fig. 6. Evolution of the generation efficiency as a function of propagation length (for perfect phasematching) in the lossless and a few lossy cases

A typical dependence of the conversion efficiency on loss is shown in Fig. 6.

The strong dependence on the fundamental and harmonic loss is clear. It limits the usable length of the waveguide to L_{opt}. It is clear that if 2 cm long waveguides are to be useful, losses below 1 dB cm^{-1} are needed at both frequencies!

Researching the literature, one may find a variety of ways to quantify the SHG efficiency. Here are some of the most widely used "figures of merit":

1) $\eta = P_2/(P_1 L)^2$ in % W^{-1} cm^{-2} normalizes the performance of a device to its length L.
2) η_0, also in % W^{-1} cm^{-2}, is the extrapolated value of η when L tends to 0. This allows the propagation losses to be removed. Therefore, it only evaluates the potential of the nonlinear guiding structure and is not a global Figure of merit that would take loss into account.
3) $\eta' = P_2/P_1^2$ is representative of the conversion efficiency dependence on input power for a given device.
4) $\eta_0' = \eta_0 L^2_{eff}$, with L_{eff} being the effective optimum length of the device that can be expected from the material parameters.
5) η_{Tmax} in % W^{-1} cm^{-2} is the idealized, but unachievable, value of η_0, assuming perfect phase-matching, constant field distributions over the area A of the waveguide and absence of losses.
6) Of course one can also give the net conversion efficiency: $\eta_{exp} = P_2/P_1$ in % for a given input power. This parameter is of minimal value since it depends on the input power.

There are some minor modifications to the above for a slab waveguide with confinement only along one dimension, for example along the y-axis. Specifically the (x,y) dependencies are replaced by (y), and the powers P_2 and P_1 are the power per unit distance along the wavefront (x-axis). Because of this different power unit, the units associated with η, η_0 and η_{Tmax} are now % W^{-1} cm^{-1}, and η' and η_0' have the units % cm W^{-1}.

2.2.2
Cerenkov Geometries

In this case the output field is a radiation wave which leaves the vicinity of the waveguide, i.e. the output field is not guided. For co-propagating fundamental beams it has been shown that the Cerenkov SHG signal grows initially quadratically and then later linearly with distance. A comparison of the growth of the SHG power for different phase-matching cases is shown in Fig. 7.

The cross-over point between the two different length dependencies for Cerenkov is typically a fraction of a millimeter and hence in most practical cases $P_2 \propto P_1^2 L$ [12]. Note that e_{2i} in the overlap integral now corresponds to a plane wave field multi-reflecting between the waveguide boundaries, and P_2 is a plane wave leaving the waveguide region. Because wavevector is conserved in the plane of the film, $k_2(2\omega)=2\beta(\omega)=n_s(2\omega)k_0(2\omega)\cos\theta_{PM}$ where θ_{PM} is the radiation angle into the substrate measured from the z-axis (Fig. 4). Here the figure

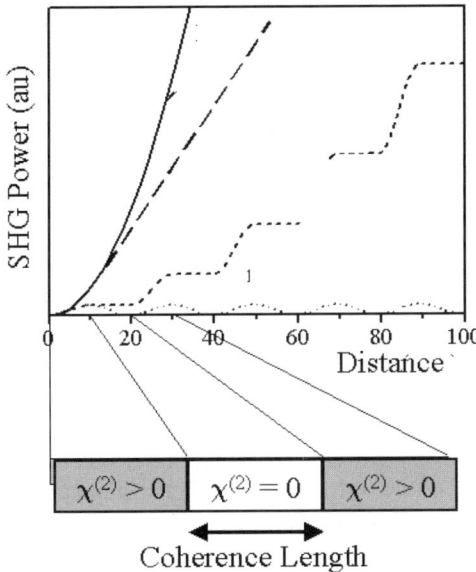

Fig. 7. The evolution of the harmonic power with propagation distance (arbitrary units) for ADM, WBM and MD phase-matching (*solid line*). The Cerenkov power changes from quadratic (*solid line*) to linear long (*dashed line*) growth. For QPM for alternate sections with no $\chi^{(2)}$ (*short dashed line*). The phase-mismatched case is also shown (*dotted line*)

of merit is $\eta = P_2/P_1^2 L$ in % $W^{-1} \cdot cm^{-2}$, and all of the other figures of merit are changed accordingly.

For the second Cerenkov case of two oppositely propagating guided waves (CP case in Fig. 4) with powers P_{1+} and P_{1-}, (in W), a sum frequency wave (usually SHG) traveling approximately (exactly) at right angles to the waveguide surface is generated. Here L is given approximately by the waveguide thickness along the radiation axis. The figure of merit used is $P_2 = A^{NL} P_{1+} P_{1-} D/W$ where D is the length of the waveguide in which the fundamental guided waves overlap and W is the width of the waveguide and the details of A^{NL} can be found in reference [13].

3
Important Physical and Structural Quantities

The first important parameter for efficient SHG that usually comes to mind is the nonlinear coefficient that is the basis for $\chi^{(2)}$ parametric interactions! But, of nearly equal importance are the losses: absorption losses of the material and more generally propagation losses of the waveguiding structure at both the fundamental and harmonic wavelength. Also governed by the material and structure is the spatio-temporal overlap between the three propagating waves, usually separately referred to and quantified in terms of a spatial overlap integral and phase-matching length. All these quantities may be combined into material and structural figures of merit.

The preceding general equation for the harmonic power was written as a product of three terms. The first concerns the local evolution of the waves, i.e. the nonlinear material coefficient and the propagation loss coefficients. A second one concerns the global effect along a given phase front, i.e. the spatial overlap integral. Last but not least, a third one deals with the accumulation of the parametric interaction effect with propagation length, i.e. the phase-matching integral. We shall now briefly discuss these three points, and their associated requirements.

3.1
Material Requirements

At the heart of any nonlinear device is the optically nonlinear material that must be able to not only produce an efficient nonlinear interaction governed by a $\chi^{(2)}$ coefficient, but should also provide low loss, preferably quasi-lossless, propagation for the interacting waves. In the case of co-propagating interacting waves, not only must the material be transparent at the fundamental wavelength λ but also at its second harmonic $\lambda/2$ as illustrated already in Fig. 6. This translates into a much more severe loss requirement than for the classical electro-optic effect with the same input wavelength. The material figure of merit in this case can be approximated by $[\chi^{(2)}/(\alpha_1 + (\alpha_2/2))]^2$ where α is the intensity absorption coefficient. Much work has been devoted to the well-known trade-off between the nonlinearity and the location λ_{max} of the absorption maximum associated with the charge transfer state. The larger the nonlinearity the larger wavelength shift of λ_{max} towards the near infrared. Because the absorption due to overtones of the C-H vibration places an upper limit of 1600 nm on transparency, the larger the nonlinearity the smaller the transparency window of the polymer. This means that for all-optical signal processing applications such as cascading at 1550 nm, the harmonic wavelength always occurs in the "red tail" of the absorption spectrum. To date little attention has been devoted to the red tail of the absorption spectrum and the impact of the host matrix on material loss there, as recently pointed out [14]. Proper choice of the host matrix can make a difference of an order of magnitude in the harmonic loss.

The nonlinear material(s) must also be combined with linear material(s) in order to form the total waveguiding structure, typically consisting of three or more layers (2 buffer layers and as many core layers as necessary, discussed later). These material systems must fulfill many technological conditions [15]. For example, they must be deposited in multi-layers, must facilitate the processing of channel waveguides, and the resulting structures must not lead to high propagation losses due to scattering. Of course the combination of materials must also allow efficient poling of the nonlinear doped polymer around its associated glass transition temperature T_g [16,17]. Furthermore, these materials and structures have to be stable with time under storage and operational conditions. Especially, the question of photostability of the chromophore-doped material, often considered as its Achilles' heel, has to be addressed [18]. For the second harmonic where the waveguided intensities are higher than for electro-optical devices, photostability is a much more difficult requirement to satisfy [19, 20].

3.2
Spatial Overlap Integral

The second integral in Eq. 1 highlights the waveguide trade-off between having multiple wavevector options for phase-matching, and the efficiency of the interaction. Assuming only one input fundamental guided wave, that field distribution appears squared and is always positive in this integral. Therefore any interference effects depend on the symmetry properties of the harmonic field. For MD phase-matching which requires a higher order harmonic mode, the integral is always reduced relative to the plane wave case and the ADM, WBM and QPM case whereas both the interacting waves can be in the lowest order mode.

However, the electric field poling technology offers opportunities to maximize this integral for MD wavevector-matching. The integrand has positive and negative regions as determined by the field structure of the harmonic. One option is to spin-coat polymers with no second order activity in the regions of opposite sign of the integrand so that interference does not occur in the integral. Another, superior option is to actually reverse the sign of the nonlinearity by using sophisticated poling techniques. This is illustrated in Fig. 8.

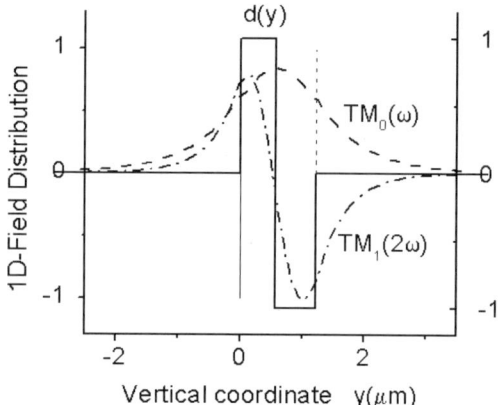

Fig. 8. An approach to increasing the overlap integral via modal dispersion phase-matching. The sign of the nonlinearity d(y) is reversed at the same point as the $TM_1(2\omega)$ field. The *solid lines* show the nonlinearity for $TM_0(\omega) \rightarrow TM_1(2\omega)$.

3.3
Phase-Matching Length (Temporal Overlap)

In any SHG device, the group velocity is different at the fundamental and harmonic wavelengths, so that temporal pulses at these wavelengths will propagate with different velocities. As a result, the two pulses will separate from each other, changing the nonlinear interaction. The principal effect will be to produce an elongated harmonic pulse of reduced intensity relative to the no temporal walk-off case. The walk-off length L_W depends on the pulse width τ, on the difference in the effective index at the two wavelengths ΔN_{eff}, and on its dispersion with wavelength:

$$L_W = \frac{c\tau}{\lambda^2}\left|\frac{\partial}{\partial\lambda}\left(\frac{\Delta N}{\lambda}\right)\right|^{-1} = \frac{c\tau}{|\Delta N - \lambda\partial\Delta N/\partial\lambda|} \xrightarrow{PM} \frac{c\tau/\lambda}{|\partial\Delta N/\partial\lambda|} \qquad (3)$$

For ps pulses in common side-chain polymers such as DANS and DR1, this length is typically a few centimeters, and this effect must be taken into account for cascading applications.

4
Experimental Realization Using Different Phase-Matching Techniques

4.1
Cerenkov Phase-Matching

We start by reviewing specific results obtained with the Cerenkov technique for which phase-matching is relatively easy, making this a very popular approach.

As discussed previously, Cerenkov SHG will occur into any waveguide bounding medium with film thicknesses for which $n_{s,c}(2\omega) > N_{eff}(\omega)$. In the example previously discussed for a slab waveguide, it was implicitly assumed that $n_s(2\omega) > N_{eff}(\omega)$. However, if $n_c(2\omega) > N_{eff}(\omega)$ also holds for the cladding, then radiation will also occur at an angle θ_{PM} into the cladding medium as long as $\cos\theta_{PM} = N_{eff}(\omega)/n_c(2\omega) < 1$. For the case of a fiber, the harmonic will emerge in a cone. For more complex waveguide geometries such as channels, radiation will be generated into every region in which the condition $n(2\omega) > N_{eff}(\omega)$ is satisfied. Although the condition for which the wavevector parallel to the waveguide axis is conserved predicts approximately the peak radiation direction, a complicated radiation pattern is produced since the source of the radiation is confined in two dimensions. That is, the channel waveguide acts as a radiation source of finite extent which will lead to a spreading diffraction pattern, and also multi-reflections and interference effects occur at the waveguide boundaries [21]. Details of the calculations of the radiation patterns can be found in [12, 22]. For example, the latter paper considered the configuration case of 2-methyl-4-nitroaniline (MNA) and 4'-nitrobenzilidene-3-acetoamino-4-metxianiline (MNBA) embedded in SF11 high index glass. The complex harmonic field patterns hinder device applications [21, 23]. However, the ease of achieving phase-matching and because the harmonic radiation leaves the waveguide region which maybe strongly absorbing at 2ω, this technique has proven useful in demonstrating the nonlinearity of many materials ranging from Langmuir-Blodgett films to organic crystal fibers, etc.

4.1.1
Langmuir Blodgett Media

There are many examples of SHG from Langmuir-Blodgett films using the Cerenkov phase-matching experimental set-up, both with the film forming the waveguide core and with a LB film deposited as a cladding layer. Early work has been already reviewed [24, 25]. In the first paper, material requirements are especially emphasized. In the second paper use of 2–docosylamino-5-nitropyridine (DCANP) which forms accordion-like layers was reported [26]. This paper also addresses the question of comparing with modal dispersion phase matching for which thickness requirements are fairly easily addressed for this type of film preparation. Many further experiments followed.

A number of other papers introduced interesting ideas. All polymeric Langmuir-Blodgett waveguide cores made from 4-(dimethylamino)-4'-(methylsulfonyl)azobenzene as the nonlinear material and poly(*tert*-butyl methacrylate) as the linear one were used [27] in order to eliminate destructive interference between different waveguide regions in a kind of transverse QPM structure. This concept was extended by Asai who inverted the nonlinearity in the thickness direction and obtained an enhancement of 20 to 30 relative to the conventional structure in DCANP [28]. Non-centrosymmetric Langmuir-Blodgett multilayers were prepared by alternate deposition of nitrophenylhydrazone (dye FA06) and trimethylsilylcellulose (TMSC) which resulted in a relatively low loss of 3.5 dBcm^{-1} at 633 nm and a reasonable $\chi^{(2)}$ of 19 pm V^{-1}. In another paper with a fundamental wavelength of 1064 nm, a conversion of 10^{-4} corresponding to a slope efficiency of 10^{-8}% W^{-1} cm^{-2} was obtained, although the reported nonlinear coefficient was about 19 pm V^{-1}. More recently, a Langmuir-Blodgett film was deposited as an overlay on the polished cladding face of a single mode fiber [29]. Using a 10 ns pulse Nd:YAG laser, in the proper temperature tuned conditions, a quadratic relationship between pump power and radiated second harmonic intensity was demonstrated.

4.1.2
Poled Polymers

As mentioned previously, the Cerenkov phase-matching technique was also used with poled polymer films. Examples include the well-known and extensively studied Disperse Red One (DR1) azobenzene chromophore in both classical organic polymers such as methacrylates [30, 31] and hybrid organic/inorganic sol-gels [32]. This technique allows up to 1% conversion efficiencies even though the second harmonic wavelength at 532 nm is near the chromophore's absorption peak at 480 nm. Great improvements were reported later by increasing the DR1 loading up to 22, and even 40 wt %, in hybrid inorganic/organic sol-gel matrices, corresponding to a d_{eff} of 75 and 110 pm V^{-1}, respectively [33]. In another paper a ridge waveguide structure was realized by photolithography and Ar sputter-etching, giving a 0.2% efficiency (62× enhancement over the slab geometry) for an average fundamental input power of 1 mW (Nd:YLF laser with 16 ns 1047 nm pulses operating at 1 kHz) and 5 mm long propagation. This specific material continues to be studied, for example an efficiency of 10^{-5} was reported in doubling a 1064 nm laser using a 0.91 µm thick corona poled DR1-methylmethacrylate-co-methylmethacrylate slab waveguide [34, 35].

Other chromophores were also investigated this way, example includes cyclobutadiene based doped polymer, of about 8 pm V^{-1} d_{33} nonlinearity with about 10 wt % loading, for green 532 nm generation and poled pNAn-PVA for blue 443 nm generation, in this case below the 480 nm cut-off wavelength of the material [36, 37].

Progress has also been reported with epoxy-based polymers. For example, bisphenol-A diglycidyl ether and a nonlinear chromophore 4-aminobenzonitrile based polymer, referred as BADGE-ABNC, had transparency down to 400 nm and an efficiency of 1.4×10^{-3}% W^{-1}cm^{-2} for doubling 1064 nm, 10 ns fundamental pulses [38]. Similarly, diglycidyl ether of bisphenol-A and 4-(4'-nitrophenylazo)phenylamine (Disperse Orange 3) based materials, referred to as DGEAB-NAC or DGEBA-DO3 C, were used to double both a pulsed Nd:YAG and a continuous wave 830 nm diode in a slab waveguide, with an efficiency of 6×10^{-3}% W^{-1}cm^{-2} (NAC compound) and 0.03% W^{-1} (DO3C) [39]. Further work with DGEAB-NAC in patterned channel waveguides by the same group yielded an efficiency 0.1% W^{-1}cm^{-2} for doubling the Nd:YAG laser [40].

A number of new ideas have been put forward recently. A ×7 enhancement of the generated second harmonic power was obtained by ultraviolet patterning to produce a chirped periodic nonlinear susceptibility on a 60 µm wide channel waveguide [41]. A complete analysis of theory and experiment is given, illustrating the potential of this new technique, very competitive with the contact electrode method developed earlier [42], and first introduced without the chirping and with an enhancement factor of about 1.5 to 2 by the same group [43, 44].

Also, a high value of 0.2% $W^{-1} cm^{-2}$ was experimentally reported using a specially designed three layer configuration optimized with respect to polymer layer thickness and substrate parameters [45, 46]. A recent paper has proposed capturing the Cerenkov radiation emitted from a sol-gel nonlinear waveguide in a secondary waveguide through a graded index layer produced by ion exchange into glass, lying on top of an ion-etched grating coupler [47]. Theoretically, more than half the harmonic energy may be recovered that way.

Using a photo-conducting polymer, self-organization of the polymer was induced. Quasi-phase-matching was achieved with a refractive index grating that was gradually formed under illumination by the guided fundamental beam [48]. The pump light generated space charges that periodically screened the poling electric field and thus reduced locally the induced reorientation of the nonlinear chromophore and hence both the nonlinear coefficient and the refractive index, which formed an antiguide structure confining the leaky lightwave [49, 50].

4.1.3
Organic Crystals

Cerenkov doubling was also investigated in nonlinear organic crystals. The best results to date have been obtained with single crystals grown inside glass fibers. For example, DMNP was used to double into the blue and figures of merit of 40% $W^{-1} cm^{-2}$ were obtained [51]. Unfortunately, the photostability proved to be the key problem.

The use of a grating to facilitate the out-coupling has been demonstrated [52]. In that work, a (methyl-3-(*p*-nitro-phenyl)carbazate, MNCZ) crystal was used as the substrate with a slab transparent waveguide made of photoresist deposited on top. The grating coupler was fabricated holographically and the SHG of a Nd:YAG 1.06 µm laser demonstrated. This nonlinear crystal has a nonlinear d_{36} coefficient of 22.7 pm V^{-1} (3 time larger than d_{24} of KTP) and a cut-off wavelength of 455 nm. Reported conversion efficiencies were at that time limited by lack of sufficient crystal flatness.

4.2
Counter-Propagating Phase-Matching

This counter-propagation Cerenkov experiment is easy to perform because one has only to launch the fundamental beams at both ends of the guiding structure [53]. This technique is very stable and puts very few constraints on the material or waveguide structure. It's efficiency can be surprisingly large for a Cerenkov process when the harmonic occurs near λ_{max} which is the same for both the maximum resonantly enhanced $\chi^{(2)}$ as well as the maximum harmonic absorption coefficient. The key is that the harmonic beam only grows a distance equal to the waveguide thickness, which is only of the order of a wavelength.

The generated harmonic represents the convolution of the two input pulses [54]. This can lead to many applications, including the measurement of the duration of short pulses. It can also serve the purpose of a digitizer [55]. In the more general case where the two fundamental beams do not have the same wavelength, the emission is tilted into the direction of the lower wavelength propagation direction. It has been suggested and demonstrated that such a configuration could be used as spectrophotometer of short pulses for potential application in WDM systems.

The initial poled polymer demonstration was with a single film of DANS doped material and it produced a figure of merit $A^{NL}=9\times10^{-9}$ W^{-1} with 100 ps, Nd:YAG pulses. Increasing the poling field to 200 V μm^{-1} later increased this to 8×10^{-7} W^{-1} [56]. Recent results using a transverse quasi-phase-matched core structure of up to 7 alternating layers of nonlinear and linear material, together with an optimized in-plane poling field in excess of 300 V μm^{-1} yielded a record SHG figure of merit efficiency of $2.3\times10^{6} W^{-1}$ [57]. This work was also performed with a multilayered structure using Langmuir-Blodgett DCANP films [58].

4.3
Anomalous Dispersion Phase-Matching

Anomalous dispersion wavevector matching (ADM) is a general technique that consists of generating a second harmonic on the low wavelength side of an absorption feature with a fundamental beam input on the long wavelength side. The standard dispersion in the refractive index caused by the absorption peak allows wavevectors to be matched for SHG. This method is well suited to polymer doped chromophores with spectrally narrow charge transfer states surrounded by wavelength regions of low absorption. Due to the proximity of the charge-transfer state, the nonlinear coefficient $\chi^{(2)}$ can be resonantly enhanced, but that same proximity implies that the linear absorption may be high. There is a direct nonlinearity-attenuation trade-off. A distinct advantage is that the lowest order mode can be used for both the fundamental and harmonic beams leading to a large overlap integral. However this technique suffers a dramatic drawback: the fundamental frequency is uniquely fixed by the material's optical index dispersion. Also very problematic is the blue side absorption at 2ω, which is a difficult to make small. A possible solution is to use a material with the large absorption in the cladding layer.

Although the basic principle dates back to the early days of nonlinear optics [59], it was only implemented three decades later with chromophores oriented by an electric field in solution [60, 61]. The chromophore was Foron Brilliant

Blue, chosen because it's absorption was peaked at 620 nm and because it had a deep absorption minimum around 450 nm. In later work, a thiobarbituric acid derivative in PMMA produced an efficiency of 39 % W^{-1} cm^{-2} over an estimated 32 μm absorption length [62]. This was followed recently with new tricyanovinylaniline doped polymers in which a net efficiency of nearly 250% W^{-1} cm^{-2} was demonstrated over 35 μm. [63] An added bonus was that this approach proved to be the stimulus for many materials to be characterized. Materials with μβ values as large as $3,800\times10^{-48}$ esu, coupled with good transparency in the 450–470 nm window, were synthesized and optimistically predicted to satisfy criteria for ADPM SHG with target efficiencies of about 2000% W^{-1} [64]!

4.4
Waveguide (and Bulk) Birefringence Phase-Matching

Waveguide birefringence wavevector matching (WBM) is a direct copy of the well-known technique for bulk materials, adapted to waveguides [65]. In Type I that requires only one fundamental input beam, the birefringence is used to generate a phase-matched harmonic in the orthogonal polarization. This approach is far less restrictive than ADPM. However, it requires off-diagonal elements of $\chi^{(2)}$ which tend to be smaller than the diagonal ones in crystals, with some notable exceptions like NPP [1]. For poled polymers deposited on substrates, it is difficult to generate a large enough birefringence just due to poling for phase-matching, especially due to the limited transparency window. Furthermore, the off-diagonal element is only one third of the diagonal one. However, it has proven possible to mechanically stretch a main chain polyurea polymer film to achieve phase-matching. [66]. The most noticeable features of this technique are that it does not require critical control of the initial dimensions of the waveguide and that the birefringence needed for phase matching can be tuned by varying the draw ratio.

4.5
Modal-Dispersion Phase-Matching

Modal-dispersion phase-matching, as discussed earlier, consists of wavevector matching from a low order fundamental spatial mode, usually the zero'th mode, to a higher order harmonic spatial mode, typically the first or second order one. This was illustrated in Fig. 5. The drawback is that the lowest order spatial mode cannot be used for the harmonic, which reduces the overlap integral due to interference effects, although the cross-sections are still larger than that for the Cerenkov technique. However, this can be overcome by suitably modifying the transverse distribution of the nonlinearity to optimize the overlap integral of Equation 1 as discussed previously. Another disadvantage is the complicated spatial structure of the harmonic beam which is not in the zero'th order waveguide mode, although this is not a disadvantage for cascading applications in which it is the fundamental beam that is the output [67].

It is with this in mind that recent experiments were conducted at telecommunication wavelengths on poled polymer channel waveguides. Early work on organics focused on single material waveguide films. The film thickness was tapered perpendicular to the propagation direction to vary the wavevector matching condition. For example on top of a tapered slab glass waveguide a thin single crystal of MNA had been grown by vapor deposition [68]. Modal-phase matching in a thick film was obtained between the first and third order spatial modes at 1.06 µm with a net efficiency of 0.5% with a fundamental irradiance of 300 kW·cm^{-2} [69]. Poled, tapered films of 2-methyl-nitroaniline (MNA) in poly(methylmethacrylate) (PMMA) gave a conversion efficiency of 7.4×10^{-3}% over 1 mm for 43 W of zero order mode 1064 nm and a third order harmonic mode [70]. In later work, a more nonlinear chromophore at 10 wt %, namely CDI, was used and a figure of merit of 9×10^{-9}% W^{-1} was obtained, higher than the preceding work and corresponding to 1.4×10^{-10}% W^{-1}, but only 1/100 of the expected value because of losses [71, 72].

The phase-matching thickness of a main chain, poled, polyarylamine polymer was also controlled by applying an electric field, tuning the thickness by 25 nm [73]. With the very reasonable optical propagation losses of 2.7 dB cm^{-1} at 633 nm, this approach should be revisited in the near future [74].

Optimizing the spatial overlap is of course the key to MD. In the case of a symmetric waveguide containing a uniform nonlinear material, this overlap integral is rigorously equal to zero if the matching is made from the fundamental zero'th spatial mode to a second harmonic odd-number spatial mode, including the first one. An option that yields a non-zero but reduced overlap integral is to use even order harmonic modes [75]. A better option is to wavevector match to the first order spatial mode with basically a different layer for each of the field lobes, using for one film a passive transparent high index material, or a nonlinear material of opposite nonlinearity to the other lobe [76]. The second case is shown by the example in Fig. 8. This requires complex poling protocols. In the first case, only half of the guiding layer is nonlinear and the overlap integral is divided by two (note that this can however be partially re-optimized by varying the index mismatch between the optical core indices). In the second case full use of the nonlinearity can be made. The easiest method to realize such a structure is through opposite poling directions for the two core layers. One may either pole the first layer, cross-link it, and then deposit and pole the second one, or use two different T_g materials. In this latter case, one then poles simultaneously both layers at the high T_g temperature, decreases the temperature to the low T_g value and then reverses the poling field direction [77]. An example of such a structure may be found on Fig. 9.

A normalized conversion efficiency of 7% W^{-1}cm^{-2} was achieved in a first experiment. Although very high propagation loss of about 100 dB/cm at the 800 nm second harmonic wavelength resulted from the repoling, these figure of merit values were further increased to 14% W^{-1}cm^{-2} for a 1 mm device [78]. It is probable that the interface between the two oppositely poled regions is the source of the loss. Another "theoretical" route that requires special materials

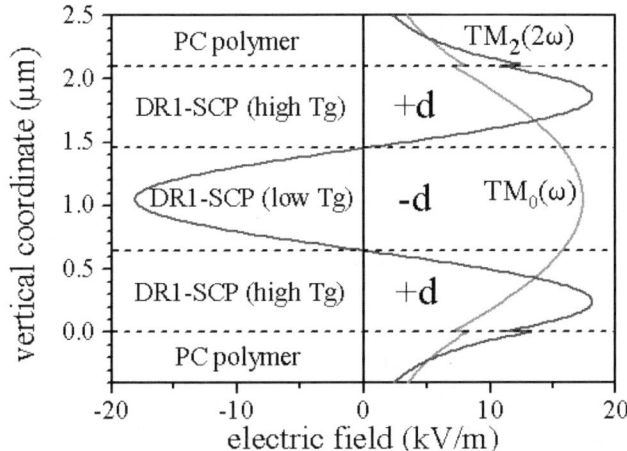

Fig. 9. An example of a structure that epitomizes the overlap integral using materials of different T_g, allowing adjacent layers to be poled in opposite directions

would be to use two materials that would give an efficient nonlinearity of opposite sign with a single poling field.

Other researchers have also used two layer cores with a polymer film to optimize the overlap integral. A two layer hybrid core containing a polymer (650 nm of spincoated MSMA, 4-alkoxy-4'-alkylsulfone stilbene side chain methyl methacrylate copolymer) and a very thin and high optical index layer (75 nm of Si_3N_4) gave a maximum conversion efficiency of 0.4% $W^{-1}cm^{-2}$ in etched channel waveguides, mainly limited by the small poling field which occurred across the nonlinear layer (d_{33} = 0.004 pm V^{-1}) [79]. When using chromophores whose absorption dominates the propagation losses at the second harmonic, a core consisting of a nonlinear and passive material works well. Clay et. al. used a thin nonlinear layer, about a quarter of the total film thickness, to overlap the central lobe of the 2nd order harmonic mode, and a transparent one to achieve guiding in essentially a transparent structure with no destructive SHG interference in the overlap integral [80]. Using a styrene chromophore, losses of 12 and 30 dB cm^{-1} were obtained for the harmonic in the 1 and 2 modes respectively. Most recently, reactive-ion etched waveguides containing DR1 produced a figure of merit of 0.1% $W^{-1}cm^{-2}$ for 2 mm long waveguide using only a modest 3 pm V^{-1} nonlinearity [81].

Langmuir-Blodgett film deposition provides very accurate control of the different film thickness, including non-centrosymmetric materials. This approach was used to generate blue light as early as 1994 [82]. The LB molecular layers were deposited to form two planar layers with oppositely directed nonlinearities and a figure of merit of 150% $W^{-1}cm^{-2}$. In a similar experiment, a fundamental at 1064 nm produced a figure of merit of 1% $W^{-1} cm^{-2}$ for a 1 mm long slab waveguide, an improvement over prior work with a film of single orientation. [83, 84].

Non-centrosymmetric Langmuir-Blodgett multilayers were also prepared by alternate deposition of nitrophenylhydrazone (dye FA06) and trimethylsilylcellulose (TMSC) [85]. Such waveguides had a relatively low loss of 3.5 dB cm^{-1} at 633 nm and the nonlinear material exhibited a reasonable $\chi^{(2)}$ of 19 pm V^{-1}. With a fundamental wavelength of 1064 nm, a conversion efficiency of 10^{-4}, corresponding to slope of the order of 10^{-8}% W^{-1}cm^{-1} was demonstrated and further work was proposed to diminish the optical losses.

The thickness requirements for this technique are very stringent, as well as for all of the other methods that incorporated guided fundamental and harmonic waves. However, careful design can reduce the critical constraints on the waveguide dimensions. This "non-critical phase-matching" (NCPM) case occurs when the two modal dispersion curves (in Fig. 5) are tangent thanks to special choices of the optical indices of the materials used [86, 87]. A particular example of this concept consisted of two inorganic core layers (Ta$_2$O$_5$–SiO$_2$ on BK7) onto which was grown a cladding of an organic IAPU crystal. The nonlinear interaction occurred in the evanescent field and the spatial overlap integral is much reduced: It can however be optimized by operating near the 2ω cut-off [88].

Another configuration was proposed using two nonlinear poled polymer layers pasted one onto another leaving a gap of air in between [89]. This extra layer, localized where the first spatial mode second harmonic electric field changes sign, reduces significantly the thickness tolerance on each layer.

Using conventional spin-coating technology, phase-matching has been demonstrated with a 2 layer core structure made of DANS doped polycarbonate and polyetherimide over 7 mm of propagation [90].

Recently modal dispersion was demonstrated with a composite structure made of a nonlinear poled polymer and an ion-exchanged glass [91]. Phase-matching was achieved by varying the ion-exchange diffusion time and, for a typical $\chi^{(2)}$ nonlinearity of about 7 pm V^{-1}, a net conversion efficiency of 0.45% was achieved using a 4 kW input average power and 1 mm propagation length. This new configuration did not take full advantage of the nonlinearity/transparency trade-off of known doped polymers and changing from PNA to another more red-shifted chromophore should further increase the efficiency. Calculations show that parametric amplification should be possible in this waveguide structure.

For blue green generation, a new polymer, poly[(4'-acetylphenylaminoformyloxy-4-styrene)-co-(4-hydroxystyrene)], was developed with an absorption cut-off wavelength of 350 nm. Using a 691 nm thick film, doubling from 1064 to 532 nm was demonstrated with an experimental figure of merit of about 4×10^{-6}% W^{-1}cm^{-2} for an interaction length of 0.2 mm [92].

4.6
Quasi-Phase-Matching

As in the preceding case, this wavevector matching technique (QPM) depends on structural modifications to waveguides to be efficient. It requires fabricating a periodic structure that contributes to wavevector matching by either modulating the refractive index or the nonlinearity along the propagation direction [93]. The highest SHG efficiency occurs when the nonlinearity is fully modulated between ±d. Nulling the nonlinearity over every half-period is 4 time less efficient. The least efficient are surface gratings. This technique, proposed in the very early days of nonlinear optics [59], has been applied with great success to ferroelectrics like $LiNbO_3$ [94]. The limiting factor is the period required for wavevector matching. It is technologically difficult to fabricate periods less than a few microns for nonlinearity modulation.

4.6.1
Nonlinearity Modulation

For organic waveguides, an early demonstration was done using liquid nitrobenzene and "live" periodic electrode poling [95]. More technology relevant is the modulation of the linear or nonlinear properties of polymer films, either by photobleaching or by periodic poling. In the first case the main advantage is the ease of fabrication, the main drawback being the induced propagation losses due to coupling to radiation fields via the associated index modulation. This drawback will be especially dramatic in the case of 2D waveguides with lateral confinement obtained by the same bleaching process. In the case of the periodic poling, a similar problem will be encountered if a single poling direction is used to obtain a +/0 type of structure. Here p/q refers to a two component core layer with nonlinearity >0 for the p'th layer and 0 for the q'th layer. The proper solution, which allows modulation of the nonlinear coefficient but not of the optical index, really consists of achieving a +/- type of structure. Such a structure requires poling by a system of at least two-interdigited periodic electrodes. It was shown that for practical reasons these electrodes should preferably be deposited on the substrate side and the flat counter-electrode on the top of the polymer stack. This design minimizes the electrostrictive problems which are encountered in the opposite geometry (interdigited electrodes on top) in which the electrodes tend to "sink" during the poling process (due to the combination of a softened material near T_g and a high electric field). Three of these different approaches, i.e. bleaching and interdigited electrodes on top or under the polymer were used with the same DANS side chain polymer, confirming the preceding comments [76].

The first QPM polymer work in 1990 used a stilbene derivative and periodic poling to demonstrate conversion from 1340 to 670 nm with an efficiency of about 0.01% W^{-1} in 5 mm long waveguides [96, 97]. Many different issues were reported, including some rare temperature tuning results. This stimulated analysis of design, synthesis and evaluation of chromophores for QPM second-harmonic generation showing that a $\chi^{(2)}$ of the order of 60 pm V^{-1} would be necessary to achieve a 10% conversion efficiency [98]. Other early work includes the use of a vinylidene cyanide/vinyl acetate copolymer to demonstrate theoretically and experimentally the enhancement factor that QPM can bring to the field of

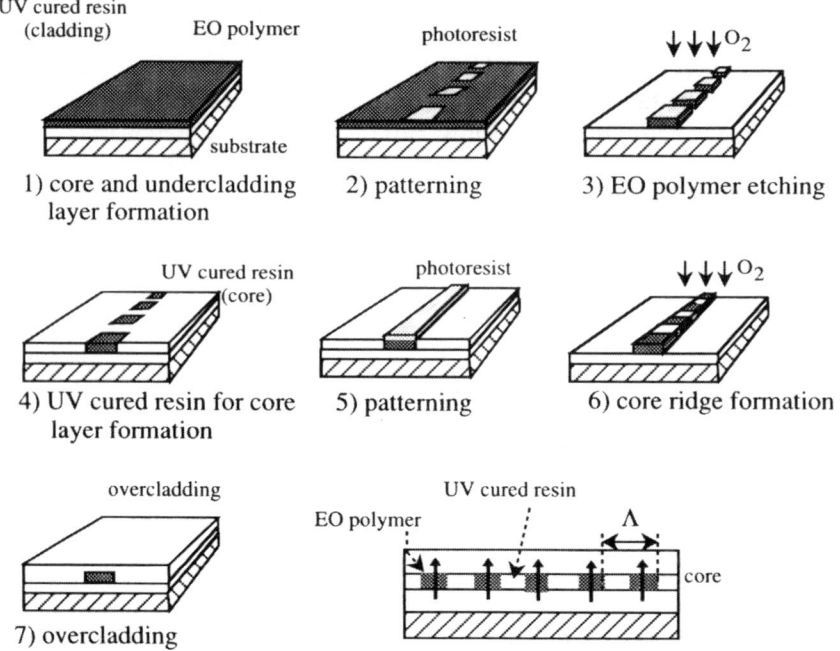

Fig. 10. A successful example of multi-step process used in fabricating a low loss +/0 QPM structure

integrated nonlinear poled polymers [99, 100]. The QPM technique was also used in a novel way for bulk SHG by fabricating stacks made from 43 layers of freestanding polymer thin films of appropriate thickness [101]. A 4-oxy-4'-nitrostilbene and methyl methacrylate copolymer was used and an efficiency of the order of 10^{-2}% was reported with a total thickness of 150 µm.

A number of experiments have been reported using azo dyes. Photobleaching of DR1 by respectively long and short exposures to UV was used to fabricate channel waveguides and $\chi^{(2)}$ gratings in both organic polymer and hybrid inorganic/organic sol-gel doped matrices [102]. Pulsed poling using interdigited electrodes, patterned on an ITO covered substrate by etching, was demonstrated using the DO3-MMA side chain polymer, resulting in a nonlinear coefficient modulated between 5 and 8 pm V^{-1} [103].

The most impressive QPM work to date was obtained by the complex procedure summarized in Fig. 10.

A photoresist pattern was deposited on a nonlinear diazo dye film and then oxygen etched to form a line of nonlinear and linear sections with a 32 µm QPM period. The upper surface was then planarized with a UV cured resin. A raised QPM +/0 channel waveguide was then formed with three more processing steps [104]. Despite the complexity of the fabrication, losses were below 3 dB cm^{-1} at 1.59 µm and a maximum SHG figure of merit of $4\times10^{-}$% W^{-1} cm^{-2} was obtained [105, 106].

4.6.2
Index Modulation Only

Although proposed a long time ago [107], it is only recently that SHG was demonstrated in a slab, spin-coated and corona poled PNA-PMMA film. An electron-beam was used to write a grating used in third order for phase-matching on a poly(methyl methacrylate)-coated glass substrate [108]. The second harmonic was generated in the counter-propagating direction from the 1.06 µm fundamental beam and the rather small efficiencies achieved, about 2×10^{-17} $W^{-1}cm^{-2}$, were attributed to the large spectral width of the laser. The role of losses was taken into account in a numerical rigorous electromagnetic theory [109]. The case where a grating structure is covered by a metal, silver in this case, has also been experimentally addressed [110].

4.7
Non-Collinear Wavevector Matching

Non-collinear wavevector matching is, in principle, a relatively easy way to fulfill the wavevector conservation conditions. However it suffers major drawbacks and it's potential in device oriented applications remains very limited.

The non-collinear phase-matching technique requires two intersecting fundamental guided waves travelling at some angle relative to each other with the harmonic being generated along the bisector of the fundamental beams. Because of the beam crossing, there is a direct trade-off between the usable interaction length and the fundamental beam transverse cross-section. As a result the conversion efficiency is small.

Only a few demonstrations have been reported with organic materials. For example, a single crystal of MNA was the substrate for a Corning glass waveguide and a SHG efficiency of 0.1% at 532 nm in a TM_3 spatial mode was obtained for two crossing, high power, TM_1 fundamentals beam at 1064 nm [111]. Others example include using stilbene functionalized polymer waveguides [112], for which modal dispersion phase-matching between fundamental TM_0 and second harmonic TM_2 spatial modes gave an efficiency of the order of 1% [113].

4.8
Self Light Organization

This is a novel technique in which the phase-matching and poling are simultaneously realized by a purely optical excitation process [114, 115]. It is based on the breaking of the centrosymmetry of an initially isotropic doped polymer medium containing azo (Disperse Red 1) chromophores. Disperse Red 1 molecules can have either a *trans* (linear) or *cis* (bent) geometry and the more stable *trans* state has a larger $\chi^{(2)}$ than the less stable *cis* state. Since the molecules have mixed parity ground and excited states, electrons can be raised from the *trans* ground state by either one or two photon absorption. When the ω $\{E_1 exp[i(Tt-$

k_1z]} and 2ω {$E_3\exp[i(2\Upsilon t-k_2z)]$} beams are co-polarized, mutually coherent, and undergo absorption simultaneously, the absorption α $C_1\cos^2\theta+C_2\cos^4\theta+$ $<E^*_1{}^2E_3>\cos^3\theta$ where θ is the polar angle defined by $E\cdot\mu$ α $\cos\theta$ and μ is the permanent dipole moment. This results in an anisotropic spatial distribution of *trans* molecules in the first *trans* excited state, i.e. more excited molecules for q=0 than for q=p. From the excited state the *trans* molecules can efficiently undergo a structural transition to the *cis* form (*trans-cis* isomerization) which is a more compact and spherical structure and hence the molecules can rotate before they reconvert back to the more stable *trans* form. As a result, for the duration of the optical fields, the *trans* population (in the ground+the excited state) along q=0 is partially depleted relative to q=p and this distribution is subsequently locked-in when the fields are turned off. Since the re-orientation of *trans* molecules in the ground state is very, very slow, thus there is a net orientation of *trans* molecules produced (more pointing along $\theta\cong\pi$), and hence a net $\chi^{(2)}$. Note that the wavevector produced by $<E^*_1{}^2E_3>$ is exactly that needed for QPM, i.e. $\chi^{(2)}\alpha$ $\exp[i(2k_1-k_3)z]$, the ideal QPM case.

By simultaneously injecting both beams into a DR1 PMMA film, within hours a self-written QPM nonlinearity of 90 pm V^{-1} was permanently inscribed, allowing further use of the film as a second harmonic generator. The limitation is that the harmonic is generated near its absorption peak and hence only very thin films with a small net conversion efficiency can be used [116]. Detuned from resonance, the nonlinearity is naturally significantly decreased. Nonetheless this method remains potentially very attractive.

5
Summary and Perspectives

Despite the fact that it dates back to the early 1990s, this is still a very young field. Many interesting "preliminary" results have been published over the past decade, and unfortunately not often followed-up by the expected impressive device demonstration. This illustrates the very difficult challenge in turning a clever configuration into a realistic, efficient, low loss structure. Furthermore, the transparency window-$\chi^{(2)}$ trade-off is a real problem in finding the right material for each projected device. At the long wavelength end, the transparency window is limited to about 1600 nm by vibrational overtones of the C-H stretch. At the short wavelength end, the larger the λ_{max}, the larger the nonlinearity. Given that harmonic loss is a limiting factor to conversion efficiency, the spectral width of the maximum absorption line and the decrease of the absorption in its tail into the infrared are critical, but essentially unaddressed issues. To be truly effective, the harmonic loss should be less than 1 dB cm^{-1}. In fact it is the loss in the tails of the dominant absorption line that ultimately limits the practicality of all-guided phase-matching techniques.

It is worth pointing out that, overall, MDPM can be more efficient than QPM. This is illustrated in the top part of Fig. 11 where the FOMs η_0, in % $W^{-1}cm^{-2}$ are represented for different structures.

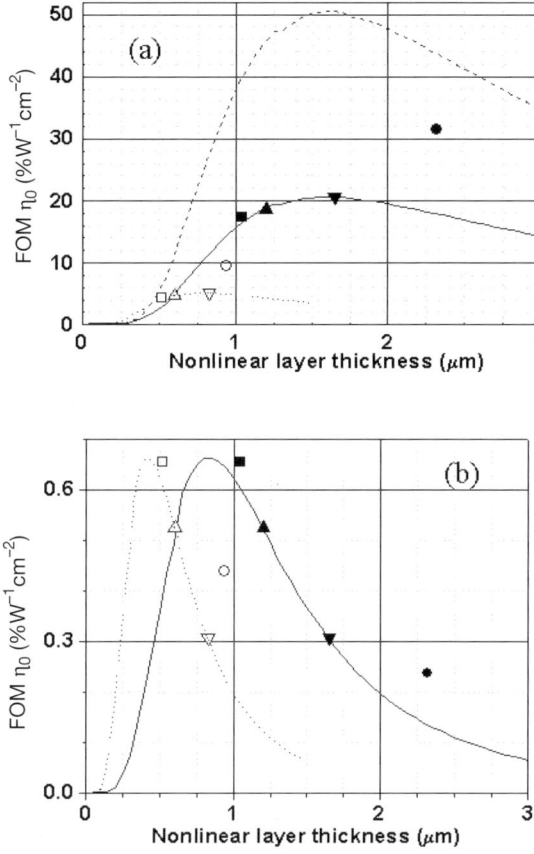

Fig. 11. (a) FOM η_0, in % $W^{-1}cm^{-2}$ and (b) conversion efficiency FOM η_0 in % $W^{-1}cm^{-2}$ as a function of core thickness with the same nonlinearity d in each case. Here the *dashed line* identifies the perfectly phasematched case, the *solid lines* identify the "+/−" and the *dotted lines* the "+/0" QPM structures respectively. The individual symbols relate to the MDPM structures with "+/−" (*solid*) and "+/0" (*open*). ■, □ MDPM, $TM_0(\omega) \rightarrow TM_1(2\omega)$; ●, ○ - MDPM, $TM_0(\omega)$ ▼, ∇ QPM maximum; and ▲, Δ QPM optimum. Calculation done for DANS $d=6$ pm V^{-1} with a second harmonic loss of 40 dB cm^{-1} and a fundamental loss of 5 dB cm^{-1} for a sample length of 2 mm. The MDPM structures are calculated assuming two layers perfectly index matched to the DANS parameters

First, the phase-matching of the fundamental modes $TM_{00}(\omega)$ and $TM_{00}(2\omega)$ are plotted as a function of core thickness, in the perfect phase matched case and in the QPM case which is smaller by a factor of $4/\pi^2$ to take into account the impact of the grating structure along the propagation axis. Points corresponding to the highest efficiency (QPM max) and to the stable (minimum) grating period (QPM opt) are represented. Second, the efficiency of the two MDPM structures matching $TM_{00}(\omega)$ to $TM_{01}(2\omega)$, and $TM_{02}(2\omega)$ respectively, are indicated for reversed poling +/− structures and one-active core layer +/0 structures. For prac-

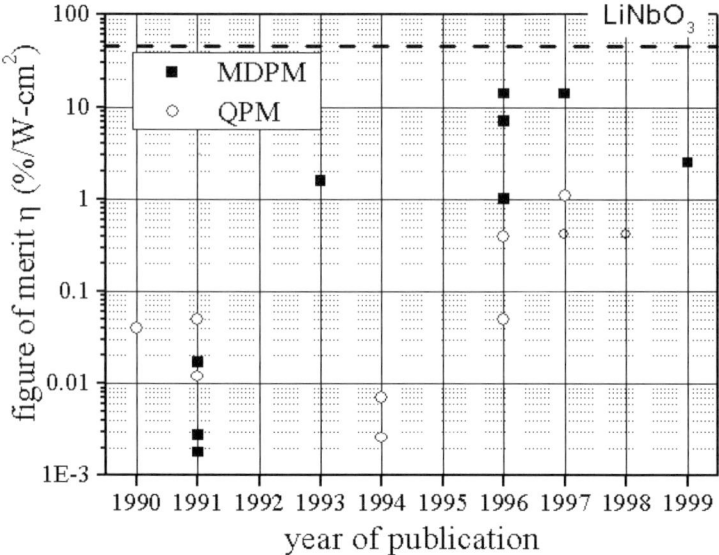

Fig. 12. SHG figure of merit η reported in poled-polymer devices using quasi-phase matching (QPM) and modal dispersion phase matching (MDPM)

tical devices, the important figure of merit is FOM η'_0, in % $W^{-1}cm^{-2}$ which takes losses into account.

There are several important conclusions to be drawn from Fig. 11. First, MDPM is as efficient as QPM; second, the highest efficiency is calculated for the "MDPM $TM_{00} \rightarrow TM_{01}$" structure; third, as long as the absorption of the nonlinear chromophores is the dominant mechanism of the beam propagation losses, it is not advantageous to use complex, oppositely poled structures. This will only increases η but not the device relevant efficiency η'_0! Structures can be made more easily with a waveguiding core of alternated nonlinear absorbing and transparent linear layers with an efficient % /W.

As indicated in Fig. 12 and Tables 1 and 2, constant progress has been made over the years. Efficiencies have been steadily increasing.

Altogether, only those wavevector matching techniques with all waves guided allow a long interaction length L and have the potential for efficient energy exchange between the waves, increasing with L^2 in the limit of low depletion. For a perfect phase-match, as in the birefringence, anomalous, or modal dispersion cases, the phase-matching integral is theoretically equal to 1. In the QPM the integral is theoretically limited to $2/\pi$ for the +/– case (or $1/\pi$ for +/0) [117] and the conversion efficiency is reduced by [$2/\pi^2$ (or $1/\pi^2$)]. But of course, the magnitude of the nonlinear coefficient must also be considered. In the poled polymer case it is the diagonal coefficient that is the largest (as it also is for the most nonlinear organic crystals). Therefore the two best options remaining are QPM and MD.

A comparison of the spatial overlap integral between the different approaches is somewhat more difficult because it depends in each individual case on the de-

Table 1. Comparison of selected quasi-phase-matched SHG devices reported in organic and some inorganic materials: $d^{(2)}$-coefficient fundamental wavelength λ, efficiency η', device length L, figure of merit η, waveguide losses $\alpha_\omega/\alpha_{2\omega}$ and waveguide structure. Note that not all of the relevant parameters were reported in die publications quoted

$d^{(2)}$ (pm V^{-1})	λ (nm)	η' (% W^{-1})	L (cm)	η (% W^{-1}cm^{-2})	$\alpha_\omega/\alpha_{2\omega}$ (dB cm^{-1})	waveguide type, QPM type material, and reference
				Poled polymers		
2	1340	0.01	0.50	0.04		slab, (+/0) polingoxynitrostilbene/MMA [Khanarian et al. 1990]
9	820				/ 12	slab, (+/0) bleaching alkoxyalkylsulfone/MMA [Rikken et al. 1990]
	1064	2×10^{-3}	0.2	0.05		slab, (+/0) poling
	2940	5×10^{-4}	0.2	0.012		VDCN/vinyl acetate [Azumai et al. 1991a]
	1064	4.0×10^{-4}	0.24	7.0×10^{-3}	30 / 30	bulk film, (+/−) poling
	2940	1.5×10^{-4}	0.24	2.6×10^{-3}	61 / 30	VDCN/vinyl acetate [Azumai et al. 1994]
25	1615	3×10^{-3}	0.25	0.05	5 /	channel, (+/0) polingDANS-polymer[Jäger et al. 1996a]
15	1590	0.1	0.50	0.4	3 /	channel, (+/0) grafting diazo dye / PMMA [Tomaru et al. 1996], [Shuto et al. 1997], [Watanabe et al. 1998]
				Other organic materials		
20	1064	0.04	0.50	0.16	14 /	channel, corrugation mNA crystal / SiN [Suhara et al. 1993]
	1064	4.4×10^{-13}	0.38	3.0×10^{-12}		slab, (+/0) ablation hemicyanine (adsorbate) [Marowsky et al. 1993]
				Inorganic crystals		
26	1545	250	2.7	43	0.6	channel, (+/−) Ti-diffusion LiNbO$_3$,[Arbore et al. 1997]
26	858	960	0.8	1500		channel, (+/−) proton exch. LiTaO$_3$, [Doumuki et al. 1994]

tails of the geometry, bounding media and there is no natural theoretical best value due to the complex field distributions. The contribution of the linear material areas to this integral is 0. This integral quantifies the useful overlap between the fundamental and harmonic spatial modes in the nonlinear region. Thus,

Table 2. Comparisen of selected waveguide SHG devices using MDPM reported in organic and some inorganic materials: $d^{(2)}$-coefficient fundamental wavelength λ, efficiency η', device length L, figure of merit η, waveguide losses $\alpha_\omega/\alpha_{2\omega}$ and waveguide structure. (ev.) – nonlinearity used via an evanescent field interaction, (+) non-optimized, single layer structure, (+/0) active/non-active structure, (+/−) inverted nonlinearity

$d^{(2)}$ (pm V^{-1})	λ (nm)	η' (% W^{-1})	L (mm)	η (% W^{-1}cm^{-2})	$\alpha_\omega/\alpha_{2\omega}$ (dB cm^{-1})	waveguide type, conversion, struct., material, & reference
				Poled polymers		
	1064	7.0×10^{-5}	2	1.8×10^{-3}	35 / 35	slab, TE$_1 \Rightarrow$ TE$_3$, (+)
	2940	1.1×10^{-4}	2	2.8×10^{-3}	61 / 30	slab, TE$_0 \Rightarrow$ TE$_1$, (+)
						VDCN/Vac copolymer [Azumai et al. 1991b]
0.71	1064	1.7×10^{-4}	1	1.7×10^{-2}		slab, TM$_0 \Rightarrow$ TM$_2$,
						MNA/PMMA (+) [Sugihara et al. 1991a]
2.0	900	0.4	5	1.6	1 /	channel, TM$_0 \Rightarrow$ TM$_1$, (ev.)
						MSMA/Si$_3$N$_4$/SiO$_2$ [Rikken et al. 1993b]
	850	(<10^{-3}%)			0.2 / 20	slab, TM$_0 \Rightarrow$ TM$_1$, (+/0), stilbene / polystyrene [Clays et al. 1994]
13	1535	0.04	2	1	5 / 50	channel, TM$_{00} \Rightarrow$ TM$_{20}$, (+) DANS-poly. [Jäger et al. 1996c]
6	1510	0.14	1	14	7 / 20	channel, TM$_{00} \Rightarrow$ TM$_{10}$, (+/0) DANS-polymer/polyetherimide [Jäger et al. 1996b]
		0.70	7	1.3		
5	1610	0.04	1	4	5 / 95	channel, TM$_{00} \Rightarrow$ TM$_{10}$, (+/−) channel, TM$_{00} \Rightarrow$ TM$_{20}$, (+/−/+) DR 1/MMA [Jäger et al. 1997b]
	1610	0.14	1	14		
3	1607	0.1	2	2.5	6 / 10	channel, TM$_{00} \Rightarrow$ TM$_{10}$, (+/0) DR1 [Cho et al. 1999]
				Other organic materials		
250	1064	5.1×10^{-5}	4	3.2×10^{-3}		slab, (+) MNA crystal [Itoh et al. 1986]
250	1064	1.2×10^{-4}	5	4.9×10^{-4}	11	slab, TE$_6 \Rightarrow$ TE$_{16}$, (+) MNA crystal [Sugihara et al. 1991c]

Table 2. (continued)

$d^{(2)}$ (pm V^{-1})	λ (nm)	η' (% W^{-1})	L (mm)	η (% W^{-1}cm^{-2})	$\alpha_\omega/\alpha_{2\omega}$ (dB cm^{-1})	waveguide type, conversion, struct., material, & reference
250	1064	3.0×10^{-4}	1	3.0×10^{-2}	15	slab, TE$_0$⇒TE$_2$, (+) MNA crystal [Sugihara et al. 1992]
13	926	0.02	2	0.5	5.5	slab, TE$_0$⇒TE$_1$, DCANP (+/0) LB-film [Flörsheimer et al. 1992]
7.8	1064	0.01	1	1.0		slab, TE$_0$⇒TE$_1$, DCANP (+/−) LB-film [Küpfer et al. 1993]
10 (8)	888 (819)	0.04	10	0.04	/ 2	slab, TM$_0$⇒TM$_1$ (TE$_0$⇒TM$_1$), (+/−) LB film [Penner et al. 1994]
35	≈900	7.4×10^{-3}	1	0.74		slab, TE$_0$⇒TE$_1$, (ev.) IAPU cryst./ Ta$_2$O$_5$ [Yamamoto et al. 1996]
0.68	1064	1.9×10^{-4}			3.7	slab, DONPU LB film, (+/−) [Fujita et al. 1996]
				Inorganic crystals		
18.5	827	160	4.1	960	<1	channel, TE$_0$⇒TE$_1$, Ta$_2$O$_5$/KTP, (ev.) [Doumuki et al. 1994]

it also takes into account how strong the waveguide structure confines the optical waves in the nonlinear area. A crude approximation normalizing to unity the QPM case matching two zero'th order spatial modes gives around one half for (+/−) MDPM for the zero order fundamental and the first spatial mode of the harmonic wave. This is insufficient for a detailed comparison of the potential of the two techniques, but does illustrate their trade-off. The net efficiencies will be comparable because of the 2/π reduction in the phase-matching integral. However, MD results in a dual maximum intensity profile, making it unattractive for any application that requires a single peak as the output. Therefore, QPM seems currently to be the most attractive option. And the reasonable losses obtained by Tomaru et al. despite a complex fabrication sequence, is very promising.

In the future, one can hope that the segmented waveguide technique will be applied to QPM with organic materials with the opposite wavevector mismatch dispersion in successive segments [118]. Thus, the phase mismatch will no longer grow between 0 and π, but will be maintained within arbitrary small boundaries around 0, only limited by the shortest segment length realizable and the associated propagation losses. Co-doping the material to control separately nonlinearity and dispersion will also certainly be experimentally investigated.

Altogether synthesis of better materials and the simultaneous use of different phase-matching techniques should lead to the lowering of the stringent requirements imposed for $\chi^{(2)}$ nonlinear interactions in integrated optics and devices should appear in the following decades following the success initiated in the electro-optic field.

Acknowledgements. This work is supported by a NSF/CNRS US/French bilateral collaboration grant. The authors would like to thank all of their colleagues who contributed to our knowledge of this field, specifically, Mathias Jaegger, Vincent Ricci, Wook-Rae Cho, Tomas Pliska, Akira Otomo, Anne-Claire Le Duff, Robert Twieg and Pong Chan.

References

1. Reviewed in Zyss J, Chemla DS (1987) Quadratic nonlinear optics and optimization of the second-order nonlinear optical response of molecular crystals. In: Chemla DS, Zyss J (eds) Nonlinear and optical properties of organic molecules and crystals. Academic Press, Orlando, p 23
2. Bosshard C, Knopfle G, Pretre P, Follonier S, Serbutoviez C, Gunter P (1995) Opt Engin 34:1951
3. Extensive discussion in Hopf F, Stegeman GI (1986) Advanced classical electrodynamics Vol II: Nonlinear optics. Wiley
4. Marder SR, Perry JW, Yakymyshyn CP (1994) Chem Mater 6:1137
5. Josse D, Dou SX, Zyss J, Andreazza P, Perigaud A (1992) Appl Phys Lett 61:121
6. Stegeman G, Seaton C (1985) J Appl Opt 58:R57
7. Ulrich D (1988) Mol Cryst Liq Cryst 160:1
8. Stegeman G, Stolen R (1989) J Opt Soc Am B 6:652
9. Rikken G, Seppen C, Venhuizen A, Nijhuis S, Staring E (1992) Philips J Res 46:215
10. Miyata S, Tao T (1996) Synth Met 81:99
11. Gase T, Karthe K (1997) Opt Commun 133:549
12. Chikuma K, Umegaki S (1990) J Opt Soc Am B 7:768; Chikuma K, Umegaki S (1992) J Opt Soc Am B 9:1083
13. Normandin R (1979) PhD thesis, University of Toronto
14. Le Duff AC, Ricci V, Pliska T, Canva M, Stegeman G, Chan K, Twieg R (2000) Appl Opt 39:947
15. Canva M, Jäger M, Stegeman G (1998) Polymer News 23:78
16. Ricci V, Stegeman GI, Chan KP (2000) J Opt Soc Am B 17:1349
17. Pliska T, Meier J, Ricci V, Le Duff AC, Canva M, Stegeman G, Raymond P, Chan K (2000) Appl Phys Lett 76:265
18. Galvan-Gonzales A, Canva M, Stegeman G, Twig R, Kowalczyk T and Lackritz H (1999) Opt Lett 24:1741
19. Zhang Q, Canva M, Stegeman G (1998) Appl Phys Lett 73:912
20. Galvan-Gonzales A, Canva M, Stegeman G, Twig R, Chan K, Kowalczyk T, Zhang X, Lackritz H, Marder S, Thayumanavan S (2000) Opt Lett 25:332
21. Tien P, Ulrich R, Martin R (1970) Appl Phys Lett 17:447
22. Hayata K, Matsumura H, Koshida K (1992) J Appl Phys 72:4514
23. Kerkoc P, Bosshard C, Arend H, Gunter P (1989) Appl Phys Lett 54:487
24. Umegaki S (1992) Nonlinear Opt 3:73
25. Bosshard C, Küpfer M, Flörsheimer M, Günter P (1992) Nonlinear Opt 3:215
26. Bosshard C, Flörsheimer M, Küpfer M, Günter P (1991) Opt Commun 3:247
27. Clays K, Armstrong N, Penner T (1993) J Opt Soc Am B 10:886
28. Asai N, Tamada H, Fujiwara I, Seto J (1992) J Appl Phys 72:4521
29. Johal S, James S, Tatam R, Ashwell (1999) G Opt Lett 24:1194
30. Sugihara O, Kunioka S, Nonaka Y, Aizawa R, Koike Y, Kinoshita T, Sasaki K (1991) J Appl Phys 70:7249

31. Sasaki K, Nonaka Y, Kinoshita T, Nihei E, Koike Y (1992) Nonlinear Opt 3:61
32. Izawa K, Okamoto N, Sugihara O (1993) Jpn J Appl Phys 32:807
33. Sugihara O, Furukubo M, Hayashi H, Okamoto N (1995) Nonlinear Opt 14:251
34. Lim E, Matsumoto S, Fejer M (1990) Appl Phys Lett 57:2294
35. Lim T, Jeong M, Cha S, Koh E (1997) J Korean Phys Soc 30:544
36. Tomono T, Nishikata Y, Pu L, Sassa T, Kinoshita T, Sasaki K (1993) Mol Cryst Liq Cryst 227:113
37. Kinoshita T, Tsuchiya K, Sasaki K, Yokoh Y, Ashitaka H, Ogata N (1994) IEICE Transact On Electron, 5:679
38. Cazeca M, Jiang X, Masse C, Kamath M, Jeng R, Kumar J, Tripathy S (1995) Opt Commun 117:127
39. Zhu X, Chen Y, Kamath M, Jeng R, Kumar J, Tripathy S (1993) Nonlinear Opt 4:175
40. Chen Y, Kamath M, Jain A, Kumar J, Tripathy S (1993) Opt Commun 101:231
41. Sato H, Matsumo M, Seo I (1998) J Opt Soc Am B 15:773
42. Sato H, Nozawa H, Azumai Y, Seo I (1995) Nonlinear Opt 10:319
43. Sato H, Azumai Y (1993) J Opt Soc Am B 10:894
44. Azumai Y, Sato H (1993) Jpn J Appl Phys 32:800
45. Onda T, Ito R (1991) Jpn J of Appl Phys 30:957
46. Schmitt K, Benecke C, Schadt M (1997) J Appl Phys 81:11
47. Montiel F, Neviere N, Peyrot P (1998) J Mod Opt 45:2169
48. Sassa T, Umegaki S (1997) Opt Lett 22:856
49. Sassa T, Umegaki S (1998) Jpn J Appl Phys 37:447
50. Sassa T, Umegaki S (1998) J Appl Phys 84:4071
51. Harada A, Okazaki Y, Kamiyama K, Umegaki S (1991) Appl Phys Lett 59:1535
52. Kim E, Kinoshita T, Sasaki K, Senoh T, Yamanaka T (1995) Nonlinear Opt 14:257
53. Normandin R, Stegeman G (1979) Opt Lett 4:58
54. Normandin R, Stegeman G (1980) Appl Phys Lett 36:253
55. Normandin R, Stegeman G (1982) Appl Phys Lett 40:759
56. Otomo A, Mittler-Neher S, Bosshard C, Stegeman G, Horsthuis W, Möhlmann G (1993) Appl Phys Lett 63:3405
57. Otomo A, Stegeman G, Flipse M, Diemeer M, Horsthuis W, Möhlmann G (1998) J Opt Soc Am B 15:759
58. Bosshard C, Otomo A, Stegeman G, Küpfer M, Flörsheimer M, Günter P (1993) Appl Phys Lett 64:2076
59. Armstrong J, Bloembergen N, Ducuing J (1962) Phys Rev 127:1918
60. Cahill P Singer K, King L (1989) Opt Lett 14:1137
61. Singer K, Kowalczyk T, Cahill P, King L (1992) Nonlinear Opt 3:3
62. Kowalczyk T, Singer K, Cahill P (1995) Opt Lett 20:2273
63. Dai T, Singer K, Twieg R, Kowalczyk T (2000) J Opt Soc Am B 17:412
64. Cross G, Bloor D, Szablewski M (1995) Nonlinear Opt 14:219
65. Extensive discussion in Hopf F, Stegeman G (1986) John Wiley and Sons
66. Tao X, Watanabe T, Zou D, Ukuda H, Miyata S (1995) Nonlinear Opt 14:225; Tao X, Watanabe T, Zou D, Ukuda H, Miyata S, (1995) Macromolecules 28:2637; Tao X, Watanabe T, Zou D, Ukuda H, Miyata S (1995) J Opt Soc Am B 12:1581
67. Stegeman G, Schiek R, Torner L, Torruellas W, Baek Y, Baboiu D, Wang Z, VanStryland E, Hagan D, Assanto G (1996) Cascading: a promising approach to nonlinear optical phenomena revisited. In: Khoo I, Simoni F (eds) Novel optical materials and applications. Wiley Interscience, New York edited by, p 49
68. Sasaki K, Kinoshita T, Karasawa N (1984) Appl Phys Lett 45:333
69. Azumai Y, Seo I, Sato H (1991) Nonlinear Opt 1:129
70. Sugihara O, Kinoshita T, Okabe M, Kunioka S, Nonaka Y, Sasaki K (1991) Appl Opt 30:2957
71. Okada A, Ishii K, Mito K, Sasaki K (1991) Nonlinear Opt 1:179
72. Tomono T, Nishikata Y, Pu L, Sassa T, Nonaka Y, Sasaki K (1992) Nonlinear Opt 3:255

73. Tao X, Watanabe T, Zou D, Shimoda S, Usui H, Sato H Miyata S (1995) J Polym Sci B 33:2205
74. Watanabe T, Zou D, Shimoda S, Tao X, Claude C, Okamoto Y (1994) Mol Cryst Liq Cryst 255:95
75. Jäger M, Stegeman G, Möhlmann G, Flipse M Diemeer M (1996) Elect Lett 32:2009
76. Jäger M, Stegeman G, Brinker W, Yilmaz S, Bauer S, Horsthuis W, Möhlmann G (1996) Appl Phys Lett 68:1183
77. Wirges W, Yilmaz S, Brinker W, Bauer-Gogonea S, Bauer S, Jäger M, Stegeman G, Ahlheim M, Stähelin M, Zysset B, Lehr F, Diemeer M, Flipse M (1997) Appl Phys Lett 70:3347
78. Jäger M, Stegeman G, Yilmaz S, Wirges W, Brinker W, Bauer-Gogonea S, Bauer S, Diemeer M, Flipse M (1998) J Opt Soc Am B 15:781
79. Rikken G, Seppen C, Staring E, Venhuizen A (1993) Appl Phys Lett 62:2483
80. Clays K, Schildkraut J, Williams D (1994) J Opt Soc Am B 11: 655
81. Cho W, Ricci V, Pliska T, Canva M, Stegeman G (1999) J Appl Phys 86:2941
82. Penner T, Motschmann H, Armstrong N, Ezenyilimba M Williams D (1994) Nature 367:49
83. Küpfer M, Flörsheimer M, Bosshard C, Günter P (1995) Nonlinear Opt 10:341
84. Flörsheimer M, Küpfer M, Bosshard C, Günter P (1992) Adv Mat 4:795
85. Hickel W, Menges B, Althoff O, Lupo D, Falk U, Scheunemann U (1994) Thin Solid Film 244:966
86. Burns W, Andrews R (1973) Appl Phys Lett 22:143
87. Burns W, Lee A (1974) Appl Phys Lett 24:222
88. Yamamoto H, Funato S, Sugiyama T, Johnson R, Norwood R, Kinoshita T, Sasaki K (1995) Nonlinear Opt 14:263
89. Watanabe T, Edel V, Tao X, Shimoda S, Miyata S (1996) Opt Commun 123:76
90. Jäger M, Stegeman G, Flipse M, Diemeer M, Möhlmann G (1996) Appl Phys Lett 69:4139
91. Alshikh Khalil M, Vitrant G, Raimond P, Chollet P, Kajzar F (1999) Opt Commun 170:281
92. Jung J, Kinoshita T (2000) J Appl Phys 87:3209
93. Suhara T, Nishihara H (1990) IEEE J Quant Electron 26:1265
94. Arbore MA, Fejer MM (1997) Opt Lett 22:151
95. Levine B, Bethea C, Logan R (1975) Appl Phys Lett 26:375
96. Khanarian G, Norwood R, Haas D, Feuer B, Karim D (1990) Appl Phys Lett 57:977
97. Norwood R, Khanarian G (1990) Elect Lett 26:2105
98. Burland D, Miller R, Reiser O, Twieg R, Walsh C (1992) J Appl Phys 71:410
99. Azumai Y, Kishimoto M, Sato H, (1992) Jpn J Appl Phys 31:1358
100. Azumai Y, Kishimoto M, Sato H (1994) IEEE J Quant Elect 30:1924
101. Mortazavi M, Khanarian G (1994) Opt Lett 19:1290
102. Nakanishi M, Sugihara O, Okamoto N, Hirota K (1998) Appl Opt 37:1068
103. Taggi V, Michelotti F, Bertolotti M, Petrocco G, Foglietti V, Donval A, Toussaere E, Zyss J (1998) Appl Phys Lett 72:2794
104. Tomaru S, Watanabe T, Hikita M, Amano M, Shuto Y, Yokohoma I, Kaino T, Asobe M (1996) Appl Phys Lett 68:1760
105. Shuto Y, Watanabe T, Tomaru S, Yokohama I, Hikita M, Amano M (1997) IEEE J Quant Elect 33:349
106. Watanabe T, Hikita M, Amano M, Shuto Y, Tomaru S (1998) J Appl Phys 83:639
107. Somekh S, Yariv A (1972) Appl Phys Lett 21:140
108. Blau G, Popov E, Kajzar F, Raimond A, Roux J, Coutaz J (1995) Opt Lett 20:1101
109. Popov E, Neviere M, Reinisch R, Coutaz J, Roux J (1995) Appl Opt 34:3398
110. Coutaz J, Blau G, Roux J, Reinisch R, Kajzar F, Raimond A, Robin P, Chastaing E, Le Barny P (1995) Nonlinear Opt 10:347
111. Sugihara O, Kai S, Uwatoko K, Kinoshita T, Sasaki K (1990) J Appl Phys 68:4990

112. Shuto Y, Takara H, Amano M, Kaino T (1989) Jpn J Appl Phys 28:2508
113. Shuto Y (1993) J Opt Soc Am B 10:1221
114. Charra F, Devaux F, Nunzi JM, Raimond P (1992) Phys Rev Lett, 68:2440
115. Fiorini C, Charra F, Nunzi JM, Raimond P (1997) J Opt Soc Am B 14:1984
116. Fiorini C, Charra F, Nunzi JM, Raimond P (1995) Nonlinear Opt 9:339
117. Somekh S, Yariv A (1972) Opt Commun 6:301
118. Nir D, Weissman Z, Rushin S, Hardy A (1994) Opt Lett 19:1732

Received August 2000

Molecular Design for Third-Order Nonlinear Optics

Ulrich Gubler, Christian Bosshard

Nonlinear Optics Laboratory, Institute of Quantum Electronics,-ETH Hönggerberg, CH-8093 Zürich, Switzerland
present adress: Centre Suisse d'Electronique et de Microtechnique (CSEM), Untere Gründlistrasse 1, CH-6055 Alpnach, Switzerland

e-mail: ulrich.gubler@csem.ch, christian.bossard@csem.ch

We discuss recent developments in the design of molecules for applications in third-order nonlinear optics with emphasis on all-optical signal processing and two-photon absorption. We especially concentrate on functional substitution patterns and conjugation length expansion. We show that low molecular symmetry with regard to the conjugation path of the delocalized electrons was found to be a good guideline towards linearly conjugated molecules with large second-order hyperpolarizabilities γ. We show that this guideline is also valid for two-dimensionally conjugated systems and that the observed effects can be explained by the symmetry of the electronic wavefunctions. We present scaling laws and critical conjugation lengths of rod-like molecules with electrons delocalized over a one-dimensional path and show that the exponent tends to be constant for various polymers in the transparency range and that the values presented here are of similar magnitude for various organic materials systems. Finally, we discuss different materials systems with regard to the figures of merit relevant for all-optical signal processing.

Keywords. Third-order nonlinear optics, Second-order hyperpolarizability, Molecular design

1	Introduction	125
2	**Third-Order Nonlinear Optics**	126
2.1	Definitions and Conventions	126
2.1.1	Definition of the Nonlinear Optical Susceptibilities and Molecular Hyperpolarizabilities	126
2.1.2	Electric Field Definition and Prefactors of Third-Order Effects . . .	128
2.1.3	Permutation and Kleinman Symmetry	129
2.1.4	Nonlinear Refractive Index and Two-Photon Absorption	130
2.2	Relations Between Microscopic and Macroscopic Coefficients . . .	132
2.3	Reference Values	134
2.4	Material Requirements for All-Optical Signal Processing	136
2.4.1	Figures of Merit	136
2.4.2	Nonlinearity and Intensity Requirements	139
2.4.3	Further Critical Factors	140

3	Measurement Techniques	141
3.1	Third-Harmonic Generation	142
3.2	Degenerate Four-Wave Mixing	145
3.3	Z-Scan	149
3.4	Two-Photon Absorption Spectroscopy	151
3.5	Fluorescence Spectroscopy	154
4	**Discussion of Molecular Properties**	155
4.1	Basic Molecular Units	155
4.1.1	Electronic States and Symmetry Considerations	159
4.1.2	Bond-Length Alternation	163
4.2	Influence of Molecular Substitution	164
4.3	1-D vs 2-D Conjugation	168
4.4	Conjugation Length Expansion	173
4.5	Two-Photon Absorption	177
5	**Concluding Remarks**	182
References		187

List of Symbols

α	linear absorption coefficient
α_2	two-photon absorption coefficient
α_3	three-photon absorption coefficient
α_{ij}	molecular polarizability
β_{ijk}	molecular first-order hyperpolarizability (or molecular second-order polarizability)
$E^\omega, E^{3\omega}$	Fourier components of the electric field strength
δ	two-photon cross-section
ε	molar extinction coefficient
f	local field factor
γ_{ijkl}	molecular second-order hyperpolarizability (or molecular third-order polarizability)
I	light intensity
m	molecular mass
μ	dipole moment
μ_{ij}	transition dipole moment
$\Delta\mu$	difference between the excited and the ground state dipole moments
n	linear refractive index
n_2	nonlinear refractive index
N_A	Avogadro's number
λ	wavelength

λ_{max} wavelength of maximum absorption
ω frequency of light
P macroscopic polarization
$p^{2\omega}, p^{3\omega}$ Fourier components of the molecular polarization
ρ density
T^{-1} two-photon figure of merit for all-optical switching
$\chi^{(n)}$ n-th order susceptibility
V^{-1} three-photon figure of merit for all-optical switching
W one-photon figure of merit for all-optical switching

1
Introduction

The influence of light by light (all-optical signal processing) is an exciting topic of basic and application related research. Various new effects have been demonstrated and many potential applications have emerged [1–3]. However, there still is a lack of materials that fulfill the requirements for applications. In this review we will discuss these requirements and the efforts in material science to develop new organic materials with the desired functionality. Materials with large nonlinearities are crucial for the development of components for of all-optical signal processing. Organic molecules with extended π-electron conjugation and electron donating and/or accepting groups are promising candidates for these nonlinear optical applications. We here focus on small molecules, that is monomers and oligomers, and we will describe new structure-property relations that yield guidelines for future material optimization. We will first describe some important basic issues such as a precise definition of nonlinear optical susceptibilities and reliable reference values that are crucial to allow a meaningful comparison between different materials. We will then discuss the material requirements for all-optical signal processing. In the following we will discuss a few measurement techniques that are useful for a characterization of the relevant materials parameters. This topic is treated since different molecular properties are probed with different experimental techniques. This section is then followed by the most important part of this review, a discussion of new molecules that have been developed in the past few years in the area of third-order nonlinear optics and the guidelines that could be developed from combined experimental and theoretical efforts. Additional information on various molecular systems can also be found in Refs. [4, 5].

We attempt to give all nonlinearities in the same units, with respect to the same reference value, and based on the same definition (see below). The various molecular systems will be compared with respect to the relevant figures of merit. Finally, conclusions will be drawn and future work will be outlined.

2
Third-Order Nonlinear Optics

2.1
Definitions and Conventions

In nonlinear optics, various definitions of the electric field and the expansion of the polarization have been introduced. Additionally, divergent values for the third-order nonlinear optical susceptibility of fused silica or other reference materials have been utilized. Moreover, most authors publish their nonlinearities in electro-static units (esu) instead of SI units. All these varying conventions complicate the comparison of nonlinear optical materials and can be quite confusing. Paying attention to the underlying conventions is inevitable when nonlinearities from different authors are compared. Unfortunately, in some publications the underlying conventions are not obvious and a quantitative classification of the measured nonlinearities is impossible. In the following we will give the definitions that are relevant for this work.

2.1.1
Definition of the Nonlinear Optical Susceptibilities and Molecular Hyperpolarizabilities

The polarization $\tilde{P}$ is generally a complicated function of an external electric field $\tilde{E}$. The real function $\tilde{P}(\tilde{E})$ can be approximated with a Taylor expansion around zero electric field strength $\tilde{E} = 0$.

$$\tilde{P}(\tilde{E}) = P(0) + \left(\frac{\partial \tilde{P}}{\partial \tilde{E}}\right)_{\tilde{E}=0} \tilde{E} + \frac{1}{2!}\left(\frac{\partial^2 \tilde{P}}{\partial \tilde{E}^2}\right)_{\tilde{E}=0} \tilde{E} \otimes \tilde{E} + \frac{1}{3!}\left(\frac{\partial^3 \tilde{P}}{\partial \tilde{E}^3}\right)_{\tilde{E}=0} \tilde{E} \otimes \tilde{E} \otimes \tilde{E} + \ldots \quad (1)$$

The operator $\otimes$ denotes the tensor product of the electric fields. The tildes above the electric field and polarization symbols portray the wave character of the fields and are used for better distinction with the field amplitudes $E(r)$ and $P(r)$ as defined below. For the notation of the polarization components follows

$$\tilde{P}_i(\tilde{E}) = P_i(0) + \left(\frac{\partial \tilde{P}_i}{\partial \tilde{E}_j}\right)_{\tilde{E}=0} \tilde{E}_j + \frac{1}{2!}\left(\frac{\partial^2 \tilde{P}_i}{\partial \tilde{E}_j \partial \tilde{E}_k}\right)_{\tilde{E}=0} \tilde{E}_j \tilde{E}_k$$

$$+ \frac{1}{3!}\left(\frac{\partial^3 \tilde{P}_i}{\partial \tilde{E}_j \partial \tilde{E}_k \partial \tilde{E}_l}\right)_{\tilde{E}=0} \tilde{E}_j \tilde{E}_k \tilde{E}_l + \ldots$$

(2)

with the Einstein summation convention of adding over common indices applied. The coefficients of the expansion orders are the linear and higher-order susceptibilities $\chi^{(j)}$

$$\chi_{ij}^{(1)} = \frac{1}{\varepsilon_0}\left(\frac{\partial \tilde{P}_i}{\partial \tilde{E}_j}\right)\bigg|_{\tilde{E}=0} \qquad (3)$$

$$\chi_{ijk}^{(2)} = \frac{1}{2\varepsilon_0}\left(\frac{\partial^2 \tilde{P}_i}{\partial \tilde{E}_j \partial \tilde{E}_k}\right)\bigg|_{\tilde{E}=0} \qquad (4)$$

$$\chi_{ijkl}^{(3)} = \frac{1}{6\varepsilon_0}\left(\frac{\partial^3 \tilde{P}_i}{\partial \tilde{E}_j \partial \tilde{E}_k \partial \tilde{E}_l}\right)\bigg|_{\tilde{E}=0} \qquad (5)$$

$$\tilde{P}_i(\tilde{E}) = P_i(0) + \varepsilon_0\left(\chi_{ij}^{(1)}\tilde{E}_j + \chi_{ijk}^{(2)}\tilde{E}_j\tilde{E}_k + \chi_{ijkl}^{(3)}\tilde{E}_j\tilde{E}_k\tilde{E}_l + ...\right) \qquad (6)$$

The dielectric constant of the vaccum ε_0 is included in the susceptibility definitions as SI units are used throughout this work. The analogous definition can be applied to the microscopic polarization $\tilde{p}$ of a molecule with the molecular dipole moment μ and the polarizabilities α, β, and γ instead of the static polarization $P_i(0)$ and the susceptibilities $\chi^{(j)}$.

$$\tilde{p}_i(\tilde{E}) = \mu_i + \varepsilon_0\left(\alpha_{ij}\tilde{E}_j + \beta_{ijk}\tilde{E}_j\tilde{E}_k + \gamma_{ijkl}\tilde{E}_j\tilde{E}_k\tilde{E}_l + ...\right) \qquad (7)$$

α_{ij} is the linear polarizability, β_{ijk} the first-order hyperpolarizability (or second-order polarizability), and γ_{ijkl} the second-order hyperpolarizability (or third-order polarizability), which is the main focus of this chapter.

In principle, the electric fields to be inserted in Eq.(7) are the electric fields at the location of the molecule. Instead of the local electric fields $\tilde{E}_{loc}$ the external fields $\tilde{E}$ are usually used. Therefore, local field correction factors have to account for the electric field screening of the surrounding material when going from the macroscopic susceptibilities to the molecular hyperpolarizabilities as shown below.

Third-order nonlinear optical effects are described by the tensors $\chi_{ijkl}^{(3)}$ and γ_{ijkl}. In contrast to second-order nonlinearities $\chi_{ijk}^{(2)}$ and γ_{ijk}, no symmetry requirements are imposed on these effects to occur. Some authors do not include the faculty denominators of the Taylor expansion in nonlinearity definitions of Eqs. (4) and (5).

$$\bar{\chi}_{ijk}^{(2)} = \frac{1}{\varepsilon_0}\left(\frac{\partial^2 \tilde{P}_i}{\partial \tilde{E}_j \partial \tilde{E}_k}\right)\bigg|_{\tilde{E}=0} \qquad (8)$$

$$\bar{\chi}_{ijkl}^{(3)} = \frac{1}{\varepsilon_0}\left(\frac{\partial^3 \tilde{P}_i}{\partial \tilde{E}_j \partial \tilde{E}_k \partial \tilde{E}_l}\right)\bigg|_{\tilde{E}=0} \qquad (9)$$

Consequently, the faculty factors appear in the power series of Eqs. (6) and (7). A correction factor of two and six has to be encountered when comparing nonlinearities with the different conventions.

$$\tilde{P}_i(E) = P_i(0) + \varepsilon_0 \left(\chi_{ij}^{(1)} \tilde{E}_j + \frac{\overline{\chi}_{ijk}^{(2)}}{2!} \tilde{E}_j \tilde{E}_k + \frac{\overline{\chi}_{ijkl}^{(3)}}{3!} \tilde{E}_j \tilde{E}_k \tilde{E}_l + \dots \right) \qquad (10)$$

$$\tilde{p}_i(\tilde{E}) = \mu_i + \varepsilon_0 \left(\alpha_{ij} \tilde{E}_j + \frac{\overline{\beta}_{ijk}}{2!} \tilde{E}_j \tilde{E}_k + \frac{\overline{\gamma}_{ijkl}}{3!} \tilde{E}_j \tilde{E}_k \tilde{E}_l + \dots \right) \qquad (11)$$

$$\chi_{ijk}^{(2)} = \overline{\chi}_{ijk}^{(2)}/2; \quad \beta_{ijk} = \overline{\beta}_{ijk}/2 \qquad (12)$$

$$\chi_{ijkl}^{(3)} = \overline{\chi}_{ijkl}^{(3)}/6; \quad \gamma_{ijkl} = \overline{\gamma}_{ijkl}/6 \qquad (13)$$

2.1.2
Electric Field Definition and Prefactors of Third-Order Effects

For the definition of the electric field and polarization amplitudes different definitions are in use. The most widely accepted convention is to define the amplitudes by

$$\tilde{E}(r,t) = \frac{1}{2} \cdot \sum_{k_p, \omega_p} \left(E^{k_p, \omega_p}(r) e^{i(k_p r - \omega_p t)} + c.c. \right) \quad (\omega_p \neq 0) \qquad (14)$$

$$\tilde{p}(r,t) = \frac{1}{2} \cdot \sum_{k'_p, \omega'_p} \left(p^{k'_p, \omega'_p}(r) e^{i(k'_p r - \omega'_p t)} + c.c. \right) \quad (\omega'_p \neq 0) \qquad (15)$$

as a superposition of monochromatic waves at the frequencies ω_p and with the wavevectors k_p. With this definition and the expansion of the polarization the Fourier components of the nonlinear polarizations for the various nonlinear optical effects can be derived.

$$p_{SHG}^{2k,2\omega} = \frac{1}{2} \varepsilon_0 \cdot \beta(-2\omega, \omega, \omega) \cdot (E^{k,\omega})^2 \qquad (16)$$

$$p_{THG}^{3k,3\omega} = \frac{1}{4} \varepsilon_0 \cdot \gamma(-3\omega, \omega, \omega, \omega) \cdot (E^{k,\omega})^3 \qquad (17)$$

$$p_{SPM}^{k,\omega} = \frac{3}{4} \varepsilon_0 \cdot \gamma(-\omega, \omega, -\omega, \omega) \cdot (E^{k,\omega})^3 \qquad (18)$$

$$p_{XPM}^{k_1,\omega_1} = \frac{3}{2} \varepsilon_0 \cdot \gamma(-\omega_1, \omega_1, -\omega_2, \omega_2) \cdot \left(E^{k_1,\omega_1} \right) \left(E^{k_2,\omega_2} \right)^2 \qquad (19)$$

$$P_{DFWM}^{k_1-k_2+k_3,\omega} = \frac{3}{2}\varepsilon_0 \cdot \gamma(-\omega,\omega,-\omega,\omega) \cdot E^{k_1,\omega} E^{k_2,\omega} E^{k_3,\omega} \qquad (20)$$

Equations (16) and (17) describe second-harmonic generation (SHG) and third-harmonic generation (THG) of one laser beam with a single polarization. Self-phase modulation (SPM) of a single laser beam is described in Eq.(18) as e.g. employed in z-scan experiments [6]. Equation (19) is the cross-phase modulation (XPM) process between two laser beams and Eq.(20) describes the four wave mixing with degenerate frequencies (DFWM).

The factors 2 and 4 in the denominators are due to the definition of the field amplitudes (Eq.(14) and Eq.(15)). In order to prevent these factors some authors drop the factor 1/2 in the definitions of the amplitudes. The disadvantage of this convention is the unusual convergence behaviour of the electric field as the frequency ω approaches zero. This different field definition of course additionally complicates the comparison of different hyperpolarizability values. The factors in the numerator arise from the different possibilities to permute the input frequencies. As an example, in self-phase modulation the three input electric fields each provide a factor 1/2 which, with the factor 1/2 from the polarization, results in a denominator of 4. The negative frequency for SPM allows three permutations yielding finally a prefactor of 3/4.

For the macroscopic polarization analogous expressions can be obtained by replacing the tensors β and γ by the corresponding macroscopic susceptibilities $\chi^{(2)}$ and $\chi^{(3)}$.

$$P_{SHG}^{2k,2\omega} = \frac{1}{2}\varepsilon_0 \cdot \chi^{(2)}(-2\omega,\omega,\omega) \cdot (E^{k,\omega})^2 \qquad (21)$$

$$P_{THG}^{3k,3\omega} = \frac{1}{4}\varepsilon_0 \cdot \chi^{(3)}(-3\omega,\omega,\omega,\omega) \cdot (E^{k,\omega})^3 \qquad (22)$$

$$P_{SPM}^{k,\omega} = \frac{3}{4}\varepsilon_0 \cdot \chi^{(3)}(\omega,\omega,-\omega,\omega) \cdot (E^{k,\omega})^3 \qquad (23)$$

$$P_{XPM}^{k_1,\omega_1} = \frac{3}{2}\varepsilon_0 \cdot \gamma^{(3)}(\omega_1,\omega_1,-\omega_2,\omega_2) \cdot (E^{k_1,\omega_1})(E^{k_2,\omega_2})^2 \qquad (24)$$

$$P_{DFWM}^{k_1-k_2+k_3,\omega} = \frac{3}{2}\varepsilon_0 \cdot \chi^{(3)}(-\omega,\omega,-\omega,\omega) \cdot E^{k_1,\omega} E^{k_2,\omega} E^{k_3,\omega} \qquad (25)$$

2.1.3
Permutation and Kleinman Symmetry

In Eqs (4) and (5) the nonlinearities are defined as the derivatives of the polarization in the electric field. Whether the derivative is performed first for E_j and then for E_k or vice versa is of no importance. Therefore, the second-order hyperpolarizabilities $\gamma_{ijkl}(-\omega_4,\omega_3,\omega_2,\omega_1)$ have a permutation symmetry in the spatial indices and the involved frequencies. If the spatial indices and the frequencies of the electric fields are subjected to the same permutation σ the tensor elements stays the same.

$$\begin{aligned}
\gamma_{ijkl}(-\omega_4,\omega_3,\omega_2,\omega_1) &= \gamma_{ijlk}(-\omega_4,\omega_3,\omega_1,\omega_2) = \\
\gamma_{ikjl}(-\omega_4,\omega_2,\omega_3,\omega_1) &= \gamma_{iklj}(-\omega_4,\omega_2,\omega_1,\omega_3) = \\
\gamma_{iljk}(-\omega_4,\omega_1,\omega_3,\omega_2) &= \gamma_{ilkj}(-\omega_4,\omega_1,\omega_2,\omega_3) = \\
\gamma_{i\sigma(jkl)}(-\omega_4,&\sigma(\omega_3,\omega_2,\omega_1))
\end{aligned} \qquad (26)$$

For special choices of the involved frequencies with respect to the resonances of the material the permutation symmetry of the nonlinearity can be further developed. These permutation symmetries are normally referred to as Kleinman relations [7]. If none of the frequencies is in resonance with a transition frequency of the material and consequently no absorption occurs, then also the spatial index i and the frequency ω_4 of the nonlinear polarization can be permuted.

$$\gamma_{ijkl}(-\omega_4,\omega_3,\omega_2,\omega_1) = \gamma_{\sigma(ijkl)}(\sigma(-\omega_4,\omega_3,\omega_2,\omega_1)) \qquad (27)$$

If all involved frequencies are far from any resonance enhancement, one sometimes assumes that the nonlinearities are free of any wavelength dispersion. In the case of the second-order hyperpolarizability this means that it is not important whether the derivative with respect to E_j is of frequency ω_1, ω_2, ω_3, or ω_4. The permutations of the spatial indices σ_s and the frequencies σ_ω are independent from one another leading to

$$\begin{aligned}
\gamma_{ijkl}(-\omega_4,\omega_3,\omega_2,\omega_1) &= \gamma_{ijkl}(-\omega_4,\omega_3,\omega_1,\omega_2) = \\
\gamma_{ijkl}(-\omega_4,\omega_1,\omega_2,\omega_3) &= \gamma_{ijkl}(\omega_1,\omega_3,\omega_2,-\omega_4) \\
&= \ldots = \gamma_{\sigma_s(ijkl)}(\sigma_\omega(-\omega_4,\omega_3,\omega_2,\omega_1))
\end{aligned} \qquad (28)$$

2.1.4
Nonlinear Refractive Index and Two-Photon Absorption

For all-optical signal processing (based on pure $\chi^{(3)}$ nonlinearities) the important quantity is the nonlinear refractive index n_2. The resulting refractive index change can be induced either by the beam at the frequency ω_1 itself (self-phase modulation) or by a beam at another frequency ω_2 (cross-phase modulation).

$$n(\omega_1) = n_0(v_1) + n_2(\omega_1) \cdot I^{\omega_1} + n_2(\omega_1,\omega_2) \cdot I^{\omega_2} \qquad (29)$$

n_0 is the normal linear refractive index at low intensities I. The relation between the nonlinear optical susceptibilities $\chi^{(3)}$ and n_2 is given by

$$n_2(\omega_1) = \frac{3}{4\varepsilon_0 c n^2} \mathrm{Re}\left\{\chi^{(3)}(-\omega_1,\omega_1,-\omega_1,\omega_1)\right\} \qquad (30)$$

$$n_2(\omega_1,\omega_2) = \frac{3}{2\varepsilon_0 c n_{\omega_1} \cdot n_{\omega_2}} \mathrm{Re}\{\chi^{(3)}(-\omega_1,\omega_1,-\omega_2,\omega_2)\} \quad (31)$$

for the self- and cross-phase modulation process. "Re" refers to the real part of the expression in the brackets and c denotes the speed of light in a vacuum. The imaginary part of the third-order susceptibility $\chi^{(3)}$ is the origin of two-photon absorption (TPA) in which two photons are simultaneously absorbed in a material. The two photons can be either of the same frequency (degenerate TPA) or originate from two different frequencies. The total absorption α is then described by

$$\alpha(\omega_1) = \alpha_0(\omega_1) + \alpha_2(\omega_1) \cdot I^{\omega_1} + \alpha_2(\omega_1,\omega_2) \cdot I^{\omega_2} \quad (32)$$

with α_0 the linear absorption coefficient.

These processes are directly proportional to the imaginary part of the third-order susceptibility $\chi^{(3)}$ of the self- and cross-phase modulation process.

$$\alpha_2(\omega_1) = \frac{3\pi}{\varepsilon_0 c \lambda n_0^2} \mathrm{Im}\{\chi^{(3)}(-\omega_1,\omega_1,-\omega_1,\omega_1)\} \quad (33)$$

$$\alpha_2(\omega_1,\omega_2) = \frac{6\pi}{\varepsilon_0 c \lambda_1 n_{\omega_1} \cdot n_{\omega_2}} \mathrm{Im}\{\chi^{(3)}(-\omega_1,\omega_1,-\omega_2,\omega_2)\} \quad (34)$$

Some authors are using the same relations between the third-order susceptibility $\chi^{(3)}$ and the nonlinear refractive index and the two-photon absorption coefficient for the self- and cross-phase modulation process.

$$n_2'(\omega_1,\omega_2) = \frac{3}{4\varepsilon_0 c n_{\varepsilon_1}^2} \mathrm{Re}\{\chi^{(3)}(-\omega_1,\omega_1,-\omega_2,\omega_2)\} \quad (35)$$

$$\alpha_2'(\omega_1,\omega_2) = \frac{3\pi}{\varepsilon_0 c \lambda_1 n_{\omega_1} \cdot n_{\omega_2}} \mathrm{Im}\{\chi^{(3)}(-\omega_1,\omega_1,-\omega_2,\omega_2)\} \quad (36)$$

Consequently, an additional factor 2 has to be encountered in the equations for the total refractive index n (Eq.(29)) and the overall absorption coefficient α (Eq.(32))

$$n(\omega_1) = n_0(\omega_1) + n_2(\omega_1) \cdot I^{\omega_1} + 2n_2'(\omega_1,\omega_2) \cdot I^{\omega_2} \quad (37)$$

$$\alpha(\omega_1) = \alpha_0(\omega_1) + \alpha_2(\omega_1) \cdot I^{\omega_1} + 2\alpha_2'(\omega_1,\omega_2) \cdot I^{\omega_2} \quad (38)$$

For many applications $\chi^{(3)}$ should be purely real in order to induce a maximum phase-shift without optical loss (see below). The imaginary part related to two-photon absorption is often considered an adverse effect, especially for all-optical signal processing. However, two-photon absorption can also be very important for e.g. two-photon excited fluorescence microscopy, optical limiting, three-dimensional optical data storage, and two-photon induced biological caging studies [8]. Furthermore two-photon absorption is a useful spectroscopic tool for the study of excited states that are forbidden in the dipole approximation. One of the main advantages of two-photon absorption is the excellent spatial resolution in three dimensions. Furthermore, the penetration into absorbing or scattering media can be greatly improved if the fundamental wave is not absorbed and absorption only takes place at the point of strong focussing.

2.2
Relations Between Microscopic and Macroscopic Coefficients

In the relations between the macroscopic susceptibilities $\chi^{(1)}$, $\chi^{(2)}$, $\chi^{(3)}$ and the microscopic or molecular properties α, β, γ, local field corrections have to be considered as explained above. The molecule experiences the external electric field $\tilde{E}$ altered by the polarization of the surrounding material leading to a local electric field $\tilde{E}_{loc}$. In the most widely used approach to approximate the local electric field the molecule sits in a spherical cavity of a homogenous media. According to Lorentz the local electric field [9] is

$$\tilde{E}_{loc} = \left(1 + \frac{\chi^{(1)}(-\omega,\omega)}{3}\right)\tilde{E} = \left(\frac{n_\omega^2 + 2}{3}\right)\tilde{E} = f^\omega \tilde{E} \tag{39}$$

with f^ω the local field correction factor. For a parallel alignment of the molecular polarizations $\tilde{p}$ and a molecular density N the macroscopic polarization is $\tilde{P} = N \cdot \tilde{p}$ and the relations between the microscopic and macroscopic properties follow as

$$\chi^{(1)}(-\omega,\omega) = N f^\omega \cdot \alpha(-\omega,\omega) \tag{40}$$

$$\chi^{(2)}(-\omega_3,\omega_2,\omega_1) = N f^{\omega_1} f^{\omega_2} f^{\omega_3} \cdot \beta(-\omega_3,\omega_2,\omega_1) \tag{41}$$

$$\chi^{(3)}(-\omega_4,\omega_3,\omega_2,\omega_1) = N f^{\omega_1} f^{\omega_2} f^{\omega_3} f^{\omega_4} \cdot \gamma(-\omega_4,\omega_3,\omega_2,\omega_1) \tag{42}$$

If the intermolecular arrangement is more complicated, the polarizabilites and hyperpolarizabilites have to be averaged over the molecular distribution. For a distribution parameter Ω with the probability $W(\Omega)$ the averaged second-order hyperpolarizability is

$$\langle \gamma \rangle = \frac{\int W(\Omega) \cdot \gamma(\Omega) d\Omega}{\int W(\Omega) d\Omega} \tag{43}$$

The definition for the averages $\langle \alpha \rangle$ and $\langle \beta \rangle$ are analogous. Most of the third-order nonlinear optical measurements are performed in solution or unpoled polymer films. For these isotropic distributions of molecules the isotropic averages $\langle \alpha \rangle$, $\langle \beta \rangle$, and $\langle \gamma \rangle$ of the molecular polarizability and hyperpolarizability are determined by integrating over all possible orientations with a constant probability $W(\Omega)$. The second rank tensor $\langle \alpha \rangle$ is diagonal with only one element α_{rot}. The rotational average α_{rot} is the normal arithmetic average of the diagonal elements α_{xx}, α_{yy}, and α_{zz} from the initial polarizability tensor.

$$\langle \alpha \rangle = \begin{pmatrix} \alpha_{rot} & 0 & 0 \\ 0 & \alpha_{rot} & 0 \\ 0 & 0 & \alpha_{rot} \end{pmatrix} \text{ with } \alpha_{rot} = \frac{\alpha_{xx} + \alpha_{yy} + \alpha_{zz}}{3} \tag{44}$$

For the isotropic average of the first-order hyperpolarizability β only elements with all three spatial indices different from one another are non zero. $\text{sgn}(\sigma)$ is the sign of the permutation $\sigma(xyz) = ijk$ of the spatial indices. For the nonlinear polarization in z-direction for example, the two elements $\langle \beta \rangle_{zxy} = +\beta_{chiral}$ and $\langle \beta \rangle_{zyx} = -\beta_{chiral}$ are different from zero. Far from any resonance Kleinman symmetry is valid ($\beta_{xyz} = \beta_{yzx} = \beta_{zxy} = \beta_{yxz} = \beta_{zyx} = \beta_{xzy}$) and the terms in the numerator of β_{chiral} cancel each other resulting in $\langle \beta \rangle_{ijk} = 0$ for all averaged tensor elements of the first-order hyperpolarizability.

$$\langle \beta \rangle_{ijk} = \begin{cases} \text{sgn}(\sigma) \cdot \beta_{chiral} & ijk = \sigma(xyz) \\ 0 & \text{else} \end{cases} \tag{45}$$

$$\beta_{chiral} = \frac{\beta_{xyz} + \beta_{yzx} + \beta_{zxy} - \beta_{yxz} - \beta_{zyx} - \beta_{xzy}}{6} \tag{46}$$

The averaged tensor elements of the second-order hyperpolarizability γ are only different from zero if at least twice the same index appears. In most experiments the polarizations of the electric fields are always kept parallel to the same direction (e.g. x). The tensor element of interest is then always $\langle \gamma \rangle_{xxxx} = \gamma_{rot}$.

$$\langle\gamma\rangle_{ijkl} = \begin{cases} \gamma_{rot} = \dfrac{1}{5}\sum_{\eta}\gamma_{\eta\eta\eta\eta} + \dfrac{1}{15}\sum_{\eta\neq\upsilon}(\gamma_{\eta\eta\upsilon\upsilon} + \gamma_{\eta\upsilon\eta\upsilon} + \gamma_{\eta\upsilon\upsilon\eta}) & i=j=k=l \\[4pt] \dfrac{1}{15}\sum_{\eta}\gamma_{\eta\eta\eta\eta} + \dfrac{2}{15}\sum_{\eta\neq\upsilon}\gamma_{\eta\eta\upsilon\upsilon} - \dfrac{1}{30}\sum_{\eta\neq\upsilon}(\gamma_{\eta\upsilon\eta\upsilon} + \gamma_{\eta\upsilon\upsilon\eta}) & i=j, k=l \\[4pt] \dfrac{1}{15}\sum_{\eta}\gamma_{\eta\eta\eta\eta} + \dfrac{2}{15}\sum_{\eta\neq\upsilon}\gamma_{\eta\upsilon\eta\upsilon} - \dfrac{1}{30}\sum_{\eta\neq\upsilon}(\gamma_{\eta\eta\upsilon\upsilon} + \gamma_{\eta\upsilon\upsilon\eta}) & i=k, j=l \\[4pt] \dfrac{1}{15}\sum_{\eta}\gamma_{\eta\eta\eta\eta} + \dfrac{2}{15}\sum_{\eta\neq\upsilon}\gamma_{\eta\upsilon\upsilon\eta} - \dfrac{1}{30}\sum_{\eta\neq\upsilon}(\gamma_{\eta\eta\upsilon\upsilon} + \gamma_{\eta\upsilon\eta\upsilon}) & i=l, j=k \\[4pt] 0 & \text{else} \end{cases} \quad (47)$$

In a molecule with a one-dimensional conjugation path for the π-electrons one can assume that the tensor element with all indices in this backbone direction (x) dominates over all other tensor elements. The rotational average for these 1D-molecules simplify to

$$\gamma_{rot} = \gamma_{xxxx}/5 \qquad (48)$$

In the case of a two-dimensionally conjugated π-electron system in e.g. the (x,y)-plane all tensor elements with the indices in the z-direction can be omitted.

$$\gamma_{rot} = \frac{1}{5}(\gamma_{xxxx} + \gamma_{yyyy}) + \frac{1}{15}(\gamma_{xxyy} + \gamma_{xyxy} + \gamma_{xyyx} + \gamma_{yyxx} + \gamma_{yxyx} + \gamma_{yxxy}) \qquad (49)$$

If some of the involved frequencies are degenerate the above expressions for 2D-molecules can be further simplified due to the permutations symmetry of the nonlinearities. The two-photon absorption is frequently characterized with a coefficient δ defined as the loss of photons through the two-photon process upon propagation along the z-direction

$$\frac{dF}{dz} = -N\delta(\omega)\cdot F^2 \qquad (50)$$

with $F = I/\hbar\omega$ the fluence of photons and N the molecular density. With the intensity relation $dI/dz = -\alpha_2(\omega)\cdot I_2$ we obtain for the degenerate case of the two-photon absorption process

$$\delta(\omega) = \frac{\hbar\omega}{N}\cdot\alpha_2(\omega) = \frac{6\pi^2\hbar}{\varepsilon_0 n^2\lambda^2}\left(\frac{n^2+2}{3}\right)^4 \mathrm{Im}\gamma(-\omega,\omega,-\omega,\omega) \qquad (51)$$

If the two photons in the absorption process are of different frequencies ω_1 and ω_2, the two-photon cross-section $\delta(\omega_1,\omega_2)$ for the photon fluence at ω_1 is

$$\delta(\omega_1,\omega_2) = \frac{\hbar\omega_1}{N} \cdot \alpha_2(\omega_1,\omega_2)$$

$$= \frac{12\pi^2 \hbar}{\varepsilon_0 n_{\omega_1} \cdot n_{\omega_2} \lambda_1^2} \left(\frac{n_{\omega_1}^2 + 2}{3}\right)^2 \left(\frac{n_{\omega_2}^2 + 2}{3}\right)^2 \tag{52}$$

$$\operatorname{Im}\gamma(-\omega_1,\omega_1,-\omega_2,\omega_2)$$

Reliable guidelines for an optimization of δ appeared only recently and are discussed in Sect 4.5.

2.3 Reference Values

On the route to all-optical signal processing the development of materials with large third-order nonlinear optical effects is of decisive importance. For the material characterization and the assessment of its usefulness for applications the absolute value of the third-order nonlinear optical susceptibility $\chi^{(3)}$ has to be known. Since most measurements are performed relative to a reference material, the establishment of a well accepted value for a standard material is important.

In the case of degenerate frequencies also the nonlinear refractive index n_2 can serve as the reference parameter as it is proportional to the real part of the third-order susceptibility $\chi^{(3)}$. In DFWM and Z-scan experiments the liquid CS_2 is most commonly used as the reference material. The main contribution to the nonlinearity of CS_2 originates in the dipolar orientation of the molecules in the electric field of the laser pulse. Therefore the dispersion of the nonlinearity ouside an electronic resonance should be flat and provide a good reference value over a broad wavelength range. The nonlinear refractive index n_2 is $(3.4\pm0.8)\times10^{-5}$ cm²/GW at the wavelength $\lambda=532$ nm and $(3.1\pm0.5)\times10^{-5}$ cm²/GW at $\lambda=1064$ nm [6]. The weighted average yields a value for the nonlinear refractive index n_2 and the third-order susceptibility $\chi^{(3)}$ for wavelengths between 500 nm and 1100 nm.

$$n_2(CS_2) = (3.2\pm0.4)\cdot 10^{-5} \text{ cm}^2/\text{GW}$$

$$\chi^{(3)}_{CS_2}(-\omega,\omega,-\omega,\omega) = (3.0\pm0.4)\cdot 10^{-20} \text{ m}^2/\text{V}^2 = (2.1\pm0.3)\cdot 10^{-12} \text{ esu}$$

For wavelengths shorter than 500 nm the electronic contribution from the resonance in the UV has growing importance and the above reference value is too small. For wavelengths further in the infrared the reference values are still adequate unless a vibrational resonance is met. At the wavelength $\lambda=10.6$ µm the nonlinear refractive index $n_2=(3.9\pm1.6)\times10^{-5}$ cm²/GW is obtained [6] in agreement with the above values at shorter wavelengths.

The orientational process of CS_2 happens on the time scale of a few picoseconds [10]. The reference value can therefore only be used for laser pulses of 10 ps or longer. For shorter laser pulses the orientational contribution is "frozen in" and only the vibrational and electronic contributions contribute to the nonlinearity. Consequently a different material or reference value has to be deployed.

The nonlinear refractive index n_2 of fused silica has been measured by different methods and a standard value has recently been reported [11] for the wavelength $\lambda=1053$ nm.

$$n_2(fs) = (2.74\pm0.17)\cdot10^{-7} \text{ cm}^2/\text{GW}$$

$$\chi^{(3)}_{fs}(-\omega,\omega,-\omega,\omega) = (2.04\pm0.13)\cdot10^{-22} \text{ m}^2/\text{V}^2 = (1.46\pm0.09)\cdot10^{-14} \text{ esu}$$

The nonlinearities of fused silica are two orders of magnitude smaller than for CS_2. They are consequently harder to measure and the application as a reference material is more difficult. However, the nonlinearity is the same for pulse lengths from nanosecond to femtosecond and therefore no problems in connection to the pulse duration appear.

In third-harmonic generation experiments the situation is somewhat different. Most people so far used the value $\chi^{(3)}(-3\omega,\omega,\omega,\omega)=3.89\cdot10^{-22}$ m²/V² ($2.79\cdot10^{-14}$ esu) of fused silica, which was obtained by cascading of second-order susceptibilities $\chi^{(2)}$ [12, 13]. The values for the second-order susceptibilities $\chi^{(2)}$, which this calibration approach is based on, changed in past years and more recent cascading experiments revealed a value for the third-order susceptibility $\chi^{(3)}(-3\omega,\omega,\omega,\omega)$ of fused silica a factor 2.4 to 3 smaller than the original one [14, 15]. These recent experiments together with very new gas-phase third-harmonic generation experiments [16] yield a new more reliable reference value for the third-order susceptibility $\chi^{(3)}(-3\omega,\omega,\omega,\omega)$ of fused silica.

$$\chi^{(3)}_{fs}(-3\omega,\omega,\omega,\omega) = (2.0\pm0.2)\cdot10^{-22} \text{ m}^2/\text{V}^2 = (1.4\pm0.15)\cdot10^{-14} \text{ esu at 1064nm}$$

$$\chi^{(3)}_{fs}(-3\omega,\omega,\omega,\omega) = (1.6\pm0.2)\cdot10^{-22} \text{ m}^2/\text{V}^2 = (1.1\pm0.15)\cdot10^{-14} \text{ esu at 1907nm}$$

The reference values discussed in this section are used throughout this book chapter.

2.4
Material Requirements for All-Optical Signal Processing

The material in consideration for all-optical signal processing can be divided in three groups: glasses, semiconductors, and organics. For comparison of the different nonlinear optical materials the definition of suitable figures of merit is helpful. However, some introductory remarks should emphasize some further differences between glasses, semiconductors and organics which have to be kept in mind when comparing the figures of merit.

For applications an integration of the elements on a wafer or some other substrate is important to allow a large scale production. Propagation distances of at most centimeters have to be sufficient to generate the required nonlinear optical phase shift. In glass fibers with among the best figures of merit (see below) interaction lengths are typically meters, disabling an integrated design. Therefore, for an integrated design semiconductors and organics are surely the better choice, although all-optical switching has been demonstrated in glass fibers with configurations as, for example, the nonlinear optical loop mirror (NOLM).

Large nonlinearities based on saturated absorption or bandfilling effects are reported for semiconductors. The response of these nonlinearities is fast but recovers only slowly due to the created excited state population. Decay times of the excited states on the order of some hundred picoseconds to nanoseconds are detrimental for all-optical switching with large repetition rates.

2.4.1
Figures of Merit

To implement an all-optical switch based on the nonlinear refractive index $n(I) = n_0 + n_2 \cdot I$ a nonlinear phase ϕ_{NL} shift of 2π or a multiple of 2π is required, depending on the geometry of the switch [1]. As the absorption of the material attenuates the intensity and consequently diminishes the nonlinear phase shift, low absorption is advantageous. To determine the net nonlinear phase shift over the propagation distance the signal depletion by absorption has to be included.

As large intensities are necessary to induce a nonlinear refractive index change, not only linear but also multiple photon absorption processes have to be considered. The intensity I and the induced nonlinear phase shift ϕ_{NL} are coupled differential equations as a function of the propagation direction z.

$$dI = -(\alpha_0 I + \alpha_2 I^2 + \alpha_3 I^3) \cdot dz \tag{53}$$

$$d\phi_{NL} = \frac{2\pi}{\lambda} \cdot n_2 I(z) \cdot dz \tag{54}$$

An analytical solution for simultaneous inclusion of one-, two-, and three-photon absorption α_0, α_2, α_3 is not possible. In reality, one absorption process is usually dominating for a chosen wavelength and the differential equations can be solved by neglecting the other absorption terms.

dominating one photon absorption ($\alpha_0 \gg \alpha_2 I, \alpha_3 I^2$):

$$I(z) = I_0 e^{-\alpha_0 z}, \quad \phi_{NL}(z) = \frac{2\pi}{\lambda} \cdot \frac{n_2 I_0}{\alpha_0} \left(1 - e^{-\alpha_0 z}\right) \tag{55}$$

dominating two-photon absorption ($\alpha_2 I \gg \alpha_0, \alpha_3 I^2$):

$$I(z) = \frac{I_0}{1 + \alpha_2 I_0 z}, \quad \phi_{NL}(z) = \frac{2\pi}{\lambda} \cdot \frac{n_2}{\alpha_2} \ln(1 + \alpha_2 I_0 z) \tag{56}$$

dominating three-photon absorption ($\alpha_3 I^2 \gg \alpha_0, \alpha_2 I$):

$$I(z) = \frac{I_0}{\sqrt{1+2\alpha_3 I_0^2 z}}, \quad \phi_{NL}(z) = \frac{2\pi}{\lambda} \cdot \frac{n_2}{\alpha_3 I_0} \left(\sqrt{1+\alpha_3 I_0^2 z} - 1 \right) \qquad (57)$$

For a dominating one-photon absorption the nonlinear phase shift grows asymptotically to a maximal value of $\phi_{NL} = 2\pi \cdot n_2 I_0 / \alpha_0 \lambda$ (Fig. 1a). In the cases of a dominating two- and three-photon absorption (Figs. 1b and 1c) the nonlinear phase shift does not saturate but deviates significantly from the linear behaviour as expected for absence of absorption (dashed line).

The typical attainable nonlinear phase shifts $\phi_{NL}(z)$ for the different absorption processes are given by the second terms in the above equations containing the nonlinear refractive indices n_2 and the absorption coefficients α_i. Figures of merit can be defined for each absorption process by demanding a typical nonlinear phase shift $\phi_{NL} > 2\pi$ [17].

$$\text{one-photon figure of merit: } W = \frac{n_2 I_0}{\alpha_0 \lambda} > 1 \qquad (58)$$

$$\text{two-photon figure of merit: } T^{-1} = \frac{n_2}{\alpha_2 \lambda} > 1 \qquad (59)$$

$$\text{three-photon figure of merit: } V^{-1} = \frac{n_2}{\alpha_3 \lambda I_0} > 1 \qquad (60)$$

The one-photon figure of merit W portrays the importance of avoiding any direct, single photon resonance. In principle, the one-photon figure of merit can be always tuned above unity by increasing the incident intensity I_0. In reality, this approach is not feasible as the energy deposited in the absorption process damages the nonlinear optical material for too large powers. It is therefore more illustrative to tabulate the necessary intensity $I_0(W>1)$ to reach a one-photon figure of merit larger than one instead of listing the figure of merit W itself.

No external parameter occurs in the two-photon figure of merit T^{-1}. It is a function of the material only without any possibility to tune the figure of merit by external means. With the nonlinear refractive index n_2 and the two-photon absorption coefficient α_2 proportional to the real and imaginary part of the complex third-order susceptibility $\chi^{(3)} = |\chi^{(3)}| \cdot e^{i\varphi^{(3)}}$ the two-photon figure of merit T^{-1} can be rewritten as a function of the nonlinearity phase $\varphi^{(3)}$.

$$T^{-1} = \frac{1}{4\pi} \cdot \frac{\text{Re}\{\chi^{(3)}\}}{\text{Im}\{\chi^{(3)}\}} > 1 \Leftrightarrow \tan\varphi^{(3)} < \frac{1}{4\pi} \Leftrightarrow \varphi^{(3)} < 4.55° \qquad (61)$$

Molecular Design for Third-Order Nonlinear Optics

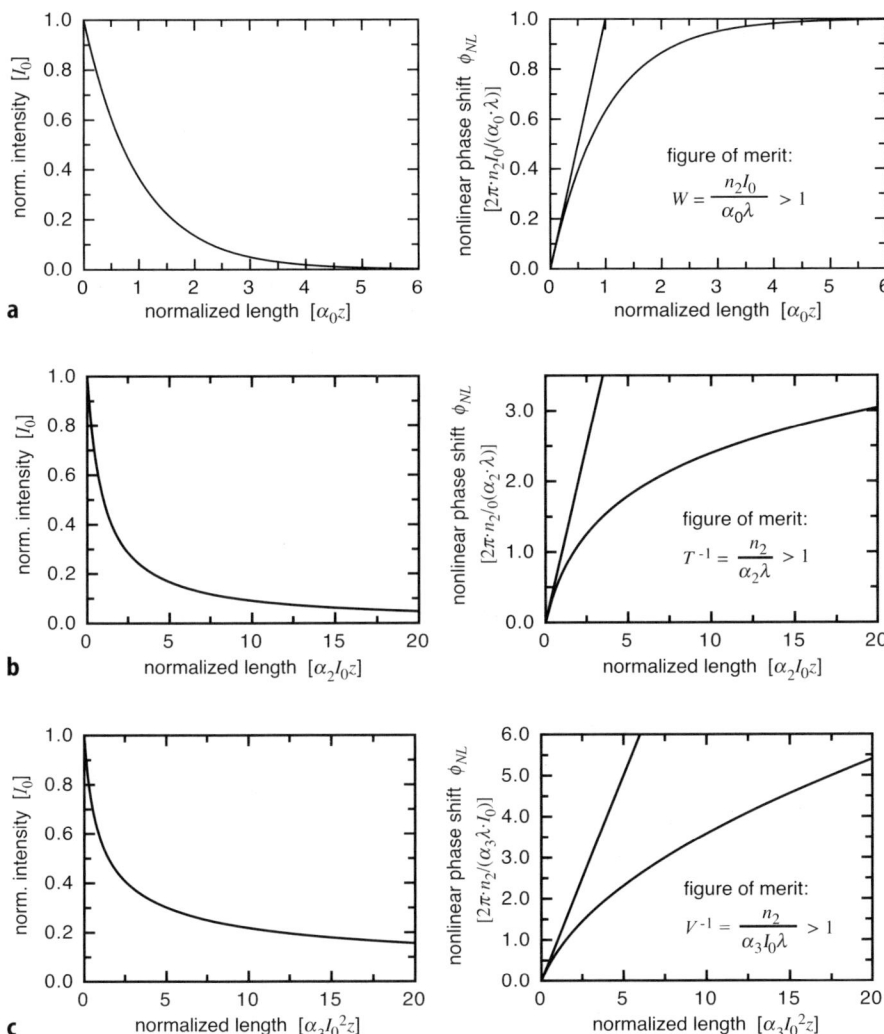

Fig. 1a–c. Absorption and nonlinear phase shift as a function of the propagation distance for a dominating **a** one-, **b** two-, and **c** three-photon absorption process. The dashed line indicates the nonlinear phase shift without any absorption

In consequence, any dominating two-photon resonance has to be avoided to maintain a small phase of the third-order susceptibility $\chi^{(3)}$.

The three-photon figure of merit V^{-1} is less important in practice and is only seldom given for nonlinear optical materials. By decreasing the incident intensity I_0 in the denominator of V^{-1} the three-photon figure of merit can be increased and brought above unity.

2.4.2
Nonlinearity and Intensity Requirements

Neglecting the absorption processes in this section the nonlinear phase shift ϕ_{NL} is proportional to the nonlinear refractive index, the intensity, and the interaction length. Keeping dimensions of centimeters in mind, either the nonlinearity n_2 or the intensity I can be increased to create the necessary nonlinear phase shift.

The intensity is, on the one hand, bound to the laser systems in telecommunication technology or the laboratory, and, on the other hand, the intensity may reach the damage threshold of the material and create irreversible degradation. In Table 1 the necessary nonlinearities n_2 to reach a 2π-phase shift over one centimeter propagation distance in a waveguide are roughly estimated for some common laser sources.

For the continuous-wave semiconductor (AlGaAs) distributed feed back lasers (DFB) used in telecommunication technology today a nonlinear refractive index $n_2=3$ cm^2/GW would be necessary for all-optical switching, which is presumably impossible to reach. Including an Erbium doped fiber amplifier (EDFA) the needed nonlinearity drops only by a factor of ten.

Moving from the single mode DFB lasers to a multimode, continuous-wave high-power pump diode as for example used to pump the EDFAs, the necessary nonlinearity drops clearly but whether the material will sustain the power of two Watts on the dimensions of micrometers is questionable.

The above discussion shows that the realization of all-optical switching with continuous wave laser sources is doubtful. Short laser pulses are a requisite as they provide large powers without damaging energy levels. For a mode-locked 10 ps Nd:YLF laser (Lightwave) with an averaged power of 500 mW and 100 MHz repetition rate a peak power in the order of 500 W is reached. In this case the necessary nonlinearity drops down to achievable values of about only twenty times fused silica. The investigation of all-optical switching with these

Table 1. Necessary nonlinearities for a nonlinear phase shift of 2π over a propagation distance of one centimeter for different laser sources. A refractive index of n =1.5 is assumed for the nonlinear material. Note that the tabulated values are only estimates and correct only on the scale of one order of magnitude. But the consequences we conclude from these estimations are nevertheless correct and important

Laser	DFB laser (cw)	DFB/EDFA (cw)	High-power cw-laser diode	Nd:YLF (10 ps pulses)
Laser power	5 mW	50 mW	2 W	500 W
Wavelength	1550 nm	1550 nm	980 nm	1047 nm
Waveguide size	3 µm	3 µm	2 µm	2 µm
n_2 [cm^2/GW]	3	0.3	2×10^{-3}	1×10^{-5}
$\chi^{(3)}$ [10^{-20}m^2/V^2]	1.5×10^5	1.5×10^4	100	0.5
$\chi^{(3)}/\chi^{(3)}_{fs}$	8×10^6	8×10^5	5000	20

laboratory lasers is possible. To introduce all-optical switching in communication systems a change to pulsed laser systems is essential as it may happen with the advent of soliton technology.

2.4.3
Further Critical Factors

The device performance of any application in the real world has to be guaranteed for a certain time without or only minor degradation. Some 10,000 hours of operation at temperatures up to 80 °C under illumination are required. Research about the lifetime of organic nonlinear optical material has only started in the past years as the chances for applications based on the second-order susceptibilities $\chi^{(2)}$ have improved. Nevertheless, only little is known about the longtime effects in nonlinear materials under real world conditions.

Concerning the chemical stability, oxidation by the ambient oxygen is of major concern [18]. Sealing and encapsulation of the material improves the lifetime. For photostability any absorption process has to be strongly avoided [19, 20]. Not only the linear absorption α_0 but also the two-photon absorption α_2 is of concern. The prevention of any absorption is therefore not only important for the figures of merit but also for the photostability.

The stability problem is currently the main disadvantage of organic materials in nonlinear optics. The firmness of glasses and semiconductors is surely superior.

An advantage of polymers is the well developed knowledge in polymer science. It is a mature technology and is compatible to other materials. Methods to produce various shapes and to structure elements are established. Specific procedures to change the mechanical or thermal properties are known.

The price of the polymer itself is not very critical as for waveguides in thin films only very small quantities are needed. Costs of 10,000 $ per gram result in only 10 cents per square inch if a two micrometer thick film and a 50% material loss in the production process are assumed. As only small quantities are needed and not tens of kilograms the possible market would be more similar to pharmaceuticals than to the classical dye market.

However, the major part of the added value is most probably not created by the material itself but in its structuring and the assembly to a device, which is enabled by the material properties.

3
Measurement Techniques

The most widely employed material characterization techniques in third-order nonlinear optics are third-harmonic generation (THG) [21], degenerate four wave-mixing (DFWM) [22], Z-scan [6], and optical limiting by direct two-photon absorption (TPA) and fluorescence spectroscopy induced by TPA [23]. All of them will be discussed in the following. Further measurement techniques such as electric-field induced second-harmonic generation (EFISH) [24], optical Kerr gating, electric field induced absorption [25], nonlinear prism coupling [26], and pump-probe schemes are described in the literature [27] and will not be covered within this chapter.

Special care has to be taken if the signal frequency is identical with one of the input frequencies as e.g. in DFWM and Z-scan. If the condition $\omega_i=\omega_j$ in the third-order susceptibility $\chi^{(3)}(\omega_4,\omega_3,\omega_2,\omega_1)$ is fulfilled, the multiplication of the electric fields with the same positive and negative frequency will lead to a time independent product which is proportional to the intensity. Consequently, nonlinear optical effects on the time scale of the employed laser pulses can additionally contribute and add to the instantaneous nonlinear response of the electrons which happens on the time scale of the inverse light frequency.

For DFWM and Z-scan with picosecond laser pulses molecular vibrations (or optical phonons in crystals) and reorientation of small molecules can add contributions to the electronic nonlinearity. For longer laser pulses even large molecules can orient and also thermal contributions can occur.

Thermal contributions to the nonlinearity can easily dominate any other effect and are normally negative in sign. To avoid them the energy deposition in the material should be kept low, which is best ensured by using low repetition rate laser pulses.

3.1
Third-Harmonic Generation

In the third-harmonic generation, the third-order susceptibility leads to a nonlinear polarization component which oscillates at the third-harmonic frequency of the incident laser beam. This leads to a light wave at the third-harmonic frequency of the fundamental wave. As optical frequencies are involved and since the output frequency is different from the input frequency only the electronic nonlinearities can participate without any contributions from thermal or orientational effects. Because one needs fast nonlinearities for all-optical signal processing, the main interest is directed towards the fast electronic nonlinearities. Therefore and also due to its simplicity, third-harmonic generation is a very attractive method to characterize newly developed materials.

The external electric field is assumed to be parallel to the x-axis. In the case of an isotropic solution only the element $\chi^{(3)}_{xxxx}(-3\omega,\omega,\omega,\omega)$ of the third-order susceptibility creates a polarization at 3ω, which is parallel to the incident electric field E^ω.

$$P^{3\omega} = \frac{\varepsilon_o}{4}\chi^{(3)}_{xxxx}(-3\omega,\omega,\omega,\omega)\left(E^\omega\right)^3 \qquad (62)$$

$$\chi^{(3)}_{xxxx}(-3\omega,\omega,\omega,\omega) = Nf_{3\omega}f_\omega^3\gamma^{THG}_{rot} \qquad (63)$$

The rotational average of the second-order hyperpolarizability in third harmonic generation is γ^{THG}_{rot}. For linear molecules the rotational average consist only of one element. For planar molecules there are four different elements.

1D-molecule: $\gamma_{rot}^{THG} = \gamma_{xxxx}/5$ (64)

2D-molecule: $\gamma_{rot}^{THG} = \frac{1}{5}\left(\gamma_{xxxx} + \gamma_{yyyy} + \gamma_{xxyy} + \gamma_{yyxx}\right)$ (65)

For third-harmonic generation measurements pulsed laser systems are used (example: wavelengths between λ=1064 nm and 2100 nm, nanosecond pulse duration, low repetition rate (e.g. 10 Hz)). Ideally, the third-harmonic signal is simultaneously measured on a sample and a reference fused silica plate (Fig. 2). By dividing the two signals the laser power fluctuations can be eliminated. The measurement beam is focused on the sample in the vacuum chamber, which is necessary as also the air can contribute to the THG signal. By the appropriate choice of the lenses in front of the chamber, the laser intensity in the windows is low enough to exclude any measurable THG signal generated from them. The focus of the laser is centered on the sample and the focal region is chosen to be much larger than the sample thickness in order to fulfill the plane wave approximation for the evaluation of the third-harmonic generation. After the sample

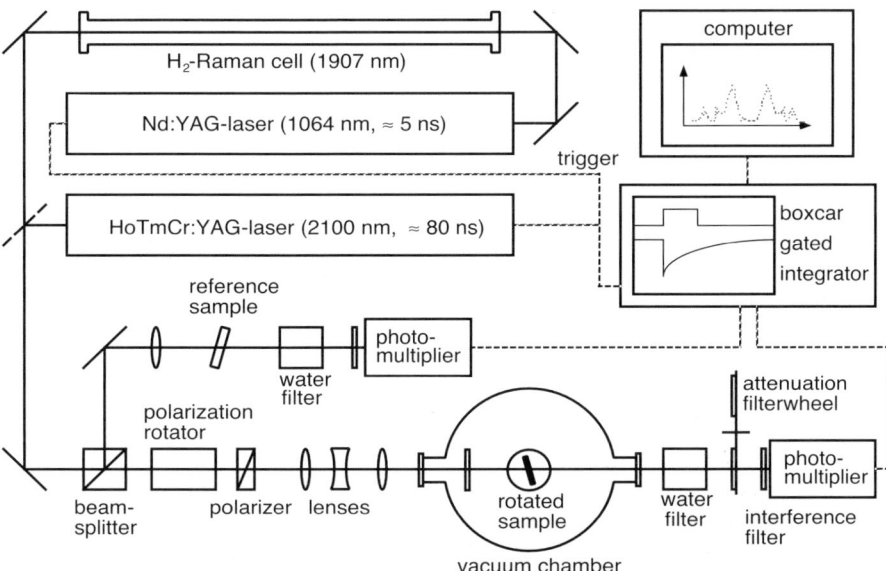

Fig. 2. Typical third-harmonic generation measurement. The beams from the laser source (here either a frequency shifted Nd:YAG or a HoTmCr:YAG laser) are split into a measurement and a reference beam. The polarization rotator and the polarizer serve as variable attenuator and yield the desired polarization of the input beam. The beam is focused on the sample in the vacuum chamber. The water filter removes the fundamental frequency and the attenuation filters limit the third-harmonic signal to the measurement range of the photomultiplier. The signal from the sample is divided by the reference signal and averaged with a boxcar gated integrator

chamber the fundamental laser frequency ω is removed with a water filter (for fundamental wavelengths between 1500 nm and 2100 nm) and the third-harmonic signal is attenuated to be within the linear range of the photomultiplier. A narrow bandpass interference filter reduces the noise from background light of the laboratory.

THG measurements can be either performed in solution or thin films. For thin films larger sample quantities are necessary (>50 mg) than for measurements in solution (1 mg), which can be of central importance for novel materials. With thin films, the THG experiments can be performed with the harmonic frequency in the absorption band and the phase of the second-order hyperpolarizability can be retrieved (see e.g. Refs.[28–30]).

We sketch in the following paragraphs the procedure to measure the second-order hyperpolarizability in solution. Figure 3 shows the situation at the sample for THG measurements in solution in more detail. The five third-harmonic contributions from the sample, the cell, and the air around the sample interfere with one another. By recording the Maker-Fringe pattern for a large range of angle, the different contributions can be separated, using the evaluation procedure of Ref. [3]. The absorption of the solution at the involved wavelengths is usually tested with a UV/VIS/NIR spectrometer (Lambda 9, Perkin-Elmer). In the case of residual absorption at the third-harmonic frequency the loss can be accounted for in the THG analysis with Beer's law [21]. If the sample is not completely dissolved an offset in the transparency region of the spectrum remains. Additionally, the sample in the THG cells should be checked under a microscope for possible inhomogeneities.

To determine the third-order susceptibility χ a step wise procedure is employed (Fig. 4). A measurement of a fused silica plate calibrates the THG setup. The measurement of the same fused silica plate in air atmosphere leads to the contribution of the air, which is a constant background reducing the signal in-

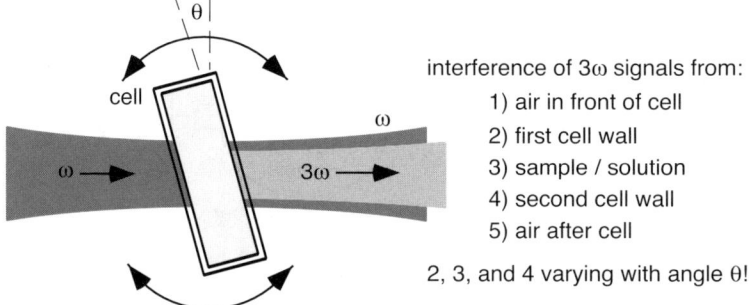

Fig. 3. Third-harmonic generation of a solution sample in a cell. The laser beam at ω creates third-harmonic signals in the air in front of the sample, in the front wall of the cell, in the solution, in the back wall, and finally in the air behind the sample. All five contribution interfere with one another. Contributions from the cell walls and the solution change differently with the rotation angle

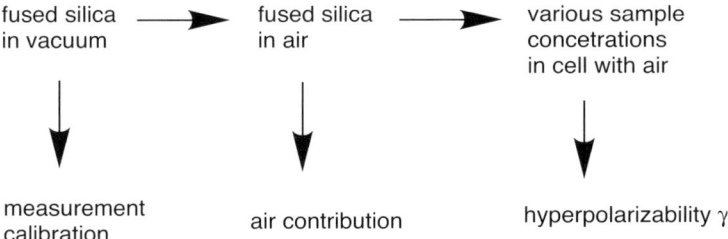

Fig. 4. Measurement sequence of third-harmonic generation in solution. A THG experiment of fused silica in vacuum yields the calibration of the whole setup. The same fused silica plate measured in air leads to the air contribution, which is afterwards taken into account in the THG measurements of the various sample concentrations

dependently of the rotation angle. The detailed derivation of the contribution of air is given in Refs.[3, 16]. In the following measurements of the sample solutions the contribution of the air is taken into account.

THG measurements are performed for different concentrations. From the third-order susceptibilities $\chi^{(3)}$ of the solution series the second-order hyperpolarizabilities γ of the molecules are determined by analyzing the data points with

$$\chi^{(3)}(-3\omega,\omega,\omega,\omega) = \frac{N_A f_\omega^3 f_{3\omega}}{1+m_m c/\rho}\left(c\gamma_{m,rot}^{THG} + \frac{\rho}{m_s}\gamma_{s,rot}^{THG}\right) \qquad (66)$$

$N_A = 6.022 \times 10^{23}$ mol^{-1} is Avogadro's number, ρ the solvent density, c the concentration of the sample in mol/l, m_m the molecular mass of the sample molecule, m_s the molecular mass of the solvent molecule, $\gamma_{s,rot}^{THG}$ the hyperpolarizability of the solvent molecule, and finally $\gamma_{m,rot}^{THG}$ the wanted hyperpolarizability of the sample molecule. Interactions between the dissolved chromophores are neglected and the density of the solution is assumed to remain constant. Unless extremely large concentrations are measured these two assumptions are fulfilled.

The third-order susceptibility of the solvent is the intersection of the concentration series with the axis at zero concentration (Fig. 5). The hyperpolarizability of the sample molecule is proportional to the slope of the concentration series. Generally, the molecular hyperpolarizability γ is retrieved with an experimental error of about ten percent.

The procedure for the THG measurements of thin films needs less steps. As the thin films can be directly measured in vacuum, the calibration of the air contribution is not required. Furthermore, the film substrate is usually fused silica that makes also the initial calibration of the THG setup with a fused silica plate superfluous. However, the film thickness and its refractive index have to be determined prior to the THG analysis.

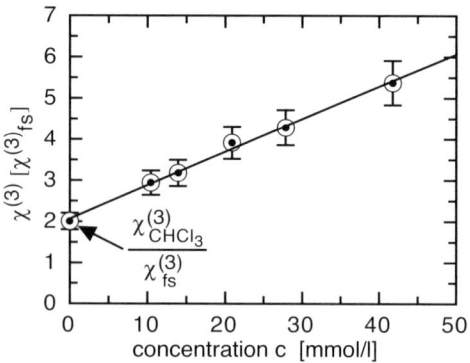

Fig. 5. Concentration series of a third-harmonic generation experiment. The third-order nonlinear optical susceptibilities of the solutions are analyzed with the solvent nonlinearity $\chi^{(3)}_{CHCl_3}$ for $c = 0$ and the slope $\partial\chi^{(3)}/\partial c$ proportional to the second-order hyperpolarizability γ of the molecule under investigation

3.2
Degenerate Four-Wave Mixing

In the degenerate four wave mixing (DFWM) experiment the third-order susceptibility $\chi^{(3)}(-\omega,\omega,-\omega,\omega)$ with degenerate frequencies can be determined [22]. This nonlinear susceptibility is directly proportional to the nonlinear refractive index n_2, which is used to describe optically induced refractive index changes. An advantage of this technique is the possibility to record the temporal shape of the third-order nonlinear optical signal.

The polarization of the incident electric fields is kept perpendicular to the plane of incidence (x-direction). In this case the nonlinear polarization is

$$P^{\omega,k_4} = \frac{3\varepsilon_0}{4}\chi^{(3)}_{xxxx}(-\omega,\omega,-\omega,\omega)\left(E^{\omega_1 k_1} E^{\omega_1 k_2} E^{\omega_1 k_3}\right) \quad (67)$$

$$\chi^{(3)}_{xxxx}(-\omega,\omega,-\omega,\omega) = N f_\omega^4 \gamma^{DFWM}_{rot} \quad (68)$$

Similar as for third-harmonic generation, in the rotational average of the second-order hyperpolarizability $\gamma^{DFWM}_{rot} = \langle\gamma(-\omega,\omega,-\omega,\omega)\rangle_{xxxx}$ the indices with the same frequency ω can be permuted. For linear molecules the rotational average consists again only of one element, planar molecules only of four different elements.

1D-molecule: $\gamma^{DFWM}_{rot} = \gamma_{xxxx}/5$ \quad (69)

2D-molecule: $\gamma^{DFWM}_{rot} = \frac{1}{5}\left(\gamma_{xxxx} + \gamma_{yyyy} + \gamma_{xxyy} + \gamma_{yyxx}\right)$ \quad (70)

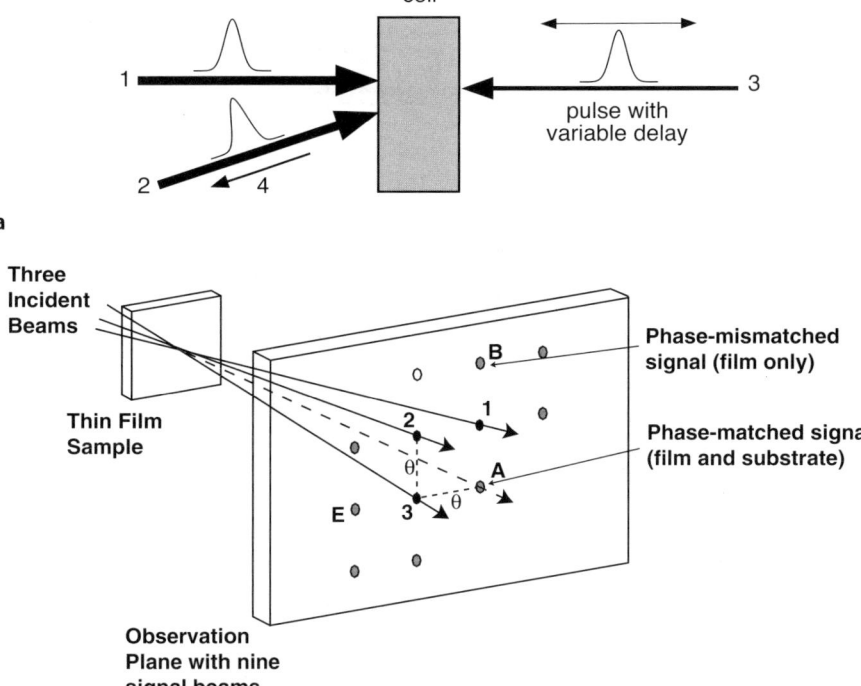

Fig. 6a,b. a Degenerate four wave mixing with counterpropagating pump beams. Beam 1 and 2 are spatially and temporally overlapping in the sample. The beam 3 has a variable delay to probe the temporal behavior of the nonlinear process. The signal with $k_4 = k_1 - k_2 + k_3$ is a phase-conjugated beam (4) in the direction opposite to beam 2. **b** Geometry for degenerate four wave mixing with copropagating pump beams. Three input beams create nine signal beams (see text for explanation)

Two different experimental geometries are used for DFWM: one with counterpropagating (Fig. 6a) and one with copropagating (Fig. 6b) laser pulses. The setup with counter propagating beams will be discussed first.

Figure 7 shows an experimental set-up based on a 10 ps Nd:YLF with 10 Hz repetition rate and a wavelength $\lambda = 1047$ nm. The initial beam is split up in three parts: two strong beams which are used as pump or writing beams and one weaker beam which is temporally delayed to read the nonlinearity induced by the writing beams. The phase conjugate signal beam is separated from the other beams and detected by a sensitive Joulemeter. To account for laser power fluctuations, the laser power is monitored with a second detector. To separate the signal from light scattered by optical elements, the distance between the sample and the detector has to be large. Spatial filtering of the signal beam and shielding of the detector from straylight improves the signal to noise ratio.

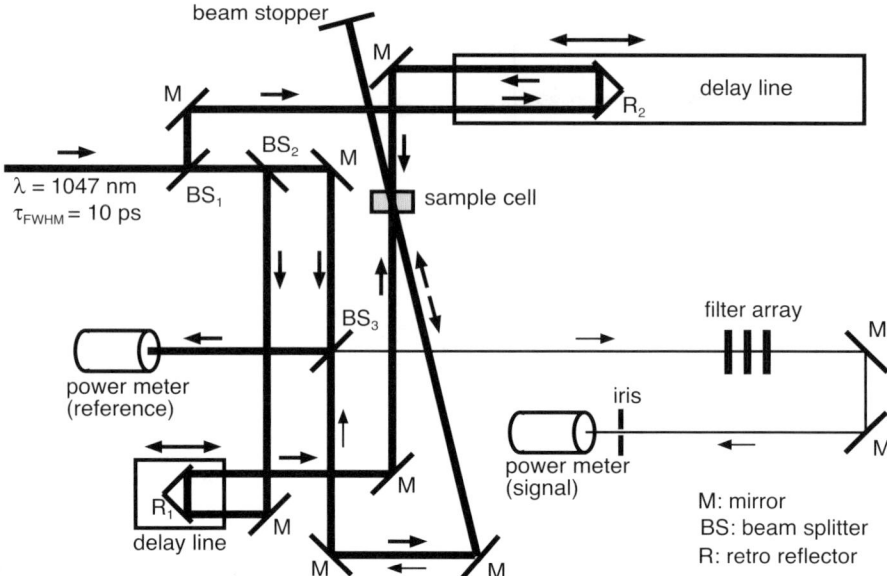

Fig. 7. Setup for the degenerate four wave mixing experiments. The input beam is split in three beams. The beam splitter BS_3 deflects a part of one of the pump beams to a power meter, which detects laser power fluctuations. The delay line with the retro reflector R_1 adjusts the temporal overlap of the two pump beams coming from the front side on the sample. The long delay line with retro reflector R_2 is moved to probe the temporal behavior of the nonlinearity in the sample. The phase conjugated signal beam propagates from the sample back to BS_3 and is then deflected through a stack of attenuation filters on a second power meter. An iris in front of the power meter increases the signal to noise ratio by removing scattered light

For a third-order susceptibility far from any resonance a signal is only detected when all three input beams overlap spatially and temporally. By delaying the reading beam the autocorrelation of the laser pulse is detected (Fig. 8). If the nonlinearity has contributions from processes leading to population changes (most often two-photon absorption in the case of DFWM) the two writing beams create a grating of molecules in the excited and ground state. The grating persists after the writing beams have passed the sample and can still be detected with the delayed reading beam. The finite lifetime of the excited state and thermal movements erase the grating on the time scale of some hundred picoseconds to some nanoseconds.

As in the case of third-harmonic generation measurements on molecules are often carried out in solution where the concentration of the molecules dissolved in the solvent is varied. The measurements are calibrated against a measurement with a reference. There is a wide choice of reference values, two of these are given below.

CS_2: $\chi^{(3)} = 3.0 \times 10^{-20} m^2/V^2 = 2.1 \times 10^{-12} esu$

$CHCl_3$: $\chi^{(3)} = 1.7 \times 10^{-21} m^2/V^2 = 1.2 \times 10^{-13} esu$

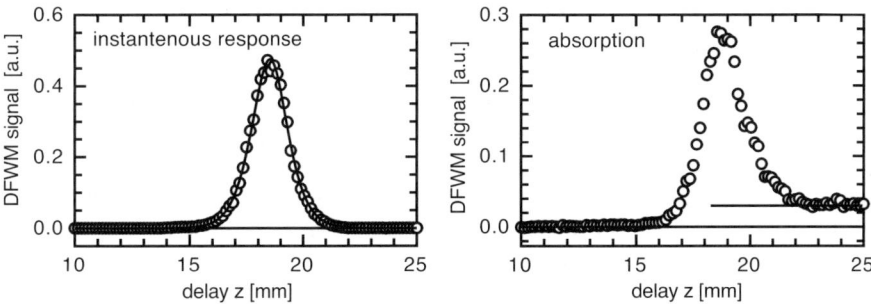

Fig. 8. Temporal scans of the phase conjugated output beam in degenerate four wave mixing experiments. On the *left side*, the nonlinearity is instantaneous with the DFWM signal being the autocorrelation of the laser pulse. On the *right side*, the DFWM signal does not vanish after the writing beams have left the sample. Two photon absorption has induced a grating of excited and ground state molecules which is read by the delayed beam and decays slowly

To be independent of the spatial beam profile the third-order susceptibility is usually measured relative to a known material. The signal from the sample is divided by the signal from the reference material I_4^{ref} and solved for $\chi^{(3)}$ to be determined.

$$\chi^{(3)} = \left(\frac{n_\omega + n_\omega^{fs}}{n_\omega^{ref} + n_\omega^{fs}}\right)^4 \frac{L^{ref}}{L} \sqrt{\frac{I_1^{ref} I_2^{ref} I_3^{ref} I_4}{I_1 I_2 I_3 I_4^{ref}}} \chi_{ref}^{(3)} \qquad (71)$$

All intensities are the external ones. n_ω, n_ω^{fs}, and n_ω^{ref} are the refractive indices of the sample, the fused silica glass windows, and the reference material, respectively, L and L^{ref} are the thicknesses of sample and reference, and the I_i are the light intensities. The analysis of the concentration series yields the molecular hyperpolarizability γ.

The susceptibilities and hyperpolarizabilities are generally complex coefficients. In the third-harmonic generation and normal degenerate four wave mixing measurement the amplitude of this complex coefficient is measured. We mentioned above that a remaining DFWM signal after the pump pulses already passed the material indicates a two-photon absorption and therefore a imaginary part of the third-order susceptibility. By an appropriate variation of the standard DFWM experiment the phase $\varphi^{(3)}$ of the third-order susceptibility $\chi^{(3)} = |\chi^{(3)}| e^{i\varphi^{(3)}}$ can be retrieved. If sufficiently large (typically above 40°) such a phase can be seen in the concentration dependence of the third-order susceptibility. Another method to determine $\varphi^{(3)}$ is the degenerate four-wave mixing interferometer [31].

The measurement of thin films is difficult as the created phase-conjugated signal is proportional to the sample length. The thick substrate with an only small nonlinearity dominates the signal unless the thin film has an extremely large third-order susceptibility. More elegant and more widely used for thin film measurements is the copropagating geometry of DFWM.

In the DFWM geometry with copropagating beams (Fig. 6b) the combination of the three incident fields creates multiple, spatially separated signal beams (see e.g. Ref. [32]). Only one of it is phase-matched and will accumulate signals from the substrate over the whole substrate thickness. In the other signal spots a combination of contributions from the substrate and the thin film are measured depending on the third-order susceptibility and the coherence length of the substrate and film. By analyzing the phase-matched and non-phase-matched signals from DFWM the nonlinearity of the film can be determined.

Due to the short interaction length in the thin film large intensities are required for a good signal to noise ratio. Most attractive for these measurments are consequently short laser pulses in the femtosecond regime.

3.3
Z-Scan

The Z-scan technique only requires a single focused beam and a circular aperture placed in the far field behind the sample [6]. It is mostly used to determine the nonlinear refractive index. The sample is translated along the optical axis through the focused laser beam (Fig. 9). The detector D1 measures the incoming intensity whereas detector D2 measures the intensity after the aperture. The translation leads to a change of the incident intensity and therefore to a variable refractive index change induced by n_2. By measuring the transmission through the circular aperture after the sample the sign as well as the magnitude of n_2 can be determined (Fig. 10a).

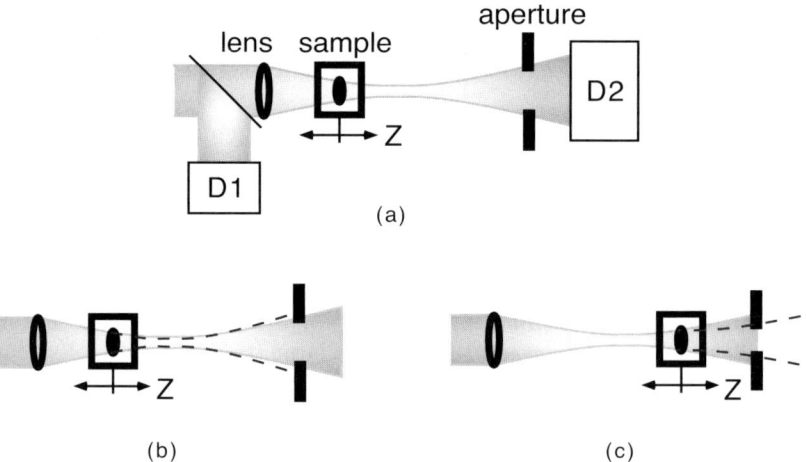

Fig. 9. Z-scan experiment. (*a*) Schematic of the set-up. (*b*) If the sample is placed to the left of the focus the transmission through the aperture is reduced as shown by the defocussing indicated by the broken line (for $n_2 > 0$). (*c*) If the sample is placed to the right of the focus the transmission through the aperture is increased as shown by the focusing indicated by the broken line (for $n_2 > 0$)

This can be understood as follows: we assume that $n_2>0$. For the sample placed far to the left of the focus the intensity and therefore also the effect of self-focusing is small. The transmission $T \propto D2/D1$ is therefore the same as the one without sample (ignoring absorption and reflection losses) and is usually normalized to one. If we move the sample towards the focus the self-focusing increases and the focus is moved towards $-z$. Therefore less energy is transmitted by the aperture. In the same way, the transmission increases if the sample is located to the right of the focal point. Finally, for large values of z the transmission becomes one again.

An analogous consideration leads to the dotted line for $n_2<0$ (Fig. 10a). An identical z-scan experiment without the aperture allows the determination of the two-photon absorption coefficient α_2 (Fig. 10b). In principle, this method also allows to determine nonlinearities of higher order (e.g. n_3 and α_3) [33].

The Z-scan theory has been described by different authors. In the thin sample limit the Z-scan measurement is described either through Fresnel integration or through a Gaussian decomposition procedure [3, 6].

The analysis of Z-scans relies on Gaussian spatial beam profiles. For short laser pulses the beam profile is frequently not completely Gaussian and the unbiased use of the standard anlaysis of Z-scan can lead to large experimental errors. For reliable Z-scan measurements the spatial beam profile has to be well

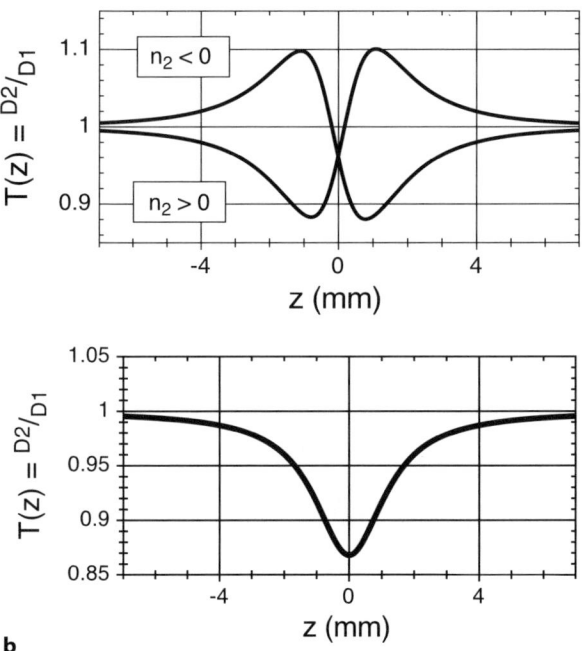

Fig. 10a,b. a Typical Z-scan curves for $n_2>0$ (*solid line*) and $n_2<0$ (*dotted line*). **b** If the aperture is removed the two-photon absorption coefficient α_2 can be determined

defined which can be quite troublesome for short pulsed laser systems. Spatial filtering in the laser cavity leads so far to the best results.

Chapple et al thourougly discussed the critical parameter of Z-scan and Mian et al. [35] showed the influence of beam ellipticity on the Z-scan measurements. A solution to overcome the troubles with non-Gaussian beams is the employment of top-hat beams [36, 37]. An aperture is placed in the expanded beam in front of the focusing lens, so that the beam profile is uniform in the aperture. The analysis follows an analogous approach as for Gaussian beams and results in similar curves but with a magnitude that is about 2.5 times larger.

Although Z-scan is a very simple experiment at first view, caution has to be taken in the interpretation of the measurement curves. By far the most common source of errors in the interpretation of Z-scan measurements is the presence of slow components in the complex susceptibility changes, especially the thermal nonlinearity.

3.4
Two-Photon Absorption Spectroscopy

In the field of optical limiting based on two-photon absorption there exists some confusion with the convention of different physical processes. We refer here to the two-photon absorption process based on $\alpha_2 \propto \text{Im}\{\chi^{(3)}(-\omega,\omega,\,\omega,\omega)\}$ and write therefore "direct" two-photon absorption. "Two-photon absorption" processes as e.g. excited state absorption or triplet state absorption are of a different physical origin and are not treated in the framework of this section.

However, if optical limiting is the goal the nonlinear absorption can be increased by making use of e.g. reverse saturable absorption originating from photogenerated transient states. Although this is a good concept to increase the effectiveness for optical limiting, no pure two-photon absorption cross-sections can be deduced this way. Therefore, one should always consider the underlying physical effect when comparing experimental values reported by different

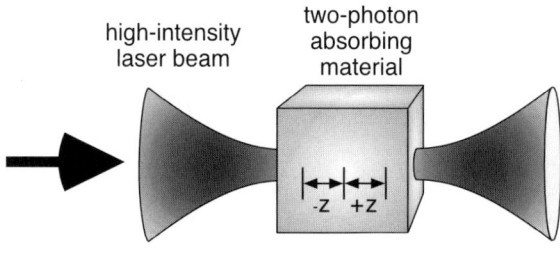

TPA ∝ (intensity)^2 → intensity ∝ z^{-2} → TPA ∝ z^{-4}

Fig. 11. Schematic of the excellent three-dimensional spatial resolution of the two-photon absorption process

groups. As already mentioned, we here concentrate on the design of efficient chromophores for direct two-photon absorption.

If a laser beam is focused in the material, the intensity required for TPA to occur will usually only be reached close to the focus. The selectivity of the absorption process in propagation direction is excellent (Fig. 11), which enables the three dimensional resolution of TPA process as mentioned above. Furthermore, the penetration into absorbing or scattering media can be greatly improved if the fundamental wave is not depleted by one-photon absorption and if the TPA only takes place at the point of strong focussing.

If intrinsic two-photon absorption cross-sections are the subject of investigation short pulses (ps or fs) should be used and two-photon fluorescence (see below) should be measured to verify the I^2-dependence of the TPA process. Most often, two-photon absorption is measured either with the Z-scan technique (see above) or with the nonlinear transmission method [38].

In nonlinear transmission experiments the transmission of a sample is measured as a function of the input intensity. At small intensities the transmission increases linearly with increasing input intensity. For large intensities the nonlinearity sets in and a deviation from the linear behavior is visible (see Fig. 12a). A theoretical analysis of such curves based on

$$I = \frac{I_o}{1 + \alpha_2 I_o L} \tag{72}$$

where I_o is the input intensity and L is the sample thickness, yields the two-photon absorption coefficient α_2 for a pure two-photon absorption process. Often, the measured data is recorded as (Fig. 12b)

$$1/T = I_o / I = 1 + \alpha_2 I_o L \tag{73}$$

since one obtains a linear relationship between $1/T$ and I_o.

3.5
Fluorescence Spectroscopy

In TPA induced fluorescence spectroscopy the excited state population for fluorescence is created by the simultaneous absorption of two photons instead of the one-photon absorption in standard fluorescence experiments. The principles of this technique are described in Ref. [23]. It allows an experimental determination of two-photon absorption cross-sections provided the material is fluorescent and that its two-photon fluorescence quantum efficiency is known (which is usually assumed to be equal to the one-photon fluorescence quantum efficiency if the same excited state is reached). In this case the method generally provides high sensitivity.

An accurate absolute determination of δ (see Eq.(51)) requires the knowledge of four parameters:

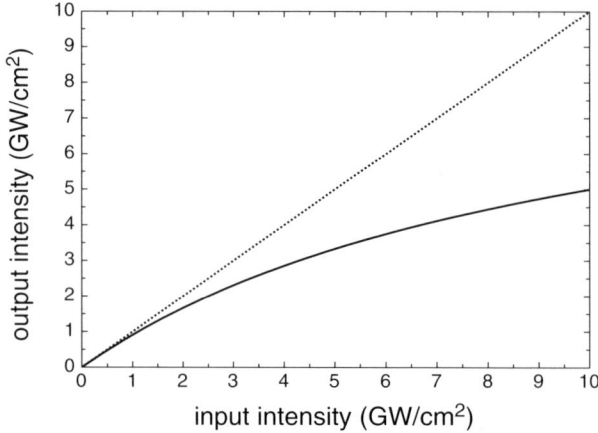

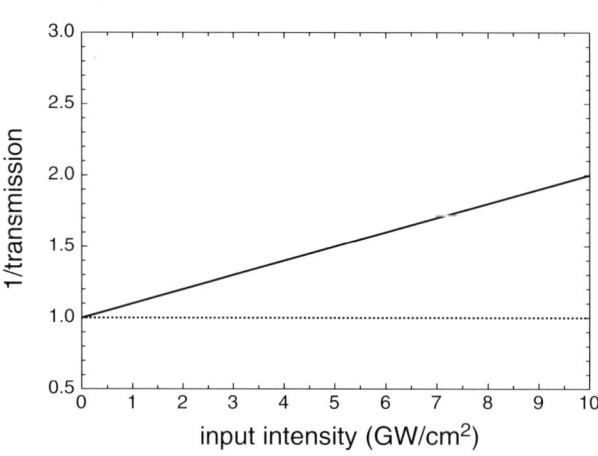

Fig. 12a,b. a Theoretically calculated output vs input intensity for $\alpha_2=1$ cm/GW, $L=0.1$ cm, input intensity in GW/cm^2 based on Eq.(72). **b** Inverse transmission versus input intensity (based on Eq.(73)) for the same parameters as in (**a**)

1. the spatial distribution of the indicent light,
2. the degree of second-order temporal coherence,
3. the fluorescence collection efficieny of the detection system, and
4. the fluorescence quantum efficiency.

All these parameters are discussed in detail in Ref. [23] and will not be described here further. If the measurements are performed relative to a material with known nonlinearities only the fluorescence quantum yield has to be known.

Molecular Design for Third-Order Nonlinear Optics

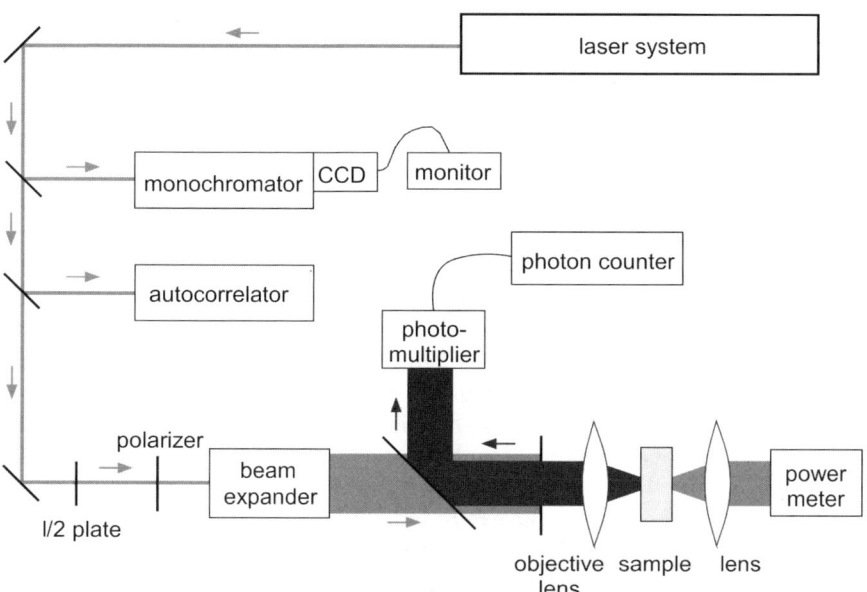

Fig. 13. Experimental set-up for fluorescence spectroscopy

In most experiments a pulsed laser is used. In the set-up of Fig. 13, fs pulses are applied that are continuously monitored with respect to pulse width, excitation wavelength, and pulse spectrum. A half-wave plate and a polarizer allow the selection of the polarization of the excitation. A combination of lenses expands the laser beam to a size larger than the aperture of the microscope objective that focuses the light on the sample. The sample thicknesses typically vary between 0.5 and 10 mm. For solution measurements sample chromophores of typically 100 µM are dissolved. The generated fluorescence is measured in the backward direction using a dichroic mirror. A liquid filter in front of the photomultiplier gets rid of the excitation wavelength. The photomultiplier is read by a photon-counter and the light transmitted by the sample is monitored by a power meter.

Note that the generated fluorescence can also be detected perpendicular to the propagation direction of the excitation laser beam [41]. In this case the sample is excited with a collimated beam. In either case one has to make sure that the intensity of the incident beam is in a range that gives a square-intensity behaviour of the measured fluorescence to be able to neglect possible higher-order contributions.

If the spectrum of the fluorescence should be measured the photomultiplier is replaced by a spectrometer. In this case the generated light is coupled into a multimode fiber and guided to the spectrometer.

4
Discussion of Molecular Properties

4.1
Basic Molecular Units

Here we discuss the basic molecular units and structures that are essential for third-order nonlinear optics in organic materials. The wide diversity of the properties of organic compounds is primarily due to the unsurpassed ability of the carbon atom to form a variety of stable bonds with itself and with many other elements. This bonding is primarily of two types, which differ considerably in the localization of the electron charge density. Whereas a two electron covalent σ C–C bond is spatially confined along the internuclear axis of the C–C bond, the π bonds have their electronic charge distributions above and below the interatomic axis and delocalize over many atoms. The electron density of π bonds is much more mobile than the one of the σ bonds. This electron distribution can be influenced by substituents; the extent of redistribution is measured by the dipole moment, and the ease of redistribution in response to an externally applied electric field is measured by the (hyper)polarizability. If the perturbation to the molecular electronic distribution caused by an intense optical field is asymmetric, a quadratic nonlinearity results. Virtually all significantly interesting nonlinear optical organic molecules exhibit π bond formation between various nuclei.

To make the picture of π-electrons more intelligible the model of linear combinations of single electron atomic orbitals to molecular orbitals is helpful (Fig. 14). In this model one concentrates only on the outermost electrons or valence orbitals. Starting from the atomic wavefunctions the s, p_x and p_y atomic orbitals are combined in the (x,y)-plane to sp and sp^2 orbitals. These sp and sp^2 orbitals of the different atoms combine to molecular orbitals, building the molecular structure framework in the (x,y)-plane. The electrons in these molecular orbitals are called σ-electrons and their wavefunctions are symmetric perpendicular to the (x,y)-plane extending only over two neighboring atoms.

The p_z orbitals have not been used so far. They are antisymmetric perpendicular to the (x,y)-plane and therefore orthogonal to the above combined sp and sp^2 orbitals. p_z orbitals from neighboring atoms combine independently to form molecular orbitals with their wavefunction delocalized over many atoms. The corresponding electrons are the π-electrons, which dictate to a large extend the linear and nonlinear optical properties.

To create a long path for the π-electrons, many atoms are connected to build a chain of p_z orbitals. On either end of this conjugated backbone electron attracting or releasing groups can modify the wavefunctions of the π-electrons. By choosing an appropriate electron acceptor or donor group the physical properties of the molecule can be tuned.

Typical dye molecules or chromophores consist of a rather short conjugated backbone with a strong donor and acceptor group on either end. Their linear

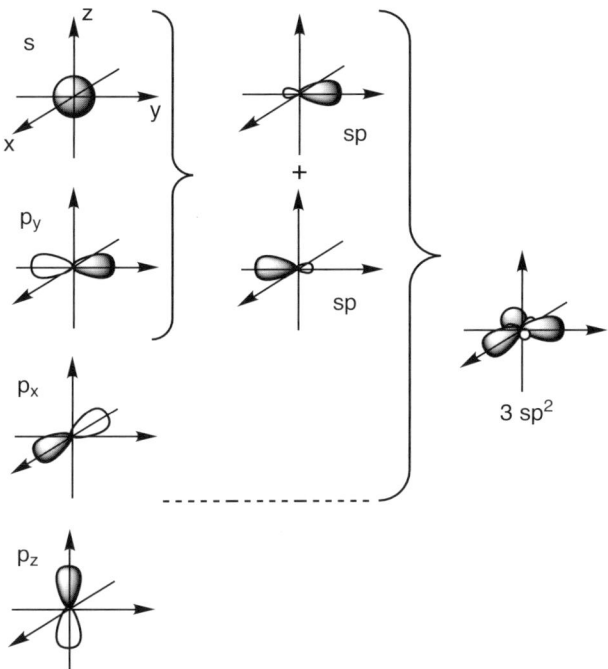

Fig. 14. Linear combination of the atomic valence orbitals s, p_x, p_y to form sp and sp^2 orbitals in the (x,y)-plane. The resulting orbitals are symmetric perpendicular to the (x,y)-plane and build the structure framework in the (x,y)-plane with the orbitals of the neighboring atoms. The remaining antisymmetric p_z orbital is orthogonal to the orbitals in the (x,y)-plane and is used for the delocalized electron orbitals

spectra usually display a strong absorption peak that is shifted to the red spectral range.

In conjugated oligomers or polymers the backbone is longer and the end groups are less important. The term "oligomers" is normally used for shorter polymers with well defined length or number of monomer units. Typical examples of conjugated polymers are shown in Fig. 15.

A special case are the two-dimensionally conjugated tetraethynylethenes (TEE). In this case the conjugation is expanded in one direction and functional groups are added along the other conjugation direction, laterally to the backbone (Fig. 16).

The development of highly active third-order nonlinear optical materials is important for all-optical signal processing. In contrast to second-order nonlinear optical molecular systems, there are few rational strategies for optimizing the third-order nonlinear optical response of molecular materials. Unlike second-order materials, there exist no molecular symmetry restrictions for the observation of a third-order nonlinear optical response. It is the instantaneous

polyene / polyacetylene (PA):

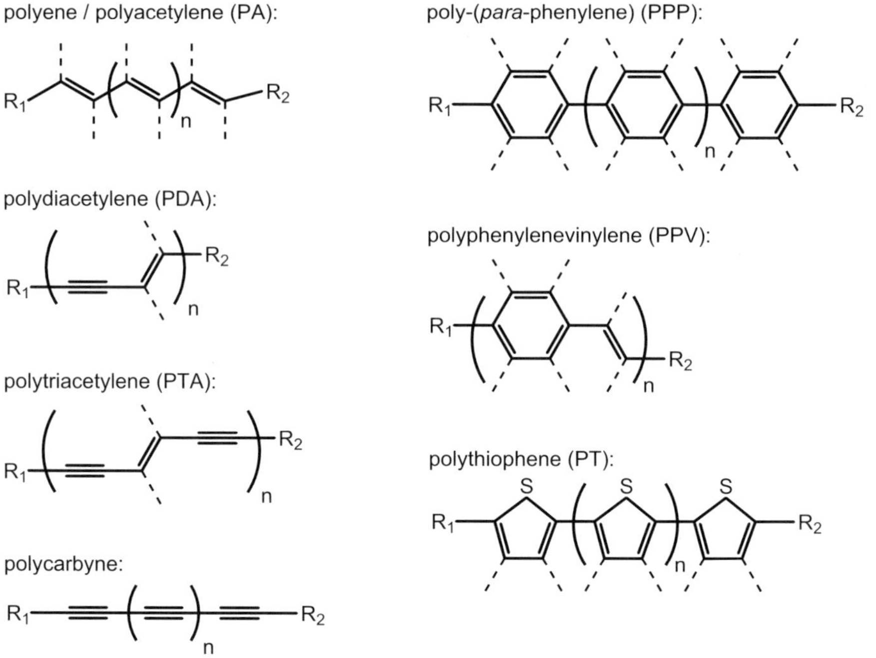

poly-(*para*-phenylene) (PPP):

polydiacetylene (PDA):

polyphenylenevinylene (PPV):

polytriacetylene (PTA):

polythiophene (PT):

polycarbyne:

Fig. 15. Examples of conjugated polymers. The *dotted lines* indicate bonds lateral to the conjugated backbone. R_1 and R_2 denote end groups. The simplest conjugated polymer is polyene with alternating single and double bonds. By introducing triple bonds, one goes from polyene over polydiacetylene and polytriacetylene to polycarbyne. The aromatic polyphenylene and polyphenylenevinylene contain additional phenyl rings. Polythiophenes additionally possess sulfur atoms

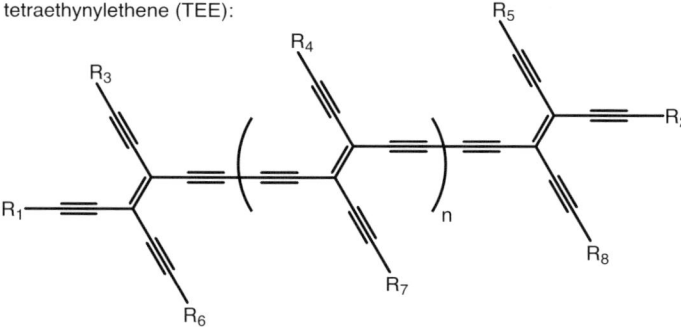

Fig. 16. Two-dimensional framework of tetraethynylethenes. One dimension can be used for the polymerization and the other for substitution with functional groups (R_i)

shift in π-electron densities over a molecule that occur on excitation which explains the large and fast polarizabilities of π-electron networks.

In quantum mechanics the definition of molecular polarizabilities is given through time-dependent perturbation theory in the electric dipole approximation. These expressions are usually given in terms of sums of transition matrix elements over energy denominators involving the full electronic structure of the molecule [42].

$$\gamma_{ijkl}(-\omega_4,\omega_3,\omega_2,\omega_1) = \frac{1}{\varepsilon_0 \hbar^3} \cdot \wp_\sigma(1,2,3) \Bigg[\sum'_{lmn} \Bigg\{$$

$$+ \frac{\mu^i_{gl}\bar\mu^j_{lm}\bar\mu^k_{mn}\mu^l_{ng}}{(\omega_{lg}-\omega_1-\omega_2-\omega_3)(\omega_{mg}-\omega_1-\omega_2)(\omega_{ng}-\omega_1)}$$

$$+ \frac{\mu^i_{gl}\bar\mu^j_{lm}\bar\mu^k_{mn}\mu^l_{ng}}{(\omega^*_{lg}+\omega_3)(\omega_{mg}-\omega_1-\omega_2)(\omega_{ng}-\omega_1)}$$

$$+ \frac{\mu^i_{gl}\bar\mu^j_{lm}\bar\mu^k_{mn}\mu^l_{ng}}{(\omega^*_{lg}+\omega_3)(\omega^*_{mg}+\omega_3+\omega_2)(\omega_{ng}-\omega_1)}$$

$$+ \frac{\mu^i_{gl}\bar\mu^j_{lm}\bar\mu^k_{mn}\mu^l_{ng}}{(\omega^*_{lg}+\omega_3)(\omega^*_{mg}+\omega_3+\omega_2)(\omega^*_{ng}+\omega_1+\omega_2+\omega_3)} \Bigg\}$$

$$- \sum'_{mn} \Bigg\{ \frac{\mu^i_{gm}\mu^j_{mg}\mu^k_{gn}\mu^l_{ng}}{(\omega_{mg}-\omega_1-\omega_2-\omega_3)(\omega_{mg}-\omega_3)(\omega_{ng}-\omega_1)}$$

$$+ \frac{\mu^i_{gm}\mu^j_{mg}\mu^k_{gn}\mu^l_{ng}}{(\omega^*_{mg}+\omega_3)(\omega_{mg}-\omega_2)(\omega_{ng}-\omega_1)}$$

$$+ \frac{\mu^i_{gm}\mu^j_{mg}\mu^k_{gn}\mu^l_{ng}}{(\omega^*_{mg}-\omega_3)(\omega^*_{mg}+\omega_2)(\omega_{ng}+\omega_1)}$$

$$+ \frac{\mu^i_{gm}\mu^j_{mg}\mu^k_{gn}\mu^l_{ng}}{(\omega^*_{mg}+\omega_1+\omega_2+\omega_3)(\omega^*_{mg}+\omega_3)(\omega^*_{ng}+\omega_1)} \Bigg\} \Bigg]$$

(74)

The sum Σ' runs over all states except the ground state g. $\wp_{\sigma(p,q,r)}$ is the average over all possible permutations of the frequencies and concurrent the spatial indices. The matrix elements $\bar\mu_{lm}=\mu_{lm}-\mu_{gg}\delta_{lm}$ include the terms with the ground state dipole moment μ_{gg}, which are omitted by the exclusion of the ground state in the sum. For $l=m$ the element $\bar\mu_{lm}$ is the difference dipole moment between the dipole moment of state l and the ground state g. For a centrosymmetric molecule no dipole moment is possible and therefore this difference dipole moment term vanishes.

Consequently, molecules with structures that lead to large transition dipole moments and small transition energies should display large second-order hyperpolarizabilities. Conjugated molecules with delocalized electronic wavefunctions enable the electrons to move over considerable distances and therefore show large transition and difference dipole moments. Furthermore, these molecules show low transition energies.

4.1.1
Electronic States and Symmetry Considerations

For an elaborated analysis of the relations between structure and hyperpolarizabilities, one has to start from the electronic wavefunctions of a molecule. By using time-dependent perturbation theory, sum-over-states expressions can be derived for the first and second-order hyperpolarizabilities β and γ. For β, a two-level model that includes the ground and one excited state has proven to be sufficient. For γ the situation is more complicated.

We first look at the electronic states of a molecule (Fig. 17) According to the overall spin of the electrons, the states are split up in singlet and triplet states. Singlets have zero spin, triplets a total spin of one. The ground state with the lowest energy is practically always a singlet state. Within the dipole interaction of the electric field only singlet states can be reached from the singlet ground state S_0. A transition to a triplet state is spin forbidden. From the first singlet excited state S_1 another photon can be absorbed in an excited state absorption process leading to a higher singlet state S_i. The electron can also go back from the excited state S_1 to the ground state S_0 by fluorescence or a non-radiative process.

Furthermore, the electron can go from S_1 to a triplet state through an intersystem crossing. From the triplet state the electron can in principle only decay by a non-radiative process back to the ground state S_0. Nevertheless, spin-orbit coupling of the electrons and magnetic interactions can allow a weak phosphorescence process from T_1 to S_0.

As resonant processes are avoided in the nonlinear optical experiments (except where two-photon absorption is optimized), no excited state molecules are created and triplet states are of no concern. For a more thorough investigation of the absorption process also the vibrations of the molecule have to be included. They are not considered here.

The symmetry of the molecules provide selection rules for the electronic transitions in the dipole approximation. A mirror plane perpendicular to a direction ξ requires the wavefunction u_n to be either symmetric $u_n(\xi)=u_n(-\xi)$ or antisymmetric $u_n(\xi)=-u_n(-\xi)$ to this mirror plane as the electron density $\rho_n^{el}(\vec{r})=|u_n(\vec{r})|^2$ is symmetric. The wavefunctions can be separated in a sym-

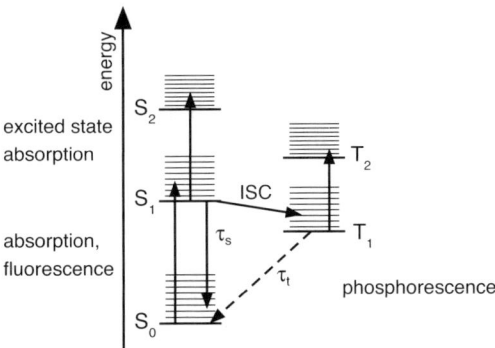

Fig. 17. Electronic states and transitions in an organic molecule. The singlet states S_i have no net electron spin and the triplet states T_i a total spin of one. The thin lines indicate higher vibrational states and ISC denotes an intersystem crossing from singlet to triplet states

metric and antisymmetric subset with respect to the direction ξ. Consequently, the transition dipole moments $\vec{\mu}_{nm} = -e\langle u_n(\vec{r})|\hat{r}|u_m(\vec{r})\rangle = e\int u_n^*(\vec{r})\vec{r}u_m(\vec{r})dV$ in such a ξ-direction are zero if the wavefunctions $u_n(\xi)$ and $u_m(\xi)$ are of the same symmetry (e:electron charge, $\hat{r}$: position operator). In a centrosymmetric molecule transitions are only symmetry allowed from a symmetric to an antisymmetric state or vice versa.

In a centrosymmetric molecule the wavefunctions can be split up in a symmetric and antisymmetric set regarding the inversion operation. The symmetric wavefunctions are labeled as nA_g, the antisymmetric ones as nB_u. The above selection rule of no transitions within the symmetry subset applies to all directions. It is impossible to reach a higher A_g state from an A_g ground state by a simple one-photon absorption.

In symmetric organic molecules the ground state S_0 is usually an A_g state. The higher states normally alternate in symmetry. There are only few molecules known, where the $2A_g$ state has a lower energy than the $1B_u$ state, as for example in elongated polyenes [43].

The above consideration influence the complex perturbation expression of the second-order hyperpolarizability γ. The number of involved terms can be reduced, leading to a sum-over-states expression of only few terms of interest.
1. With the transition energies in the denominator the magnitude of terms for higher states drops quickly. Therefore only few excited states have to be considered. Already two excited states yield a model which often describes the second-order hyperpolarizability sufficiently.
2. The conjugated molecules are elongated with a much better electron mobility in the backbone direction than perpendicular to it. The dipole and transition dipole moments are pointing in this direction (x) with $\mu_{nm}^x \gg \mu_{nm}^y, \mu_{nm}^z$. Consequently, the second-order hyperpolarizability tensor γ is dominated by

the tensor element γ_{xxxx}. In the rotational average of one-dimensional (1D) molecules tensor elements besides γ_{xxxx} can be omitted.

3. For the two-dimensional (2D) conjugated molecules the second-order hyperpolarizability tensor elements γ_{ijkl} with an index in z direction (perpendicular to the conjugation plane) can be neglected with the same arguments as for the 1D molecules.

4. For a centrosymmetric molecule the terms with a difference dipole moment $\Delta\mu$ vanish, as no dipole moment can occur. The symmetry selection rule between the A_g and the B_u states reduces the sum-over-states expression additionally. For an organic molecule the ground state can be assumed to be symmetric and the two lowest excited states to be of different symmetry. In the following the ground state $1A_g$ is always labeled $g = 0$, the excited $1B_u$ state with $n=1$, and the excited $2A_g$ state with $n=2$. The numbering n does not imply the energy ordering of the excited states. Consequently, the transition dipole moment μ_{02} is zero and also these terms do not appear in the sum-over-states expression.

5. For nonlinear optical molecules with a centrosymmetric backbone but an asymmetric substitution of the end groups the transition dipole moment μ_{02} is different from zero. The substitution can be regarded as a distortion of the initial backbone wavefunction. If the distortion is not too large, the transition dipole moment to the former $2A_g$ state is still significantly smaller than to the former $1B_u$ state: $|\mu_{02}| \ll |\mu_{01}|$. The μ_{02} terms can be neglected with respect to the μ_{01} terms. The terms with the difference dipole moments $\Delta\mu$ cannot be omitted.

For a molecule without symmetry reduction the sum-over-states expression for the hyperpolarizability γ can often be modeled within the three level approximation (the ground state (index 0) and two excited states (indices 1,2 – not necessarily indicating the hierarchy of the energy eigenvalues)). If we assume $|\mu_{02}| \ll |\mu_{01}|$ for substituted molecules with a centrosymmetric backbone as described above, the number of terms reduces leaving only three terms to consider: the negative term N_i, the dipolar term D_i, and the two-photon term TP_i, which includes the second excited state $2A_g$. In the static limit one obtains for a one-dimensionally conjugated molecule

$$\gamma_{xxxx} = \frac{4}{\varepsilon_o} \left\{ -\frac{\left(\mu_{01}^x\right)^4}{E_{01}^3} + \frac{\left(\mu_{01}^x\right)^2 \left(\Delta\mu_1^x\right)^2}{E_{01}^3} + \frac{\left(\mu_{01}^x\right)^2 \left(\mu_{12}^x\right)^2}{E_{01}^2 E_{02}} \right\} \tag{75}$$

$$= N_{xxxx} \qquad = D_{xxxx} \qquad = TP_{xxxx}$$

Molecular Design for Third-Order Nonlinear Optics

with the figures depicting the involved states and where $\Delta\mu$ is the difference of the dipole moments in the state $n=1$ and the ground state.

For a two-dimensionally conjugated molecule in solution two different notations are possible. In the rotational average both notation are equivalent. In the first one, the permutation symmetry of the spatial indices is sacrificed in order to retrieve simpler expressions of the tensor elements.

$$Y_{xxyy} = Y_{xyxy} = Y_{yyxx} = Y_{yxyx} = \frac{4}{\varepsilon_o}\left\{-\frac{\left(\mu_{01}^x\right)^2\left(\mu_{01}^y\right)^2}{E_{01}^3} + \frac{\mu_{01}^x \Delta\mu_1^x \Delta\mu_1^y \mu_{01}^y}{E_{01}^3} + \frac{\mu_{01}^x \mu_{12}^x \mu_{12}^y \mu_{01}^y}{E_{01}^2 E_{02}}\right\} \tag{76}$$

$$Y_{xyyx} = \frac{4}{\varepsilon_o}\left\{-\frac{\left(\mu_{01}^x\right)^2\left(\mu_{01}^y\right)^2}{E_{01}^3} + \frac{\left(\mu_{01}^x\right)^2\left(\Delta\mu_1^y\right)^2}{E_{01}^3} + \frac{\left(\mu_{01}^x\right)^2\left(\mu_{12}^y\right)^2}{E_{01}^2 E_{02}}\right\} \tag{77}$$

$$Y_{yxxy} = \frac{4}{E_o}\left\{-\frac{\left(\mu_{01}^x\right)^2\left(\mu_{01}^y\right)^2}{E_{01}^3} + \frac{\left(\mu_{01}^y\right)^2\left(\Delta\mu_1^x\right)^2}{E_{01}^3} + \frac{\left(\mu_{01}^y\right)^2\left(\mu_{12}^x\right)^2}{E_{01}^2 E_{02}}\right\} \tag{78}$$

In the second notation the permutation symmetry is maintained, resulting in

$$Y_{xxyy} = Y_{xyxy} = Y_{xyyx} = Y_{yyxx} = Y_{yxyx} = Y_{yxxy}$$

$$= \frac{4}{\varepsilon_o}\left\{-\frac{\left(\mu_{01}^x\right)^2\left(\mu_{01}^y\right)^2}{E_{01}^3} + \frac{1}{6}\frac{\left(\mu_{01}^x\right)^2\left(\Delta\mu_1^y\right)^2}{E_{01}^3} + \frac{1}{6}\frac{\left(\mu_{01}^y\right)^2\left(\Delta\mu_1^x\right)^2}{E_{01}^3} + \frac{2}{3}\frac{\mu_{01}^x \Delta\mu_1^x \Delta\mu_1^y \mu_{01}^y}{E_{01}^3}\right.$$

$$\left. + \frac{1}{6}\frac{\left(\mu_{01}^x\right)^2\left(\mu_{12}^y\right)^2}{E_{01}^2 E_{02}} + \frac{1}{6}\frac{\left(\mu_{01}^y\right)^2\left(\mu_{12}^x\right)^2}{E_{01}^2 E_{02}} + \frac{2}{3}\frac{\mu_{01}^x \mu_{12}^x \mu_{12}^y \mu_{01}^y}{E_{01}^2 E_{02}}\right\} \tag{79}$$

In the above expressions the permutation symmetry in all four spatial indices is used as feasible outside a resonance and without wavelength dispersion. This approach is strictly valid in the static limit and a good approximation in the transparancy range of the nonlinearity.

If the molecule has full centrosymmetry also the dipolar term D_x vanishes. The discussion of the molecules discussed in this work will rely to a large extent on the above equations of the three level model.

4.1.2
Bond-Length Alternation

Although not further discussed below, bond-length alternation (BLA) will be shortly described in the following. BLA has recently been identified as one useful structural parameter to predict trends for the second-order hyperpolarizability γ (Fig. 18) [44–46]. It is defined as the difference between the neighboring carbon-carbon bonds in a molecule. For polyenes with alternating single and double bonds, the bond length alternation can be expressed by a single parameter (BLA). In the following, the hyperpolarizabilities are optimized in respect to this parameter (BLA). The dependence of α, β, and γ on the bond length alternation (BLA) is schematically shown in Fig. 18 [44]. The whole curve could be also experimentally confirmed using different solvents and several different molecules for the first-order hyperpolarizability β and in part for the second-order hyperpolarizability γ [46]. The evolutions of β and γ with BLA are first- and second-order derivative-like with respect to the evolution of α. In principle, the largest values of γ can be obtained for a BLA of 0. Such materials have not yet been identified except may-be for squaraine dyes [47, 48]. However, there are regions in which an increase of β is associated with an increase of γ.

The bond length alternation approach describes only the position of the molecular nonlinearities on the BLA axis, but does not give an absolute magnitude of the curves in Fig. 18. Schemes as conjugation length expansion (see below) are believed to increase the magnitude of the curves in the bond length alternation picture.

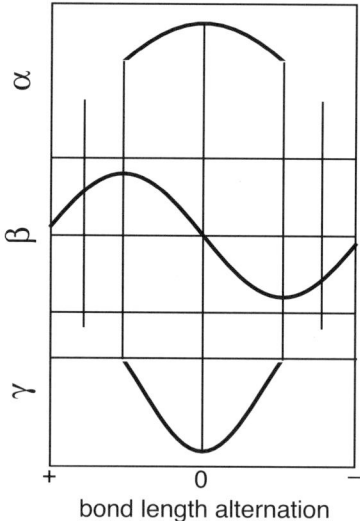

Fig. 18. Schematic illustration of the dependence of α, β, and γ on the bond length alternation. The dashed lines indicate maximum values of β and γ

An analytical structure-(hyper)polarizability relationship based on a two-state description has also been derived [49]. In this model a parameter MIX is introduced that describes the mixture between the neutral and charge-separated resonance forms of donor-acceptor substituted conjugated molecules. This parameter can be directly related to BLA and can explain solvent effects on the molecular hyperpolarizabilities. NMR studies in solution (e.g. in $CDCl_3$) can give an estimate of the BLA and therefore allow a direct correlation with the nonlinear optical experiments. A similar model introducing a resonance parameter c that can be related to the MIX parameter was also introduced to classify nonlinear optical molecular systems [50, 51].

4.2
Influence of Molecular Substitution

In second-order nonlinear optics, the relationships between the molecular structure and large first-order hyperpolarizabilities β have been well established: ideal molecules contain a conjugated backbone with an asymmetric substitution pattern consisting of a strong electron donor and a strong electron acceptor endgroup [24, 52, 53]. Further enhancement of the first-order hyperpolarizability β has been achieved via replacement of phenyl rings in the conjugated backbone by heteroaromatic spacers such as thienyl rings [54, 55].

In third-order nonlinear optics, guidelines for the optimization of the second-order hyperpolarizability γ of a molecule have been steadily improving, but the understanding is far less developed than for β. Experimental and theoretical observations have shown that for molecules with an extended conjugation in a single direction, asymmetric substitution yields the largest second-order hyperpolarizabilities γ [56, 57] similar to the case of β in several systems. This issue will be discussed in this Section for a selection of molecules that is by no means complete.

In polytriacetylenes (Fig. 19) the conjugation across the backbone is weaker than for polyenes. Conversely, the PTAs are stable under normal laboratory condition and show no degradation over months, which makes them an interesting candidate for applications. Furthermore, the chemical synthesis allows to have a well defined number of monomer units and to attach definite functional groups enabling the study of the structure-property relationships in nonlinear optics.

The PTA oligomers have been substituted with the electron donating dimethylaniline and electron accepting nitrophenyl either in symmetric or asymmetric geometry (Fig. 19). In the short substituted PTAs the donor and acceptor groups have a significant influence on the electronic wavefunctions on the backbone which is portrayed by the shift of about 100 nm of the absorption spectra to the red compared to the pure PTA.

The symmetry dependence of the second-order hyperpolarizabilities of PTAs was first assessed on monomers and dimers with different substitution patterns (Table 2). The difference of asymmetric and symmetric geometries can be seen in the three level model Eq.(80). The dipolar term $D = \mu_{01}^2 \Delta \mu_1^2 / E_{01}^3$ only occurs

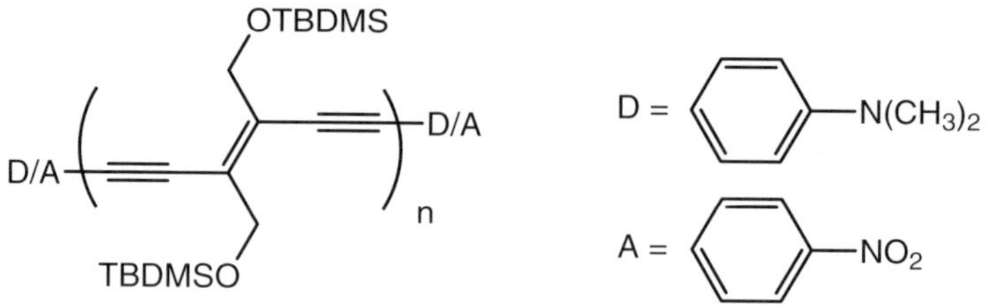

Fig. 19. Molecular structure of polytriacetylenes (PTA). The conjugation ends are substituted by donor (dimethylaniline) and acceptor (nitrophenyl) groups to modify the nonlinear optical properties. The lateral OTBDMS ((*tert*-butyl)dimethylsilyloxyl) groups serve for improved solubility in the solvent (typically chloroform)

Table 2. THG measurements of the second-order hyperpolarizabilities $\gamma(-3\omega, \omega, \omega, \omega)$ for substituted monomers and dimers of PTA at a fundamental wavelength of $\lambda = 1907$ nm. ε is the extinction coefficient at the harmonic wavelength of the THG experiment (reference $\chi^{(3)}_{fs} = 1.6 \times 10^{-22}$ m^2/V^2, 10% experimental error)

Substitution	n	λ_{00} [nm]	$\varepsilon(3\omega)$ [l/(cm·mol)]	$\|\gamma^{THG}_{rot}\|$ [10^{-36} esu]	$\|\gamma^{THG}_{rot}\|$ [10^{-48} m^5/V^2]
TMS / TMS	1	296.4±0.1	0	9.2	0.13
TMS / A	1	376.6±1.6	0	51	0.71
TMS / D	1	379.1±0.1	0	49	0.68
A / A	1	402.9±0.5	0	88	1.23
A / A (*cis*)	1	404.3±0.5	0	96	1.34
D / D	1	408.6±0.1	0	123	1.72
D / A	1	450.4±0.9	0	250	3.5
A / A	2	404.3±0.5	0	220	3.1
D / D	2	454.8±0.3	16	430	6.0

for asymmetric molecules and vanishes for centrosymmetric substitution [59]. Therefore the approach to increase the second-order hyperpolarizability is to break the symmetry by a donor-acceptor substitution.

$$\gamma = \frac{4}{\varepsilon_o}\left(-\frac{\mu_{01}^4}{E_{01}^3} + \frac{\mu_{01}^2 \Delta\mu_1^2}{E_{01}^3} + \frac{\mu_{01}^2 \mu_{12}^2}{E_{01}^2 E_{02}}\right) \tag{80}$$

The substitution by electron donating and accepting groups clearly enhances the nonlinearity of the PTA monomers. The single donor and single acceptor substituted PTA reveal the same second-order hyperpolarizabilities with an improvement compared to the unsubstituted monomer of a factor of about five.

From the double acceptor substituted monomer two isomers could be separated: the normal *trans* AA-PTA and the *cis* AA-PTA. Neither in the absorption wavelength nor in the hyperpolarizability a significant difference can be seen. The DD-PTA exhibits a nonlinearity little larger than the AA-PTAs. The asymmetrically substituted DA-PTA has a second-order hyperpolarizability about twice the size of the symmetrically substituted monomers. Concluding, for monomers the DD-substitution is somewhat better than the AA-substitution but clearly inferior to the asymmetric AD-geometry. The reason for this trend is the dipolar term of the three level model in Eq.(80).

In the dimers the DD-PTA reveals a significantly improved second-order hyperpolarizability compared to the AA-substitution and nearly matches the γ value of the asymmetric AD-PTA. The established guideline of asymmetric substitution for large nonlinearities almost failed for the PTA dimer. The validity of this guideline for even longer PTA oligomers can be questioned. The negative term $N = \mu_{01}^4 / E_{01}^3$ and the two-photon term $TP = \mu_{01}^2 \mu_{12}^2 / \left(E_{01}^2 E_{02} \right)$ in the three level model seem to play a more important role for long molecules compared to the dipolar term in short molecules. Unfortunately, the synthesis of the asymmetric trimer has not been possible and a probable failure of the guideline could not be verified. In contrast, from the double donor DD-PTAs and double acceptor AA-PTAs oligomers up to the hexamer and two polymer samples could be investigated both by THG (λ=1907 nm) and DFWM (λ=1047 nm).

The second-order hyperpolarizabilities generally start at larger values for short oligomers and disclose a peak of the nonlinearity per length at intermediate backbone lengths. In THG the hyperpolarizabilities γ for long PTAs tend to go to the same saturation as the pure ones (see below). In contrast, in DFWM the nonlinearities of the symmetrically substituted polymeric PTAs are larger than for pure ones, which can be explained by the closeness of the two-photon resonance. The magnitude of the nonlinearity for short substituted PTAs is enhanced in DFWM when compared to THG, which also indicates an already starting enhancement by the two-photon resonance. However, the phase-sensitive DFWM experiments could not disclose a phase angle of the complex nonlinear optical coefficient above the experimental uncertainties.

In conclusion, in short PTAs, $\Delta\mu$ in the dipolar term of the three level model optimizes the value of the second-order hyperpolarizability.

Unlike classical, elongated π-conjugated chromophores hydrazone derivatives can basically be divided into two groups since the hydrazone skeleton is not symmetrical. (Fig. 20). Type I hydrazones are shown in Fig. 20. If donor (D) and acceptor (A) are exchanged we obtain different molecules with different properties, type II hydrazone derivatives. These chromophores, which possess a bent hydrazone skeleton due to the non-rigid nitrogen-nitrogen single bond can be considered being formed by two dipolar chromophores e.g. (Donor–Ar–CH=N—) and (—NH-Ar-Acceptor) connected in a head-to-tail fashion. However, it was both experimentally and theoretically shown that the charge transfer process extends from the donor to the acceptor going through the entire hydrazone skeleton.

Type I hydrazones can be described as a bis-chromophoric system D-A'-D'-A where A' is equal to the azomethine double bond (C=N) and D' is equal to the central amino group which suggests lower values of β in comparison to type II hydrazones (see below). Also, one expects increasing values of γ with increasing

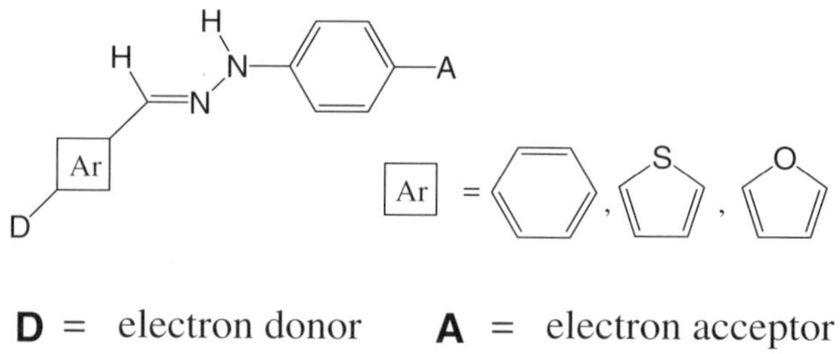

Fig. 20. Molecular structure of type I hydrazone derivatives. If donor (D) and acceptor (A) are exchanged we obtain different molecules, type II hydrazone derivatives

Table 3. THG measurements of the second-order hyperpolarizabilities γ for substituted Type I and Type II hydrazone derivatives in the solvent DMF at a fundamental wavelength of $\lambda = 1907$ nm. The aromatic ring Ar is either benzene or thiophene (see Fig. 20). ε is the extinction coefficient at the harmonic wavelength of the THG experiment (reference $\chi^{(3)}_{fs} = 1.6\times10^{-22}$ m^2/V^2, 10% experimental error)

Substitution R$_1$/R$_2$	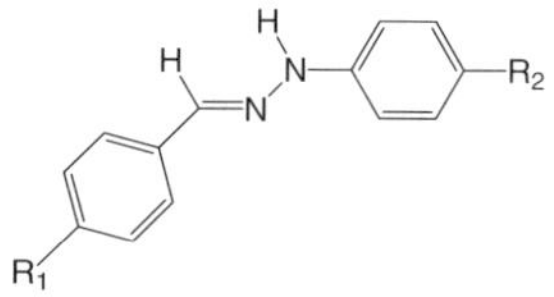 γ^{THG}_{rot} [10^{-48} m^5/V^2]	ε [l/(cm mol)]	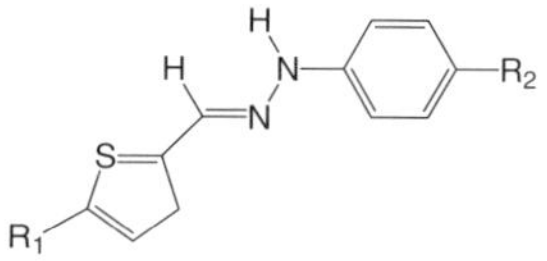 γ^{THG}_{rot} [10^{-48} m^5/V^2]	ε [l/(cm mol)]
CH$_3$/CH$_3$	0.26	0.0	0.21	0.0
OCH$_3$/OCH$_3$	0.23	0.0	0.21	–
NO$_2$/NO$_2$	1.3	–	1.8	–
CH$_3$/NO$_3$	0.87	2.3±0.3	0.95	1.2±0.1
OCH$_3$/NO$_2$	0.70	2.1±0.2	0.83	6.0±0.5
NO$_2$/CH$_3$	1.9	1.1±0.2	2.8	155±4
NO$_2$/OCH$_3$	1.9	15.5±0.6	–	1040±50

values of β. This behaviour is fully confirmed by the experimental results (Table 3) that show larger values of γ for the Type II hydrazones in comparison to the Type I hydrazones: again the asymmetric substitution enhances $\Delta\mu$ in the dipolar term of the three level model and therefore γ.

The same strategy to take advantage of the dipolar term was employed in a series of carotenoids (Fig. 21). Combined third-harmonic generation experiments and Stark spectroscopic measurements showed that a large difference dipole moment $\Delta\mu$ between ground state and excited state leads to large values of γ [57]. Considerable enhancement in γ could be achieved in this way without large increases in molecular length or volume. This investigations again show the im-

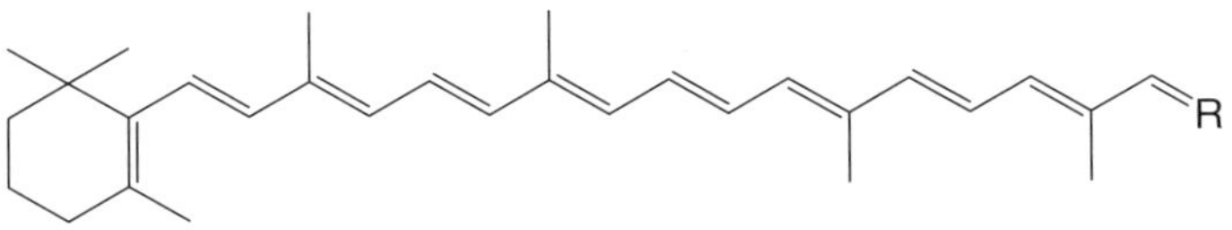

Fig. 21. Structure of the polarized carotenoids where R represents the substitution

be found in Refs. [61, 62]. Moreover, in bithiophene derivatives the same trend was experimentally deduced [63].

4.3
1-D vs 2-D Conjugation

Only little has been reported on second-order hyperpolarizabilities γ in two-dimensionally conjugated molecules. Planar systems as e.g. phthalocyanines have been studied for two photon absorption which is proportional to the imaginary part of the nonlinearity γ. For planar molecules with a three-fold symmetry, the importance of charge transfer from the periphery to the center of the molecule in order to realize large nonlinearities γ was reported [65]. Off-resonant DFWM experiments revealed promising third-order nonlinearities in two-dimensional phenylethynyl substituted benzene derivatives [66]. Recently, the advantage of two-dimensional conjugation to increase the values of the first-order hyperpolarizability β has also been pointed out [67–69].

Using two-dimensional conjugation the backbone can e.g. be conjugated in one dimension and the functional groups grafted along another conjugation direction laterally to the backbone. The increased freedom to substitute electron donating and accepting groups is a fascinating possibility to design tailor-made molecules [56, 71–72].

As an example we can look at tetraethynylethenes in which different conjugation paths can be distinguished (Fig. 22). Conjugation is either possible directly through the central double bond or indirectly through cross-conjugation with two neighboring single bonds. The interactions from the end group R_1 to R_3 and R_4 are more efficient than from R_1 to R_2. The substitution along the direct conjugation paths should therefore have much more influence on the physical properties than along the cross-conjugation path.

From TEEs unsubstituted, mono-, di-, tri-, and tetrasubstituted compounds were synthesized and investigated with third-harmonic generation [56] as well as electric field induced second-harmonic generation [67]. We here concentrate on the discussion of monomers of the most intriguing tetrasubstituted molecules and the symmetry dependence of their third-order nonlinearities [72]. Taking two donors and two acceptors three possible geometries can be realized (Fig. 23). The DDAA(*cross*)-TEE molecule has a mirror plane perpendicular to the y-axis, DDAA(*cis*)-TEE a mirror plane perpendicular to the x-axis, and DDAA(*trans*)-TEE a two-fold rotation axis along z.

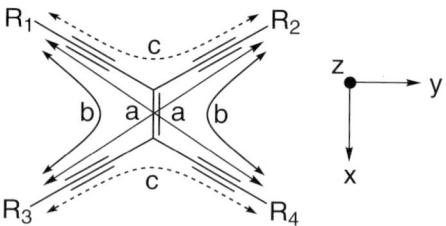

Fig. 22. Conjugation paths for tetraethynylethene molecules. Whereas paths (*a*) are in the (x,y)-plane paths (*b*) are only along the x-direction, providing a total of four conjugation paths along the x-direction but only two along the y-direction. The conjugation along the so-called cross-conjugation paths (*c*) is weak relative to paths (*a*) and (*b*)

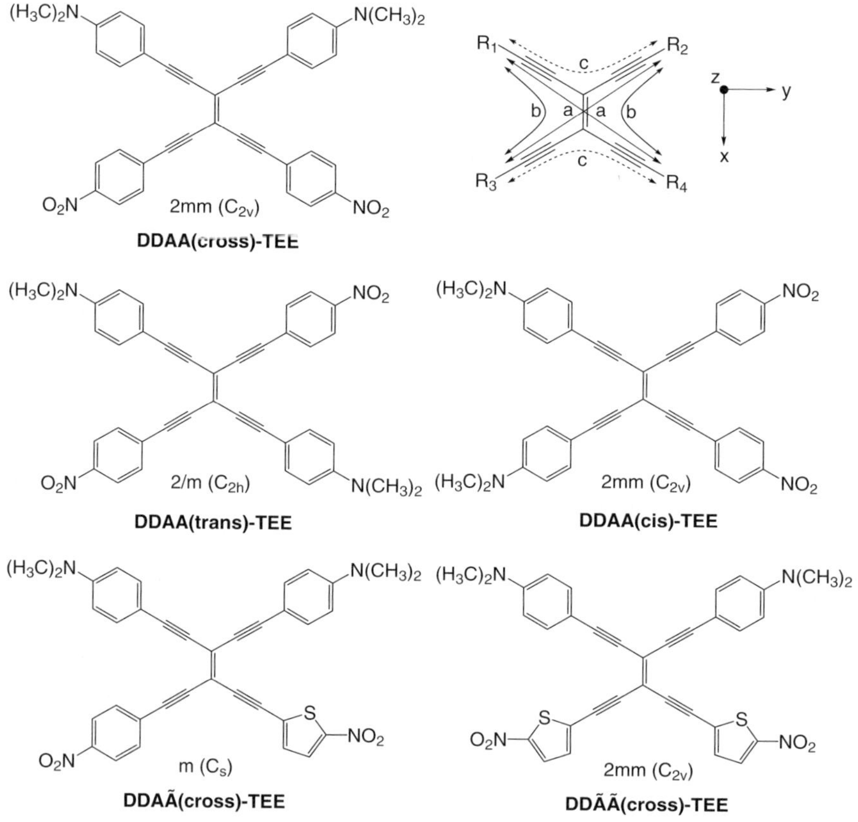

Fig. 23. Structures of the tetrasubstituted TEEs. Below the molecules the symmetry is given in international and Schönflies notation. In the upper right corner the different conjugation paths and the coordinate system are depicted as used in the derivation of the symmetry relations

The molecule DDAÃ(*cross*)-TEE with one phenyl ring of the acceptor replaced by a thienyl ring has a completely broken symmetry in the (x,y)-plane. The incorporation of a thienyl ring also on the second acceptor yields again the same symmetry for the DDÃÃ(*cross*)-TEE as for the DDAA(*cross*)-TEE with a mirror plane perpendicular to the y-axis.

Due to the strongly red shifted absorption spectra the THG experiments had to performed at the fundamental wavelength of γ=2100 nm. Only minor absorptions at the harmonic wavelength occur and the resonance enhancement is still weak and of comparable size for all five tetrasubstituted TEEs. For the DDAA(*cross*)-TEE with the absorption band at the shortest wavelength also a THG experiment at γ=1907 nm has been performed. In this case the harmonic wavelength is only little absorbed with an extinction coefficient ε=430/(mol·cm). The nonlinearity $\gamma_{rot}^{THG} = 18.6 \cdot 10^{-48}\ m^5/V^2\ \left(1330 \cdot 10^{-36}\ \text{esu}\right)$ for the DDAA(*cross*)-TEE is obtained.

The THG measurement results are summarized in Table 4 together with the wavelengths of maximal absorption λ_{max}. The values of the nonlinearities cannot be explained by increasing resonance enhancement: for the molecules with the phenyl ring on the acceptor, the largest nonlinearity is obtained for DDAA(*cross*)-TEE, which has the smallest wavelength of maximal absorption λ_{max} and, conversely, DDAA(*trans*)-TEE with the largest value of λ_{max} shows the smallest nonlinearity γ. The origin of this at first unexpected trend for the second-order hyperpolarizability lies in the electronic wavefunctions and their symmetries what is shown below.

The exchange of a first phenyl ring for a thienyl ring in going from DDAA(*cross*)-TEE to DDAÃ(*cross*)-TEE, while maintaining the same relative donor-acceptor geometry, increases the non linearity, concurrent with a red shift of the absorption spectrum. Replacing the second phenyl ring to afford DDÃÃ(*cross*)-TEE has a detrimental effect on γ, despite a further red shift of the absorption spectrum. The decreased nonlinearity for the higher symmetry molecule DDÃÃ(*cross*)-TEE, in comparison that of the lower symmetry DDAÃ(*cross*)-TEE, convincingly portrays the importance of lowered symmetry for large second-order hyperpolarizabilities γ.

Table 4. Second-order hyperpolarizabilities γ(-3ω, ω, ω, ω,) of tetrasubstituted TEE monomers measured by THG at a fundamental wavelength of λ =2100 nm. λ_{max} is the wavelength of maximal absorption and ε the extinction coefficient at the harmonic wavelength of the THG experiment (reference $\chi_{fs}^{(3)} = 1.6 \cdot 10^{-22} m^2/V^2$, experimental error 10%)

Molecule	λ_{max} [nm]	ε (3ω) [l/(cm·mol)]	$\|\gamma_{rot}^{THG}\|$ [10^{-36} esu]	$\|\gamma_{rot}^{THG}\|$ [10^{-48} m^5/V^2]
DDAA(*cross*)-TEE	486	90	830	11.6
DDAA(*cis*)-TEE	533	120	660	9.2
DDAA(*trans*)-TEE	520	40	370	5.2
DDAÃ(*cross*)-TEE	544	120	1060	14.8
DDÃÃ(*cross*)-TEE	570	310	910	12.8

The second-order hyperpolarizability of the tetrasubstituted TEEs are analyzed with the three-level model of the sum-over-states expression. The rotational average of the second-order hyperpolarizability γ_{rot}^{THG} (Eq.(49)) in the case of THG of a two-dimensional, planar molecule is

$$\gamma_{rot}^{THG} = \frac{1}{5}\left(\gamma_{xxxx} + \gamma_{yyyy} + \gamma_{xxyy} + \gamma_{yyxx}\right) \tag{81}$$

$$\gamma_{xxxx} = \frac{4}{\varepsilon_o}\left\{-\frac{\left(\mu_{01}^x\right)^4}{E_{01}^3} + \frac{\left(\mu_{01}^x\right)^2\left(\Delta\mu_1^x\right)^2}{E_{01}^3} + \frac{\left(\mu_{01}^x\right)^2\left(\mu_{12}^x\right)^2}{E_{01}^2 E_{02}}\right\} \tag{82}$$

$$\phantom{\gamma_{xxxx}} = N_{xxxx} = D_{xxxx} = TP_{xxxx}$$

γ_{yyyy} is analogous

$$\gamma_{xyyx} = \frac{4}{\varepsilon_o}\left\{-\frac{\left(\mu_{01}^x\right)^2\left(\mu_{01}^y\right)^2}{E_{01}^3} + \frac{\left(\mu_{01}^x\right)^2\left(\Delta\mu_1^y\right)^2}{E_{01}^3} + \frac{\left(\mu_{01}^x\right)^2\left(\mu_{12}^y\right)^2}{E_{01}^2 E_{02}}\right\} \tag{83}$$

For TEE molecules the dipolar terms in the three level model are of special interest as they contain the difference dipole moment $\Delta\mu_1$ which vanish perpendicular to a mirror plane. The dipolar contribution γ_D to the overall nonlinearity is

$$\gamma_D = \frac{4}{5\varepsilon_o E_{01}^3}\left\{\begin{array}{l}\left(\mu_{01}^x\right)^2\left(\Delta\mu_1^x\right)^2 + \left(\mu_{01}^y\right)^2\left(\Delta\mu_1^y\right)^2 \\ + \frac{1}{3}\left[\left(\mu_{01}^x\right)^2\left(\Delta\mu_1^y\right)^2 + \left(\mu_{01}^y\right)^2\left(\Delta\mu_1^x\right)^2 + 4\mu_{01}^x\Delta\mu_1^x\Delta\mu_1^y\mu_{01}^y\right]\end{array}\right\} \tag{84}$$

For a forbidden dipole moment in x or y-direction the dipolar contribution is reduced to

$$\gamma_D\left(\Delta\mu_1^x = 0\right) = \frac{4}{5\varepsilon_o E_{01}^3}\left\{\left(\mu_{01}^y\right)^2\left(\Delta\mu_1^y\right)^2 + \frac{1}{3}\left(\mu_{01}^x\right)^2\left(\Delta\mu_1^y\right)^2\right\} \tag{85}$$

$$\gamma_D\left(\Delta\mu_1^y = 0\right) = \frac{4}{5\varepsilon_o E_{01}^3}\left\{\left(\mu_{01}^x\right)^2\left(\Delta\mu_1^x\right)^2 + \frac{1}{3}\left(\mu_{01}^y\right)^2\left(\Delta\mu_1^x\right)^2\right\} \tag{86}$$

and can be used to rationalize the observed trend in the above series.

In Table 5 the allowed dipole moments in the (x,y)-plane are listed and the consequences for the dipolar contribution to the overall nonlinearity. For the DDAA(*cross*)-TEE the dipoles in the y-direction are symmetry forbidden and only the $\Delta\mu^x$ terms can contribute. In the case of DDAA(*cis*)-TEE only the terms with the $\Delta\mu^y$ are different from zero. As only two direct conjugation paths lead

Table 5. Symmetry allowed dipole moment components $\Delta\mu$ in the (x,y)-plane for the tetrasubstituted TEEs. The number of terms in the numerator of the dipolar contribution γ_D is consequently reduced

Molecule	Allowed dipole	Terms in dipolar contribution γ_D
DDAA(*cross*)-TEE	$\Delta\mu^x$	$(\mu_{01}^x)^2 (\Delta\mu_1^x)^2, (\mu_{01}^y)^2 (\Delta\mu_1^x)^2$
DDAA(*cis*)-TEE	$\Delta\mu^y$	$(\mu_{01}^y)^2 (\Delta\mu_1^y)^2, (\mu_{01}^x)^2 (\Delta\mu_1^y)^2$
DDAA(*trans*)-TEE	–	–
DDÃA(*cross*)-TEE	$\Delta\mu^x, \Delta\mu^y$	all
DDÃÃ(*cross*)-TEE	$\Delta\mu^x$	$(\mu_{01}^x)^2 (\Delta\mu_1^x)^2, (\mu_{01}^y)^2 (\Delta\mu_1^x)^2$

in the y-direction compared to four in the x-direction the dipolar contribution in the DDAA(*cross*)-TEE is larger than in DDAA(*cis*)-TEE, explaining the larger second-order hyperpolarizability of the *cross* TEE.

The two-fold rotation axis perpendicular to the (x,y)-plane of the DDAA(*trans*)-TEE prevents any dipole moment in the plane. Consequently the dipolar contribution vanishes totally and the nonlinearity of DDAA(*trans*)-TEE is the smallest although it shows the largest value of λ_{max} (Table 4).

By introducing the thienyl ring instead of the phenyl ring for the DDÃA(*cross*)-TEE the symmetry in the (x,y)-plane is completely broken and all terms in the numerator of Eq.(81) occur. The dipolar contribution is not symmetry reduced and therefore DDÃA(*cross*)-TEE exhibits the largest nonlinearity (Table 4).

The second thienyl ring in the DDÃÃ(*cross*)-TEE again introduces the mirror plane perpendicular to the y-direction and only $\Delta\mu^x$ contributes to the dipolar term. The overall second-order hyperpolarizability is consequently decreased compared to DDÃA(*cross*)-TEE but nevertheless larger than for the DDAA(*cross*)-TEE due to the superior influence of the thienyl moiety on the nonlinearity compared to phenyl (Table 4).

With the above symmetry considerations one can explain the unexpected trend in the tetrasubstituted TEE series by the molecular symmetry elements without knowing the wavefunctions themselves. It was assumed that the dipolar term dominates and that the negative and two-photon term are of secondary importance. As all the nonlinearities of the tetrasubstituted TEEs are positive the negative term can assumed to be small.

For a small molecule with strong donor-acceptor substitution and charge transfer character, the difference dipole moment $\Delta\mu$ is large and the dipolar contribution can be assumed to dominate the two-photon term.

In conclusion, for small TEE molecules low two-dimensional symmetry enhances the second-order hyperpolarizability without a concurrent redshift of the absorption spectra. The lowering of symmetry is not only important for one-dimensional molecules but also for two-dimensional ones.

4.4
Conjugation Length Expansion

Another issue under investigation is the evolution of the second-order hyperpolarizability upon elongation of the π-electron path in a molecular backbone. Measurements and quantum-chemical calculations provide an empirical power law of the second-order hyperpolarizability γ versus conjugation length (or the number of monomer units n) in short oligomers.

$$\gamma \propto n^a \qquad (87)$$

For longer chain lengths, this power law saturates and reaches a linear increase versus elongation.

In the following we will first discuss examples of polytriacetylenes (Fig. 19) and then compare these results with the ones of other molecular structures. Monomers of polytriacetylenes with the three neutral end groups TMS (trimethylsilyl), TES (triethylsilyl), and TIPS (triisopropylsilyl) have been investigated. By deconvolution of the absorption spectra assuming Gaussian line shapes the wavelengths λ_{00} for the absorption process from the ground state to the zero vibrational mode of the lowest excited state are obtained as described below.

The three monomers yield similar small values of γ with a deviation that is nevertheless larger than the experimental 10% error bar. The increase in the nonlinearity is concurrent with a shift in the absorption spectra to slightly longer wavelengths. The end groups can act as a very weak donor, with the donating effects smaller for TMS than TES and TIPS.

The pure TMS and TES polytriacetylene oligomer series were investigated including a polymer sample. The different end groups in the polymer have a negligible influence on the molecular property as the PTA backbone is long enough to dominate the molecular properties.

In pure PTA, the absorption bands are shifted to longer wavelengths and the extinction coefficients are increased (Fig. 24).

To determine the excitation energy of the lowest electronic level the contributions in the absorption spectrum from transitions to different vibrational modes of the excited state have to be separated. In the deconvolution of the absorption spectra Gaussian line shapes are assumed for the transitions to the different vibrational levels. The analysis leads to the transition wavelength λ_{00} for the excitation from the ground state to the zero vibrational level of the excited state.

The transition energies $E_{01}=hc/\lambda_{00}$ decrease inversely proportional to the number of monomer units as depicted in Fig. 25. For the long polymer samples the $1/n$ behavior levels off with a saturation at $E_{01}= 2.58$ eV. The critical conjugation length for saturation is between eight and ten monomer units.

The THG experiments of the pure PTA (TMS) series show a power law increase of the second-order hyperpolarizability with an exponent $a=2.52\pm0.10$ for the oligomers $n=1$ to 6. The polydisperse PTA polymer samples exhibit a constant second-order hyperpolarizability per monomer unit of $\gamma/n=4.1\times10^{-48}$ m^5/V^2 (300×10^{-36} esu) indicating an only linear increase for longer PTAs. The

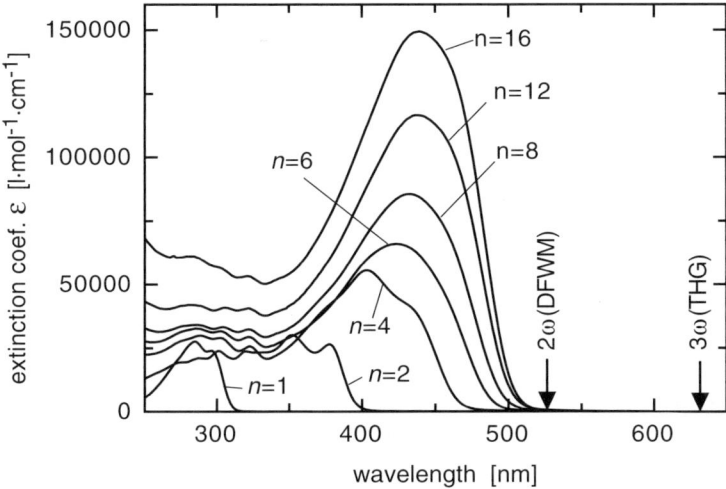

Fig. 24. Absorption spectra of the pure PTA series with TES end groups. The spectra shift with increasing conjugation length to the red part of the spectrum until around ten monomer units with a peak wavelength around 440 nm. The DFWM and THG experiment are performed outside the resonance

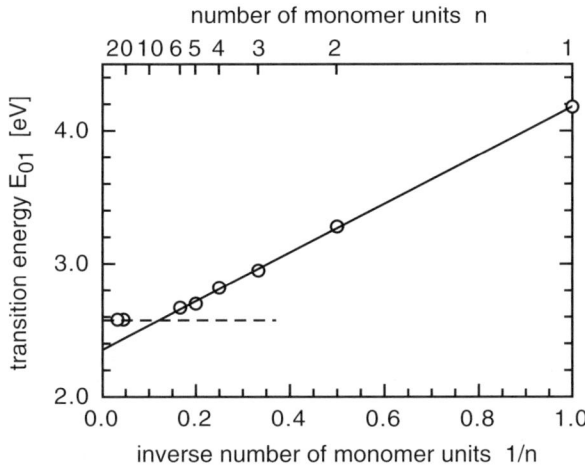

Fig. 25. Transition energy E_{01} from the ground state to the lowest excited state vs. the inverse number of monomer units. The transition energies E_{01} drop with $1/n$ until a saturation level is reached (*dashed line*). The critical conjugation length for the saturation is between eight and ten monomer units

change of the power law to the linear increase is around ten monomer units. With this pure PTA series it was not yet possible to investigate in detail the region of the actual saturation.

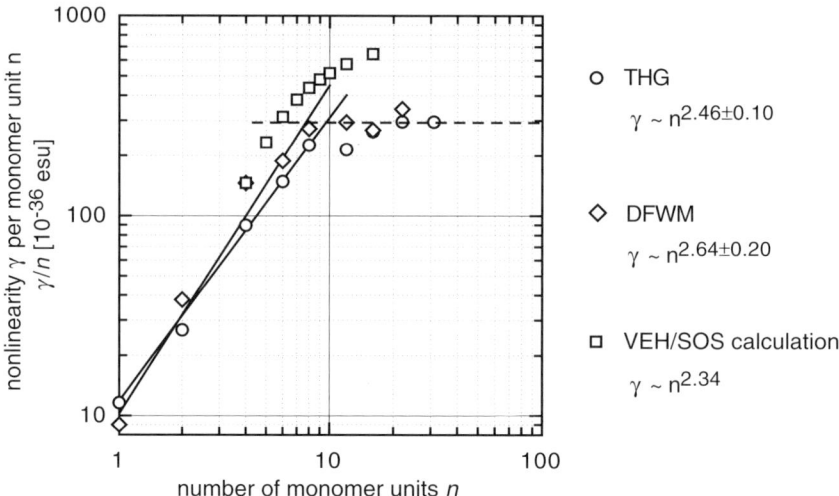

Fig. 26. Power law scaling of the second-order hyperpolarizabilities as a function of the number of monomer units for the pure (TES) PTAs. Data from THG, DFWM, and quantum chemical calculations (VEH/SOS) are given and follow the same general trend with the power law exponents close together

With the second pure PTA (TES) series longer oligomers with n=8, 12, 16 could be measured and the onset of the saturation covered (Fig. 26) [74]. The power law for short oligomer shows an exponent of a=2.46±0.10 in good agreement with the first pure PTA (TMS) series. The saturation is again around ten monomer units. The change from the power law to the saturated nonlinearity per number of monomers reveals to be smooth.

The nonlinearities measured by DFWM reveal a smooth saturation of the second-order hyperpolarizability around ten monomer units as well (Fig. 26). Below the saturation, the experiments show a power law dependence of the second-order hyperpolarizabilities with an exponent a=2.64±0.20. Comparing the exponents for DFWM and THG, the difference is small and within the experimental error. The absolute values of the second-order hyperpolarizabilities are similar with an increasing deviation for longer PTAs.

The slightly larger exponent in the power law for DFWM compared to THG can be explained by the fact that the resonance is closer in DFWM than in THG (Fig. 24). This also agrees with the tendency of yielding larger differences in the hyperpolarizabilities measured by the two techniques for longer PTA oligomers.

In DFWM vibrational contribution can add to the electronic part of the second-order hyperpolarizability measured by THG. For the measurements of the short PTAs far from any resonance the same values by both techniques were obtained which would indicate that this vibrational contribution is small relative to the electronic one.

The increase of second-order hyperpolarizabilities upon backbone elongation has also been evaluated by quantum chemical means by the Brédas group [74]. With a valence effective Hamiltonian approach (VEH/SOS) the parameters in the sum-over-states expression are evaluated leading to the second-order hyperpolarizabilities γ of the molecules. With the VEH/SOS approach the description of larger molecules is feasible, which means in the case of PTA molecules longer than the tetramer.

The conjugation length evolution of the VEH/SOS-calculated static γ values can be described by a power law with an exponent $a=2.34$ for PTA oligomers containing 5 to 8 monomer units. The calculated nonlinearities deviate from the power law beyond n=9–10, thus indicating the onset of a saturation regime. The quantum chemical calculations are in good agreement with the measurements and support the experimental findings.

We can compare these results with previous experiments where information of the saturation regime itself is, however, missing. Experiments on unsubstituted polyenes give an exponent of $a=2.5$ from THG [75] and $a=3.0$ from electric field induced second-harmonic generation (EFISH) measurements [76]. The saturation regime was published first to occur at 120 carbon–carbon bonds [75] but was lately corrected to about 60 bonds [77], which is in good agreement with the conjugation limit for PTAs.

The second-order hyperpolarizabilities in oligothiophenes measured by THG showed an exponent $a=2.8$ [78]. DFWM and EFISH data revealed an exponent $a=4.05$ [79] and $a=4.6$ [80], respectively, for these oligomers. These larger exponents, however, can be explained by a two-photon resonance enhancement in the DFWM and EFISH experiments.

The second-order hyperpolarizabilities of a series of oligothienyleneethynylenes were measured by THG at two different wavelengths. At the fundamental wavelength $\lambda=1064$ nm an exponent $a=3.66$ [81] was found, with the harmonic wavelength in resonance. At $\lambda=1907$ nm a smaller exponent $a=2.4$ [82] was determined. The saturation regime was reached for these oligomers at 60 linearly conjugated carbon-carbon bonds.

We can conclude that the exponent a thus tends to be around 2.5 for various oligomers in the transparency range. In the presence of two- or three-photon resonances, the measured exponent increases. The critical conjugation length of 60 bonds seems to be a collective limit for conjugated polymers. In polyenes the conjugation is believed to be much more efficient but nevertheless the saturation occurs in the same elongation range as for PTA, polythiophenes and other polymers. Saturation of γ was also observed in a series of symmetric cyanines [83].

About the physical origin of the exponent 2.5 and the collective critical conjugation limit of 60 bonds can be speculated. The electronic wavefunctions of the π-electrons seems to extend over maximal 60 bonds. The reason for the loss of electron coherence from one end of the backbone to the other for long molecules might be electron-phonon scattering on the molecular vibrations [84]. For metals electron mean free paths on the same length scale as our critical conjugation length are obtained at room temperature which may be supporting the above hypothesis.

4.5
Two-Photon Absorption

Up to now we have concentrated on the real part of the third-order susceptibility which is related to the nonlinear refractive index. Until recently, no guidelines for an optimization of the two-photon absorption cross-section δ (Eq.(51)) were available. Thorough investigations both theoretically and experimentally then lead to important design guidelines. Summaries of various approaches can be found in Refs. [8, 41, 85–90]. In the case of efficient two-photon absorption at λ= 800 nm the following general optimization rules emerged [85]:
- the conjugation length should be extended
- the planarity of the chromophores is important
- the π-donor strength should be increased
- the conjugation bridge should be modified
- more polarizable double bonds should be used

Prototype examples of molecules that were developed based on these criteria are shown in Fig. 27. Huge effective two-photon absorption cross-sections with values close to $12,000 \times 10^{-50}$ cm^{4}s/photon were obtained with ns pulses. Since the two-photon absorption cross-section was determined at a single wavelength and only with ns pulses excited state absorption may also have contributed to the observed large effects. Detailed studies on heterocyclic chromophores revealed that the molecular environment also plays a role on the two-photon absorption properties if excited state absorption is present [91].

In the following we will specify similar guidelines that were developed by other researchers in more detail since the results are related to the work discussed above. Just as in the case of the real part of $\chi^{(3)}$ it was realized that functionalization with electron donors and acceptors greatly improves δ [41]. Several strategies to enhance δ appeared both from theoretical modeling and detailed experimental investigations and are summarized below:
- A symmetric charge transfer and a change of the quadrupole moment appear to be important.
- An increase in conjugation length leads to improved values of δ.

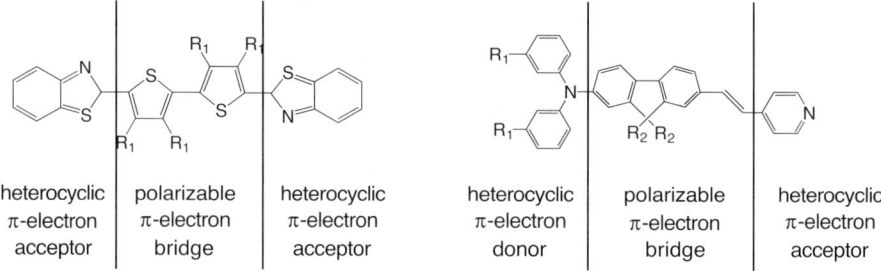

| heterocyclic π-electron acceptor | polarizable π-electron bridge | heterocyclic π-electron acceptor | heterocyclic π-electron donor | polarizable π-electron bridge | heterocyclic π-electron acceptor |

Fig. 27. Prototype examples of highly two-photon active dyes. R_1–R_3 represent molecular units suitable for substitution

Table 6. Experimentally determined two-photon absorption cross-sections δ, the imaginary part of the second-order susceptibility γ, single photon absorption maxima λ_{max}, and the wavelength of maximum two-photon absorption λ_{max}^{TPA}. Unless indicated otherwise, the measurements were performed in toluene. For the calculation of $Im\ \gamma(-\omega,\omega,-\omega,\omega)$ from δ (see Eq.(51)) a refractive index of $n=1.5$ was taken. All data originate from Refs.[8] and [41].

Compound	Compound number	λ_{max} (nm)	λ_{max}^{TPA} (nm)	δ (10^{-50} cm^4 s/photon)	$Im\ \gamma(-\omega,\omega,-\omega,\omega)$ (10^{-48} m^5/V^2)	λ (nm)
	1	–	514	12	0.26	–
	2	374	605	210	6.1	–
	3	408	730 (~725)	995 (635)	42.2	–
	4	428	730 (~725)	900 (680)	38.2	–
	5	456	775 (~750)	1250 (1270)	59.7	–
	6	472	835 (810)	1940 (3670)	92.7	–

Table 6. (continued)

Compound	Compound number	λ_{max} (nm)	λ_{max}^{TPA} (nm)	δ (10^{-50} cm^4 s/photon)	$Im\,\gamma(-\omega,\omega,-\omega,\omega)$ (10^{-48} m^5/V^2)	λ (nm)
	7	513	825 940 (815) (910)	480 620 (650) (470)	26.0 43.8	–
	8	554	970	1750	131	–
	9	618	975 (945)	4400 (large error) (3700)	333 (large error)	–
	10	424	~800	450	22.9	–

- Donor-acceptor-donor and acceptor-donor-acceptor compounds are promising for large values of δ.
- Noncentrosymmetric molecules with strong donors and acceptors can also lead to large values of δ.

It was found that the maximum of two-photon absorption shifts to longer wavelengths with increasing conjugation length and symmetric charge transfer. If the molecules are also fluorescent they become of great interest as fluorescent probes for two-photon microscopy. Selected molecules are given in Table 6. All compounds are from Refs. [8] and [41]. The two-photon absorption cross sections of these molecules were measured using the two-photon fluorescence excitation method with nanosecond and femtosecond laser pulses. In both cases, measurements were performed using fluorophore reference standards. Generally, the ns measurements were in good agreement with the ones using fs pulses. δ can be significantly increased by conjugation length expansion (compare molecules 3, 4, and 5 with 2 (Table 6)). Donor-acceptor-donor and acceptor-donor-acceptor compounds have enhanced values of δ (compare bis-1,4-(2-methylstyryl)benzene with δ of 55×10^{-50} cm^{4}s/photon with molecules 6 to 9 (Table 6)). The peak position of the two-photon absorption is significantly shifted to longer wavelengths on conjugation length expansion and an increase of the symmetric charge transfer. Some of the molecules in Table 6 also have large fluorescence quantum yields and might therefore be of large interest as fluorescent probes. Molecule 10 is the results of an attempt to introduce heavy atoms (here bromine) with a large spin–orbit coupling to allow more efficient intersystem crossing from the S_1 state to the lowest excited triplet state, T_1, therefore creating two-photon absorbing molecules that could be used as efficient triplet sensitizers. Its fluorescence quantum efficiency indicates strong intersystem crossing.

It was also noted that the two-photon absorption peak generelly occurs at shorter wavelengths than twice the one-photon absorption peak. This observation indicates that the two-photon state is energetically above the lowest one-photon state.

A further approach to enhance two-photon absorption deals with donor-acceptor substituted compounds [87] (Fig. 28). An optimization can be achieved if the chromophore is designed for maximum values of the first-order hyperpolarizability β and for two-photon absorption into the lowest excited state S_1. A second possibility is to consider two-photon absorption into the second excited state S_2 and to optimize the molecule for large linear absorption at the same

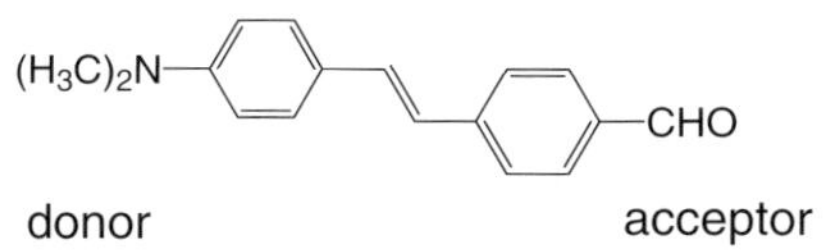

Fig. 28. Structure of the noncentrosymmetric molecule that was theoretically analyzed for the optimization of the two-photon absorption cross-section δ

time. This double resonance potentially leads to huge values of δ with the obvious drawback of substantial linear absorption although it was noted that this drawback may be circumvented by the use of cyanine molecules with their extremely sharp absorption lines [8]. Detailed experimental analysis of noncentrosymmmetric dyes have also been performed with emphasis on the spectral dispersion of two-photon absorption. For the case of e.g. Disperse Red 1 a two-level model was shown to be sufficient to describe the dispersion [92].

Independently, PPV oligomers were also studied by Spangler and coworkers with the aim to develop highly efficient chromophores for optical limiting (Fig. 29). Huge values of effective two-photon absorption cross-sections exceeding $8,000 \times 10^{-50}$ cm^{4}s/photon were obtained with ns pulses making use of 'bi-mechanistic' optical limiting (two-photon absorption and reversible saturable absorption).

Recent theoretical work on two-dimensional charge transfer molecules containing cumulene moieties indicate that just as in the case of the real part of the third-order susceptibilities (see Sect. 4.3) two-dimensionality can also increase the imaginary part [89] and lead to large two-photon absorption in the visible spectral region (Fig. 30). Donor-acceptor substituents further increase δ in the visible range.

In a further study the same authors investigated two-photon absorption in five-membered heteoaromatic oligomers with special emphasis on conjugation length expansion [88, 90]. The calculations reveal structure-property relations with a power law dependence with respect to the molecular length of δ prop. L^a and a varying between 4 and 7 for the investigated molecules. For longer oligomers this power law changes towards a linear increase. The general behaviour is in agreement with the conjugation length expansion results of Sect. 4.4. The larger values of a found in the two-photon absorption also confirms that the increase of a that is experimentally found in third-harmonic generation when two-photon absorption becomes important is due to that effect.

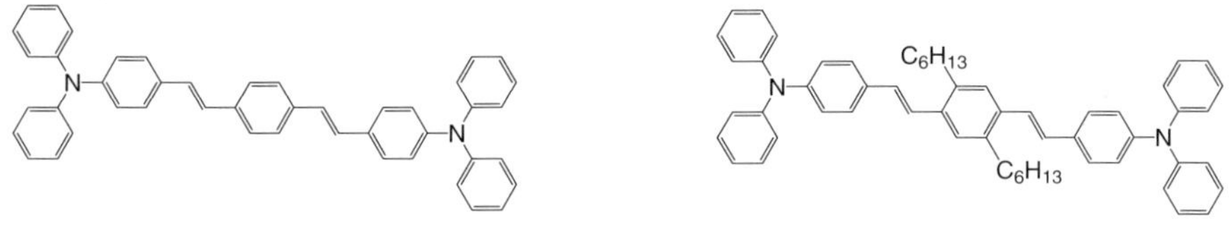

Fig. 29. Two molecules with excellent effective two-photon absorption cross-sections. The increased values (6670×10^{-50} cm^{4}s/photon for the molecule on the left and 8180×10^{-50} cm^{4}s/photon for the one on the right) in comparison with similar molecules of Table 6 indicates that they are enhanced by excited state absorption. The molecule on the right has a large potential for optical limiting in the wavelength range from 675 to 800 nm

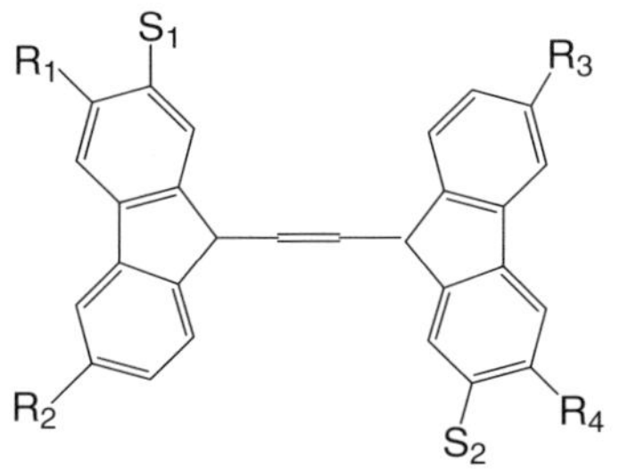

Fig. 30. Structure of donor-acceptor substituted molecules with two-dimensional charge transfer. R and S denote the positions of the substituents which are NH_2 for the donor and NO_2 for the acceptor

5
Concluding Remarks

We have discussed recent developments in the molecular design for third-order nonlinear optical applications. The following optimization guidelines can be derived from the above discussions.
1. For small molecules the asymmetric substitution with strong donors and acceptors leads to the largest nonlinearities. This applies to one- and two-dimensionally conjugated molecules.
2. The conjugation length expansion increases the second-order hyperpolarizability as good as donor-acceptor substitution with the advantage of better transparency in the visible and infrared spectra.
3. Two-dimensional conjugation is superior to the one-dimensional one and has the additional advantage of increased freedom to attach functional groups.
4. For any application, resonant contributions to the second-order hyperpolarizability have to be strongly avoided. It is possible to retrieve huge values in the two-photon resonances but they are useless if all-optical signal processing applications are in mind.

In the following we will discuss the materials properties with respect to the relevant figures of merit.

The third-order susceptibilities of glasses are small but due to their excellent transparency the figures of merit are nevertheless fulfilled (Table 7) and are actually superior to semiconductors and organics. The main drawback are large interaction lengths, which are necessary for a nonlinear phase shift of the order of 2π. Fiber loops of typically tens of centimeters to meters are required, making an integrated design impossible and limiting the application possibilities of glasses.

Normal fused silica fibers show among the best figures of merit but need long fiber loops for switching. The incorporation of heavy ions (e.g. lead silicate) increases the nonlinearity but also the absorption. The figure of merits are worse

Table 7. Figures of merit for glasses and semiconductors

Glasses and semiconductors	λ_{gap} [nm]	λ_{laser} [nm]	n_2 [cm²/GW]	α_0 [1/cm]	α_2 [cm/GW]	$I(W>1)$ [GW/cm²]	T^{-1}	Ref.
SiO_2	–	1064	2.74×10^{-7}	5×10^{-6}	–	0.002	>>1	[11]
Lead silicate, $Pb:SiO_2$	–	1064	2.2×10^{-6}	5×10^{-3}	7.2×10^{-4}	0.2	30	[99]
Pb/Bi/Ga-glass, RN (Corning)	–	1064	1.25×10^{-5}	0.012	–	0.1	–	[100]
Chalcogenide, $As_{0.38}S_{0.62}$	–	1303	4.2×10^{-5}	0.015	<0.16	0.05	>2	[101]
		1552	9.3×10^{-6}	0.002	<0.16	0.03	>0.4	
GaAs	870	1064	-3×10^{-4}	1	25	0.4	0.1	[102]
$Al_{0.12}Ga_{0.88}As$	790	810	-3×10^{-3}	16	20	0.45	2	[103], [104]
		850	-3×10^{-4}	0.5	20	0.15	0.2	
$Al_{0.18}Ga_{0.82}As$	750	1555	1×10^{-4}	0.1	<0.1	0.15	>6	[105], [106]
$In_{0.77}Ga_{0.23}As$	–	1530	-3.2×10^{-3}	active material		–	–	[107]

than for fused silica but still sufficient. The fiber lengths can be reduced from meters to centimeters.

The lead-bismuth-gallate glass RN (40% PbO, 35% $BiO_{1.5}$, 25% $GaO_{1.5}$) exhibits a large nonlinear refractive index. An even larger nonlinearity exhibit the chalcogenide glasses, which have the absorption band shifted far to the red (~ 650 nm). In order to keep out of the two-photon resonance, the measurements have to be performed at longer wavelengths. As all-optical switching application are most interesting for the telecommunication wavelength regions at 1300 nm and 1550 nm, the red shifted spectrum is not disturbing. The two-photon figure of merit T^{-1} is only slightly larger than one for 1300 nm and already below one at 1550 nm, questioning the suitability of chalcogenide glasses for real devices.

In semiconductors, nonlinearities induced by absorption as band filling effects and exciton bleaching lead to large negative nonlinear refractive indices for photon energies just below the energy band gap E_{gap} (Table 7). The concurrent linear absorption is large, depleting the light in a waveguide strongly. For linear absorption coefficients α_0 of the order of 10 cm^{-1} the laser power is absorbed in the first millimeter of propagation, leading to heating of the waveguide material unless very short pulses are used. The use of semiconductors just below the bandgap without amplification is therefore doubtful.

A further disadvantage of using absorptive effects in semiconductors is the recovery time of the excited electrons which are responsible for the large nonlinearity. Recombination times of typically 100 ps to some nanoseconds limit the repetition rate of the switching severely.

Going to longer wavelengths further away from the band gap, the two-photon figure of merit T^{-1} drops below unity until the photon energy is half the band gap. Below half the band gap, the two-photon figure of merit T^{-1} remains larger than one, providing an interesting spectral region for all-optical switching. For even longer wavelengths the three-photon figure of merit can also play a role.

The semiconductor $Al_{0.18}Ga_{0.82}As$ with a band gap λ_{gap} =750 nm is designed for operation at the telecommunication wavelengths around 1550 nm. The two-photon figure of merit is fulfilled and the intensity requirement $I(W >1)=0.15$ GW/cm^2 from the one-photon figure of merit is reasonable. Operation of all-optical switching based on the approach to work below half the band gap have been demonstrated .

A possible way to become independent of the figures of merit W, T^{-1} related to absorption is the application of the semiconductors not only as a nonlinear material but also as an amplifier. The depletion of the laser beam is compensated by the active material, which is pumped either optically or electronically. The large nonlinear refractive indices shown in this approach are quite promising for all-optical signal processing ($In_{0.77}Ga_{0.23}As$ in Table 7).

Similar to semiconductors, polymers show the best figures of merit below half the absorption edge. At shorter wavelengths the two-photon absorption is too strong resulting in a two-photon figure of merit T^{-1} close to or below unity. With the exception of the values at 800 nm for PPV (Fig. 15) and at the 1064 nm for DANS (4-dialkylamino-4'-nitro-stilbene) and DAN2 (4-dialkylamino-4'-nitro-diphenylbutadiene), all materials in Table 8 were obtained below half the absorption edge.

The polymers can be divided in two subgroups: the pure polymers with the backbone as the active unit itself and the chromophore polymers with the active molecule as a guest in the host matrix or laterally attached to the polymer backbone.

Polydiacetylene (PDA, Fig. 15) can be processed in different forms. In the PTS polydiacetylene crystal, the crystallized monomers are polymerized in the solid state by UV-light. The two-photon figure of merit is larger than for any other polymer or semiconductor. A Mach-Zehnder type all-optical switch with a nonlinear phase shift larger than π has been demonstrated in a two millimeter long waveguide [94]. The main problem to overcome are the scattering centers and cracks induced by the solid state polymerization. Unless a reproducible technique to create low scattering waveguides is found, the application of PTS single crystals is questionable. The solution processed 4BCMU polydiacetylene exhibits forty times smaller nonlinearities than PTS. This can be explained either by the different side groups, the intermolecular ordering or defects that arise from the polymerization of 4BCMU. The figures of merits with T^{-1} close to unity are similar to other polymer systems. The waveguide production is easier than for PTS but the linear absorption is nevertheless dominated by scattering from waveguide imperfection [95].

The most advanced polymer with respect to structuring is polyphenylenevinylene (PPV, Fig. 15) which exhibits suitable figures of merit just below half the absorption edge. The waveguide production is well established and the interesting wavelength region compatible with standard semiconductors lasers. PPV (e.g. DPOP) with different side groups have shown sufficient figures of merit and also copolymers (e.g. RCo 52) with other monomer units have satisfactory properties.

The side-chain polymers DANS and DAN2 exhibit two-photon figures of merit T^{-1} below unity at 1064 nm as the nonlinearity is still in a two-photon resonance. At 1319 nm the two-photon resonance is left for DANS and the figures of merit are fulfilled. A nonlinear phase shift of $\phi_{NL}=0.5\pi$ in a 1.65 cm long Mach-Zehnder interferometer was demonstrated [96].

Waveguides of the chromophore main chain polymer PSTF66 have been measured, disclosing also a too large two-photon absorption for applications.

For the comparison with the materials discussed just now the nonlinear refractive indices of PTA and TEE can be estimated. These estimations lead to values comparable with PPV and other polymers measured outside resonance result (Table 8).

Since we have focussed on direct third-order nonlinearities the interesting concept of cascading of second-order nonlinearities to generate nonlinear phase-shifts [2, 3, 97, 98] are not discussed in this review.

In the last few years major contributions to the understanding of the relationships between the molecular structure and the third-order nonlinear optical

Table 8. Figures of merit of polymers.([#]assuming linear absorption for a typical polymeric waveguide outside resonance)

Polymer	λ_{laser} [nm]	n_2 [cm^2/GW]	α_0 [1/cm]	α_2 [cm/GW]	$I(W>1)$ [GW/cm^2]	T^{-1}	Ref.
PTS (PDA)	1600	2.2×10^{-3}	<0.8	<0.5	0.06	>30	[108]
4BCMU (PDA)	1320	4.8×10^{-5}	1.8	<0.25	5	>1.5	[95]
PPV	800	10^{-2}	1[#]	80	0.008	1.6	[109],
PPV (DPOP)	880–920	1×10^{-5}	<0.35	0.07	3.2	1.6	[110]
	970	1×10^{-5}	<0.35	0.01	3.4	10	
PPV (RCo 52)	880–920	2×10^{-5}	<0.35	0.15	1.6	1.5	
	970	2×10^{-5}	<0.35	0.06	1.7	3.4	
PSTF66	1550	2.8×10^{-5}	1.5	0.81	8	0.2	[111]
DANS	1064	7×10^{-5}	2.5	1.5	4	0.4	[112],
	1319	8×10^{-5}	0.4	0.08	0.7	7.5	[96]
DAN2	1064	1.9×10^{-4}	4	2	2.2	0.9	[112]
PTA	1907	2×10^{-5}	1[#]	-	10	-	this work
TEE	1907	1.5×10^{-5}	1[#]	-	13	-	this work

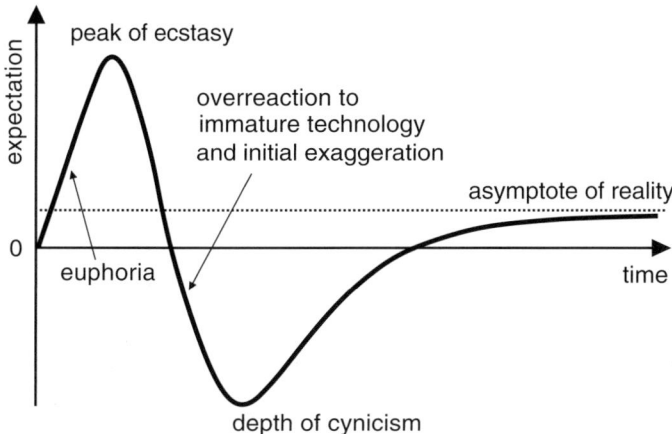

Fig. 31. Expectations in a scientific topic over time. After an initial euphoria and large excitement the exaggerations of this first period and the technological challenges lead to a disillusionment. Persistent scientific work solves and overcomes a part of the problems resulting in an asymptotical rapprochement to reality. No time scale is given and the asymptote or reality is always above the origin

properties of organic molecules have been achieved. However, how far are we on the road to an organic all-optical switch?

The state-of-the-art can best be discussed by a general "scaling law of expectation", which is observed in many scientific areas with a technological focus (Fig. 31).

In the mid 1980s, conjugated polymers were a hot topic in nonlinear optics. Many people promoted third-order nonlinear optics based on conjugated polymers and predicted their incorporation in future all-optical applications. However, the disillusionment came during the beginning of nineties when it was realized that it will not be so easy to have all required parameters under control. The initial hopes did not fulfill on the short time scale and quite some problems were left to be tackled.

In recent years, with the developments in laser technology and the advanced skills and tools in synthetic organic chemistry new approaches and experiments allowed to steadily improve the knowledge about organic molecules in nonlinear optics. The slope of the improvements is not as steep as in the beginning but definitely positive.

With today's continuous-wave modulation techniques in telecommunications the necessary nonlinearities are very difficult if not impossible to achieve. However, for other transmission schemes (e.g. solitons) and new potential concepts the development of organic materials fulfilling the requirements imposed by applications may is possible.

To further improve the materials the collaboration between physics, chemistry, and material science is of central importance and only a mutual understanding of the limiting factors in each field will drive the topic closer to possible applications. The possible multi-functionality of polymers is after all one of the exciting advantages that organic materials have to offer.

References

1. Stegeman GI, Wright EM (1990) Optical and Quantum Electronics 22:95
2. Stegeman GI, Hagan DJ, Torner L (1996) J Optical and Quant Electron 28:1691
3. Bosshard C (2000) Third-order nonlinear optics in polar materials. In: Günter P (ed) Nonlinear Optical Effects and Materials. Springer Verlag, Berlin Heidelberg New York, p. 7
4. Kuzyk MG, Dirk C (ed) (1998) Characterization Techniques and Tabulations for Organic Nonlinear Optical Materials, Marcel Dekker,
5. Dick B, Stegeman G, Twieg J, Zyss J (eds) (1999) Molecular Nonlinear Optics: Materials, Phenomena, Devices, Special issue of Chem Phys 245
6. Sheik-Bahae M, Said AA, Wei T-H, Hagan DJ, Van Stryland EW (1990) IEEE J Quantum Electron 26:760
7. Kleinman DA (1962) Phys Rev 126:1977
8. Marder SR, Perry JW, Brédas JL, McCord-Maughon D, Dickinson ME, Fraser SE, Beljonne D, Kogej T (1999) Design and applications of two-photon absorbing organic chromophores. In: Brédas JL (ed) Conjugated Oligomers, Polymers, and Dendrimers: from Polydiactylene to DNA. De Boeck & Larcier, s.a., Bruxelles, p. 395
9. Boyd RW (1992) Nonlinear Optics, Academic Press, San Diego
10. Butcher PN, Cotter D (1990) The elements of nonlinear optics, Cambridge University Press, Cambridge
11. Milam D (1998) Appl Opt 37:546
12. Meredith GR (1981) Phys Rev B 24:5522
13. Buchalter B, Meredith GR (1982) Appl Opt 21:3221
14. Mito A, Hagimoto K, Takahashi C (1995) Nonlinear Opt 13:3
15. Bosshard C, Gubler U, Kaatz P, Mazerant W, Meier U (2000) Phys Rev B 61:10688
16. Gubler U, Bosshard C (2000) Phys Rev B 61:10702
17. Stegeman GI (1993) In: Proceedings of Nonlinear optical properties of advanced materials, SPIE 1852, edited by Etemad S, SPIE-The International Society for Optical Engineering, Bellingham, Washington , p. 75
18. Galvan-Gonzales A, Canva M, Stegeman G, Twieg R, Kowalczyk T, Lackritz H (1999) Opt Lett 24:1747
19. Zhang Q, Canva M, Stegeman G (1998) Appl Phys Lett 73:912
20. Galvan-Gonzales A, Canva M, Stegeman G, Twieg R, Kowalczyk T, Lackritz H (1999) Appl Phys Lett 75:3306
21. Kajzar F, Messier J (1985) Phys Rev A 32:2352
22. Hellwarth RW (1977) Prog Quant Electr 5:1
23. Xu C, Webb WW (1992) J Opt Soc Am B 13:481
24. Bosshard C, Bösch M, Liakatas I, Jäger M, Günter P (2000) Second-order nonlinear optical organic materials: recent developments. In: Günter P (ed) Nonlinear Optical Effects and Materials. Springer Verlag, p. 163
25. Wortmann R, Elich K, Lebus S, Liptay W, Borowicz P, Grabowska A (1992) J Phys Chem 96:9724
26. Ueberhofen K, Deutesfeld A, Koynov K, Bubeck C (1999) J Opt Soc Am B 16:1921
27. Miyata S, Nalwa HS (eds) (1997) Nonlinear Optics of Organic Molecular and Polymeric Materials, CRC Press, Inc., Boca Raton
28. Kajzar F, Messier J, Rosilio C (1986) J Appl Phys 60:3040

29. Neher D, Wolf A, Bubeck C, Wegner G (1989) Chem Phys Lett 163:116
30. Miyano K, Nishiwaki T, Tomioka A (1992) Opt Commun 91:501
31. Tang N, Partanen JP (1996) Opt Lett 21:1108
32. Drenser KA, Larsen RJ, Strohkendl FP, Dalton LR (1999) J Phys Chem A 103:2290
33. Lawrence B, Torruellas WE, Cha M, Sundheimer ML, Stegeman GI, Meth J, Etemad S, Baker G (1994) Phys Rev Lett 73:597
34. Chapple PB, Staromlynska J, Hermann JA, McKay TJ, McDuff RG (1997) Journal of Nonlinear Optical Physics and Materials 6:251
35. Mian SM, Taheri B, Wicksted JP (1996) J Opt Soc Am B 13:856
36. Zhao W, Pallfy-Muhoray P (1993) Appl Phys Lett 63:1613
37. Zhao W, Palffy-Muhoray P (1994) Appl Phys Lett 65:673
38. Feneyrou P, Doclot O, Block D, Baldeck PL, Delysse S, Nunzi JM (1997) Opt Lett 22:1132
39. Miniewicz A, Delysse S, Nunzi J-M, Kajzar F (1998) Chem Phys Lett 287:17
40. Ehrlich JE, Wu XL, Lee I-YS, Hu Z-Y, Röckel H, Marder SR, Perry JW (1997) Opt Lett 22:1843
41. Albota M, Beljonne D, Brédas JL, Ehrlich JE, Fu J-Y, Heikal AA, Hess S, Kogej T, Levin MD, Marder SR, McCord-Maughon D, Perry JW, Röckel H, M. R, Subramaniam G, Webb WW, Xu C (1998) Science 281:1653
42. Orr BJ, Ward JF (1971) Mol Phys 20:513
43. Brédas JL, Adant C, Tackx P, Persoons A (1994) Chem Rev 94:243
44. Meyers F, Marder SR, Pierce BM, Brédas JL (1994) J Am Chem Soc 116:10703
45. Marder SR, Perry JW, Bourhill G, Gorman CB, Tiemann BG, Mansour K (1993) Science 261:186
46. Marder SR, Gorman CB, Meyers F, Perry JW, Bourhill G, Brédas JL, Pierce BM (1994) Science 265:632
47. Andrews JH, Khaydarov JDV, Singer KD, Hull DL, Chuang KC (1995) J Opt Soc Am B 12:2360
48. Kuzyk MG, Paek UC, Dirk CW (1991) Appl Phys Lett 59:902
49. Barzoukas M, Runser C, Fort A, Blanchard-Desce M (1996) Chem Phys Lett 257:531
50. Wortmann R, Poga C, Twieg RJ, Geletneky C, Moylan CR, Lundquist PM, DeVoe RG, Cotts PM, Horn H, Rice J, Burland DM (1996) J Chem Phys 105:10637
51. Maslak P, Chopra A, Moylan CR, Wortmann R, Lebus S, Rheingold AL, Yap GPA (1996) J Am Chem Soc 118:1471
52. Zyss J (ed) (1994) Molecular Nonlinear Optics: Materials, Physics, Devices, Academic Press, Boston
53. Bosshard C, Sutter K, Prêtre P, Hulliger J, Flörsheimer M, Kaatz P, Günter P (1995) Organic Nonlinear Optical Materials, Gordon and Breach Science Publishers, Amsterdam
54. Cheng LT, Tam W, Marder SR, Stiegman AE, Rikken G, Spangler CW (1991) J Phys Chem 95:10643
55. Wong MS, Meier U, Pan F, Gramlich V, Bosshard C, Günter P (1996) Adv Mater 8:416
56. Bosshard C, Spreiter R, Günter P, Tykwinski RR, Schreiber M, Diederich F (1996) Adv Mater 8:231
57. Marder SR, Torruellas WE, Blanchard-Desce M, Ricci V, Stegeman GI, Gilmour S, Brédas J-L, Li J, Bublitz GU, Boxer SG (1997) Science 276:1233
58. Martin RE (1998) Monodisperse poly(triacetylene) oligomers, PhD thesis ETH Zürich, No. 12821
59. Garito AF, Heflin JR, Wong KY, Zamani-Khamiri O (1989) In: Proceedings of Organic Materials for Non-linear Optics:Royal Society of Chemistry Special Publication No. 69, edited by Hann RA, Bloor D, The Royal Society of Chemistry, Burlington House, London, p. 16
60. Martin RE, Gubler U, Boudon C, Bosshard C, Gisselbrecht J-P, Günter P, Gross M, Diederich F (2000) Chem Eur J (in print)

61. Alain V, Thouin L, Blanchard-Desce M, Gubler U, Bosshard C, Günter P, Muller J, Fort A, Barzoukas M (1999) Adv Mater 11:1210
62. Alain V, Rédoglia S, Blanchard-Desce M, Lebus S, Lukaszuk K, Wortmann R, Gubler U, Bosshard C, Günter P (1999) Chem Phys 245:51
63. Steybe F, Effenberger F, Gubler U, Bosshard C, Günter P (1998) Tetrahedron 54:8469
64. Perry JW, Mansour K, Lee I-YS, Wu X-L, Bedworth PV, Chen C-T, Ng D, Marder SR, Miles P, Tian M, Sasabe H (1996) Science 273:1533
65. Greve DR, Schougaard SB, Geisler T, Petersen JC, Bjornholm TB (1997) Adv Mater 9:1113
66. Kondo K, Yasuda S, Sakaguchi T, Miya M (1995) J Chem Soc, Chem Commun 1995:55
67. Spreiter R, Bosshard C, Knöpfle G, Günter P, Tykwinski RR, Schreiber M, Diederich F (1998) J Phys Chem B 101:29
68. Wolff J, J,, Längle D, Hillenbrand D, Wortmann R, Matschiner R, Glania C, Krämer P (1997) Adv Mater 9:138
69. Nalwa HS (1995) Adv Mater 7:754
70. Tykwinski RR, Gubler U, Martin RE, Diederich F, Bosshard C, Günter P (1998) J Phys Chem B 102:4451
71. Schreiber M, Tykwinski RR, Diederich F, Spreiter R, Gubler U, Bosshard C, Poberaj I, Günter P, Boudon C, Gisselbrecht J-P, Gross M, Jonas U, Ringsdorf H (1997) Adv Mater 9:339
72. Gubler U, Spreiter R, Bosshard C, Günter P, Tykwinski RR, Diederich F (1998) Appl Phys Lett 73:2396
73. Martin RE, Gubler U, Boudon C, Gramlich V, Bosshard C, Gisselbrecht J-P, Günter P, Gross M, Diederich F (1997) Chem Eur J 3:1505
74. Gubler U, Bosshard C, Günter P, Balakina MY, Cornil J, Brédas JL, Martin RE, Diederich F (1999) Opt Lett 24:1599
75. Samuel IDW, Ledoux I, Dhenaut C, Zyss J, Fox HH, Schrock RR, Silbey RJ (1994) Science 265:1070
76. Craig GSW, Cohen RE, Schrock RR, Silbey RJ, Puccetti G, Ledoux I, Zyss J (1993) J Am Chem Soc 115:860
77. Ledoux I, Samuel IDW, Zyss J, Yaliraki SN, Schattenmann FJ, Schrock RR, Silbey RJ (1999) Chem Phys 245:1
78. Cheng L-T, Tour JM, Wu R, Bedworth PV (1993) Nonlinear Opt 6:87
79. Zhao M-T, Singh BP, Prasad PN (1988) J Phys Chem 89:5535
80. Thienpont H, Rikken GLJA, Meijer EW, ten Hoeve W, Wynberg H (1990) Phys Rev Lett 65:2141
81. Geisler T, Petersen JC, Bjornholm T, Fischer E, Larsen J, Dehu C, Brédas J-L, Tormos GV, Nugara PN, Cava MP, Metzger M (1994) J Phys Chem 98:10102
82. Samuel IDW, Ledoux I, Delporte C, Pearson DL, Tour JM (1996) Chem Mater 8:819
83. Werncke W, Pfeiffer M, Johr T, Lau A, Grahn W, Johannes H-H, Dähne L (1997) Chem Phys 216:337
84. Ashcroft NW, Mermin D, N. (1976) Solide state physics, International Edition W. B. Saunders Company, London
85. Reinhardt BA, Brott LL, Clarson SJ, Dillard AG, Bhatt JC, Kannan R, Yuan LX, He GS, Prasad PN (1998) Chem Mater 10:1863
86. Spangler CW (1999) J Mater Chem 9:2013
87. Kogej T, Beljonne D, Meyers F, Perry JW, Marder SR, Bredas JL (1998) Chem Phys Lett 298:1
88. Norman P, Luo Y, Agren H (1998) Chem Phys Lett 296:8
89. Norman P, Luo Y, Agren H (1999) J Chem Phys 111:7758
90. Norman P, Luo Y, Agren H (1999) Opt Commun 168:297
91. Baur JW, Alexander MD, Banach M, Denny LR, Reinhardt BA, Vaia RA, Fleitz PA, Kirkpatrick SM (1999) Chem Mater 11:2899
92. Delysse S, Raimond P, Nunzi J-M (1997) Chem Phys 219:341

93. Stegeman GI, Villeneuve A, Kang J, Aitchison JS, Ironside CN, Al-hemyari K, Yang CC, Lin C-H, Lin H-H, Kennedy GT, Grant RS, Sibbett W (1994) Int J Nonlinear Opt Phys 3:347
94. Thakur M, Krol DM (1990) Appl Phys Lett 56:1213
95. Rochford K, Zanoni R, Stegeman GI, Krug W, Miao E, Beranek MW (1991) Appl Phys Lett 58:13
96. Kim DY, Sundheimer M, Otomo A, Stegeman G, Horsthuis WHG, Möhlmann GR (1993) Appl Phys Lett 63:290
97. Bosshard C (1996) Adv Mater 8:385
98. Schiek R, Baek Y, Stegeman GI (1996) Phys Rev A 53:1138
99. Newhouse MA, Weidman DL, Hall DW (1990) Opt Lett 15:1185
100. Hall DW, Newhouse MA, Borrelli NF, Dumbauh WH, Weidman DL (1989) Appl Phys Lett 54:1293
101. Asobe M, Kobayashi H, Itoh H, Kanamori T (1993) Opt Lett 18:1056
102. Aitchison SJ, Oliver MK, Kapon E, Colas E, Smith PWE (1990) Appl Phys Lett 56:1305
103. LaGasse MJ, Anderson KK, Wang CA, Haus JA, Fujimoto JG (1990) Appl Phys Lett 56:417
104. Anderson KK, LaGasse MJ, Wang CA, Fujimoto JG, Haus HA (1990) Appl Phys Lett 56:1834
105. Aitchison JS, Villeneuve A, Stegeman GI (1993) Opt Lett 18:1153
106. Villeneuve A, Aitchison JS, Yang CC, Wigley PJ, Ironside CN, Stegeman GI (1992) Appl Phys Lett 56:147
107. Hall KL, Darwish AM, Ippen EP, Koren U, Raybon G (1993) Appl Phys Lett 62:1320
108. Lawrence BL, Cha M, Kang JU, Toruellas W, Stegeman G, Baker G, Meth J, Etemad S (1994) Electron Lett 30:447
109. Samoc A, Samoc M, Woodruff M, Luther-Davies B (1995) Opt Lett 20:1241
110. Gabler T, Waldhäusl R, Bräuer A, Bartuch U, Stockmann R, Hörhold H-H (1997) Opt Commun 137:31
111. Asobe M, Yokohama I, Kaino T, Tomaru S, Kurihara T (1995) Appl Phys Lett 67:891
112. Marques MB, Assanto G, Stegeman GI, Möhlmann GR, Erdhuisen EWP, Horsthuis WHG (1991) Appl Phys Lett 58:2613

Received August 2000

Light-Emitting Characteristics of Conjugated Polymers

Hong-Ku Shim[1] and Jung-Il Jin[2]

[1] Department of Chemistry, Center for Advanced Functional Polymers,
Korea Advanced Institute of Science and Technology (KAIST), Taejon, Korea
e-mail: hkshim@sorak.kaist.ac.kr

[2] Division of Chemistry and the Center for Electro- and Photo-Responsive Molecules,
Korea University, Seoul, Korea
e-mail: jijin@mail.korea.ac.kr

Abstract. This article reviews mainly the results of our recent research on the relationship between the structure and the luminescence properties of PPV derivatives. PPV derivatives are particularly useful in an effort toward the establishment of such relationship because their chemical structures can be manipulated very systematically. Attachment of a wide variety of substituents, inclusion of kinky structural units, modification of main chain structures by inclusion of hole- and/or electron-transferring structures, and blending of polymers having different optical and electronic properties are representative approaches. The device characteristics of the light-emitting diodes (LEDs) fabricated from these polymers are discussed in relation to their structures. In certain cases, their photoluminescence (PL) properties are compared with their electroluminescence (EL) properties.

Keywords. Light-emitting diodes, Conjugated polymers, Energy transfer, Electroluminescence, Quantum efficiency

1	Introduction	194
1.1	Synthesis of Simple PPV Derivatives	195
2	Optical Properties of Simple PPV Derivatives	197
2.1	UV-vis Absorptions	197
2.2	Photoluminescence (PL)	199
2.3	Electroluminescence (EL)	201
3	PPV Derivatives Bearing Anthracene, Carbazole, and Oxadiazole Pendants	203
3.1	Synthesis	203
3.2	UV-vis Absorption and PL Spectra of PPV Derivatives with Anthracene Pendants	205
3.3	EL Characteristics of PPV Derivatives with Anthracene Pendants	207
3.4	UV-vis Absorption and PL Spectra of PPV Derivatives Bearing Carbazole and BPD Pendants	211
3.5	EL Properties of PPV Derivatives with Cz and PBD Pendants	213
3.6	Examples of Highly Efficient Light-Emitting Devices	215

4	Kinked Conjugated Polymers	217
4.1	Synthesis and Characterization	218
4.2	Optical and PL Properties	218
4.3	EL Properties and Current-Voltage-Luminance (I-V-L) Characteristics	221
5	Conjugation Polymers with High Electron Affinity	223
5.1	Synthesis and Characterization	223
5.2	Optical and PL Properties	226
5.3	Current-Voltage-EL Power (I-V-L) Characteristics	226
6	Light-emitting Properties of Blend Polymers	227
6.1	Conjugated-Nonconjugated Multiblock Copolymers (CNMBC)	228
6.1.1	Preparation of Blend Polymers	228
6.1.2	Optical and PL Properties of the Blend Polymers	228
6.1.3	LED Device Characteristics of the Blend Polymers	230
6.2	Conjugated-Conjugated Polymer Blends	232
6.2.1	Preparation of Blend Polymers	232
6.2.2	Absorption and Emission Properties of the Blend Polymers	232
6.2.3	LED Device Characteristics of the Blend Polymers	234
7	Other Conjugated Polymers	236
8	Conclusions and Outlook	239
	References	241

1
Introduction

The potential for making large area multicolor displays from easily processible polymers has driven much of the recent research in the area of polymer light-emitting diodes (LEDs) [1–3]. The polymer LEDs have many advantages compared with inorganic LEDs. One of the important advantages over inorganic LED lies in the easy construction of polymer LEDs enabling the fabrication of displays in a variety of unusual shapes. Other advantages include the ability to provide colors which span the visible spectrum by altering the π-π^* energy gap through controlled change of molecular structure or introduction of various substituents such as electron-donating and electron-withdrawing groups [4–8].

Since the first report of the polymer LED based on poly(1,4-phenylenevinylene) (PPV) by the Cambridge group [2,9], a number of different polymers have been synthesized and extended efforts have been made to obtain high-performance devices from polymeric materials. Polymer LEDs have a low operating

Light-Emitting Characteristics of Conjugated Polymers

voltage, relatively high quantum efficiency, mechanical strength, amorphous properties and color tunability. Color tuning for multicolor applications can be achieved by modifying chemical structures by varying synthetic routes [10,11]. The processibility of polymers for the fabrication of devices also gives an advantage over inorganic semiconductors. With soluble conjugated polymers in organic solvents, direct casting or spin coating is possible, which simplifies device fabrication. In addition, established techniques for the solution deposition of polymer thin films can be exploited to develop methods for manufacturing large area information displays based on this technology.

In this article, we focus briefly on the topics related to light-emitting properties using PPV and its derivatives including related conjugated polymers. The contents consist of the synthetic chemistry of the several important conjugated polymers used for EL generation and their light-emitting characteristics-polymer structure relationships. Working principles and device structural design aspects of the LEDs made of conjugated polymers will be discussed, and finally the outlook in this area is mentioned in the last section.

1.1
Synthesis of PPV Derivatives [12, 13]

PPV derivatives can be prepared by water-soluble precursor route [14–16] involving sulfonium polyelectrolyte or organic-soluble precursor route [3, 17, 18] involving alkoxy substituted non-ionic form, respectively. The representative synthetic pathways for PPV derivatives are shown in Scheme 1.

PPV derivatives synthesized by the precursor route show very high molecular weight, but have problems of insolubility. Therefore, a nonconjugated polymer with good solubility and processing properties is used as an intermediate or pre-

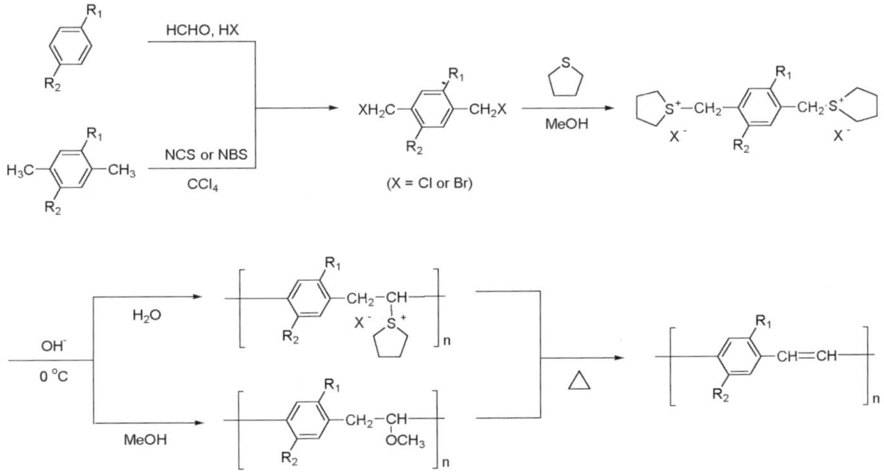

Scheme 1. Synthetic routes of PPV derivatives

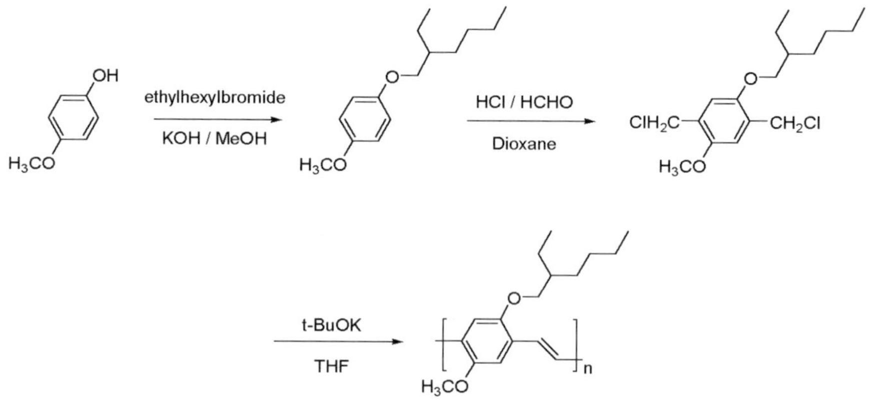

Scheme 2. The monomer synthesis and polymer structure of MEH-PPV by dehydrohalogenation reaction

cursor. The precursor polymer can be purified, characterized in terms of molecular weight, and fabricated into the desired forms prior to conversion to the final structure. In particular, fully dense films with various thicknesses that are suitable for most experiments may be obtained readily either in free-standing films or spin-coated forms on a suitable substrate. And then, the coated film can be easily transformed into a fully conjugated polymer by thermal elimination technique.

Generally, PPV and its derivatives are not soluble in polyconjugated form because of their high rigidity, so there have been many efforts to develop organic-soluble conjugated polymers in the polyconjugated forms. In 1992, Heeger's group reported synthesis of organic soluble poly[2-methoxy-5-(2-ethylhexyloxy)-1,4-phenylenevinylene] (MEH-PPV) [3]. Polyconjugated MEH-PPV is soluble in common organic solvents such as methylene chloride, THF and cyclohexanone, etc. Therefore, MEH-PPV is much more easily processible into thin-films. The monomer synthesis and the polymer structure are shown in Scheme 2.

Another famous synthetic method of conjugated polymers is Wittig condensation reaction. Even though the molecular weights of the polymers synthesized by Wittig polycondensation reaction are not high (10,000 more or less), many conjugated polymers are prepared by this method because organic-soluble polymers can be readily obtained by reacting bisphosphonium salt with dialdehyde monomer. Scheme 3 shows the preparation method of PPV derivatives by this method.

And also, conjugated polymers can be prepared through Heck's coupling between dihalide monomer and divinylbenzene monomer. This coupling reaction is very useful when the synthesis of dialdehyde monomer and bissulfonium or bisphosphonium salt monomer is difficult or impossible. But Heck's coupling has a problem in that very often the preparation of divinylbenzene monomers is

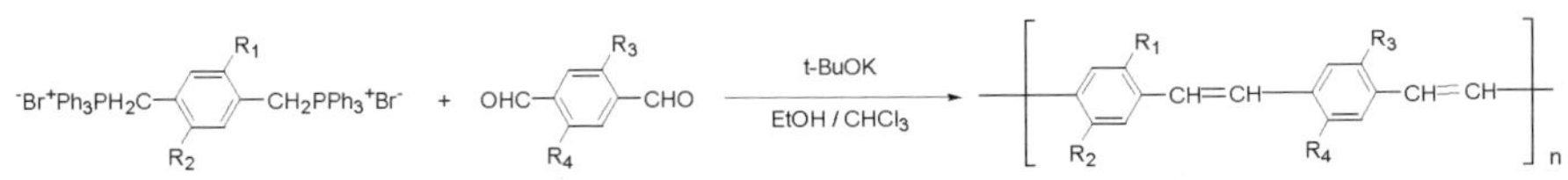

Light-Emitting Characteristics of Conjugated Polymers

Scheme 3. Synthetic route of PPV derivatives by Wittig condensation

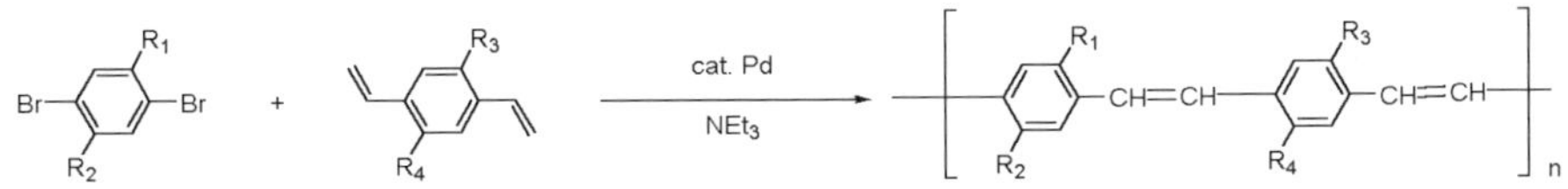

Scheme 4. Synthetic route of PPV derivatives by Heck's coupling

quite difficult. Scheme 4 shows a chemical equation for the preparation of conjugated polymer.

2
Optical Properties of Simple PPV Derivatives

2.1
UV-vis Absorptions

Typical absorption spectra of some of the PPV derivatives are shown in Fig. 1. The absorption maximum of the fully converted PPV film is red shifted from that of precursor polymer. In comparison with the theoretical energy dispersion based on the π-band structure, the π-π^* transition corresponding to the band gap is assigned at 420 nm for PPV. The shape and peak position of the lowest en-

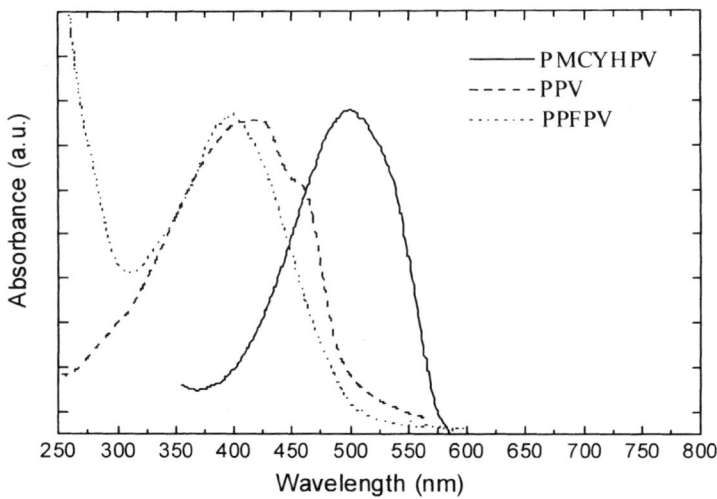

Fig. 1. Absorption spectrum of PPV and its derivatives (see Table 1 for acronyms).

Table 1. UV-visible absorption maxima and edges of PPV derivatives

PPV derivatives	Name	Absorption maxima (nm)	Absorption edges (nm)
	PPV	420	520
	PMPPV	450	540
	MEH-PPV	500	580
	PMCYH-PV	500	590
	PMEHT-PV	430	530
	TMS-PPV	410	510
	PFPV	410	520
	PPFPV	400	520

ergy absorption will be determined by the distribution of conjugation length in the polymer as well as the electronic effect of the substituents on the phenylene ring. The substituents modify the density of states and band gap of the polymers according to the degree of electron donation or attraction. When electron-donating substituents such as alkoxy groups are attached to the conjugated polymer chain, the absorption peak is red-shifted. For example, dialkoxy-substituted PMCYH-PV [19] shows red-shifted absorption maximum and edge at 500 nm and 590 nm, respectively.

The UV-visible absorption maxima and edges of the various PPV derivatives are summarized in Table 1. PFPV [7] with electron-withdrawing substituent shows a blue-shifted absorption at 410 nm and perfluorinated biphenyl substituted PPFPV [20] shows more blue-shifted absorption at 400 nm. Benjamin et al. [21] have reported that a copolymer composed of the PPV part and the 2,3,5,6-tetrafluoro-PPV part showed a blue-shifted absorption compared with that of unsubstituted PPV. These shifts are explained as being a result of the inductive electron withdrawing properties of the fluorine substitution leading to reduced electron density in the conjugated polymers and leading to an increase in the HOMO-LUMO band gap.

As mentioned before, strong electron-donating alkoxy-substituted PPVs show red-shifted absorptions and edges. Dialkoxy-substituted MEH-PPV and PMCYH-PV showed about 80 nm red-shifted absorption maxima compared to unsubstituted PPV. Monoalkoxy-substituted PMPPV showed the absorption and edge in between PPV and MEH-PPV [22]. In the case of PMEHT-PV [23], which contains sulfur atom, the absorption and edge are much more blue-shifted than those of MEH-PPV. This indicates that the bulky alkylthio group diminishes the conjugation length along the polymer chain rather than donates electrons to the backbone. The effect of silicon substitution on the optical properties are of interest. Trimethylsilyl group shows neither electron donating nor withdrawing effects, just gives a steric effect. The absorption and edge of the TMS-PPV [24] are 410 nm and 510 nm, respectively. These positions are a little bit blue-shifted compared with those of PPV, presumably owing to the steric effect of the bulky trimethylsilyl group.

2.2
Photoluminescence (PL)

Photoluminescence is a luminescence from a material after it absorbs light of energy larger than the band gap. Photoexcitation of an electron from the HOMO to the LUMO generates a singlet exciton that can decay radiatively with emission of light at a longer wavelength than that absorbed. PL spectra of PPV, PMCYH-PV and PPFPV are shown in Fig. 2.

The PL from PPV is quite strong, with a quantum yield in the range from a couple of percent to several tens of percent [25, 26], and there is a clear evidence of vibronic structure, with the zero-phonon transition at 515 nm. The efficiency of the photoluminescence is larger when the lifetime of the singlet exciton is

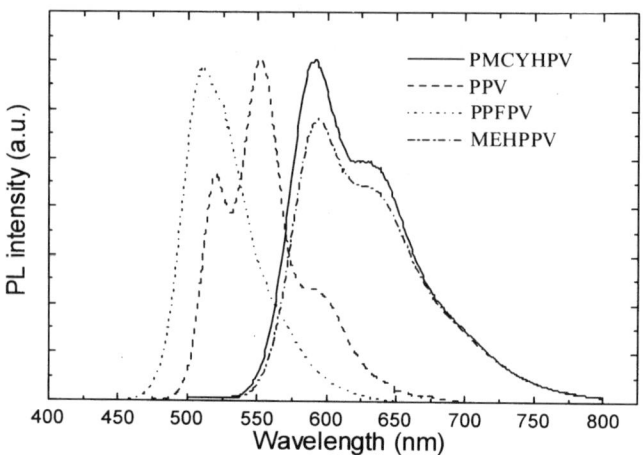

Fig. 2. PL spectra from PMCYH-PV, PPV, PPFPV and MEH-PPV

longer. As with absorption, the peak position of PL can be controlled by the conversion condition, conjugation length and variation of the substituents. The direction of shift of the PL peak position follows that of absorption. Especially, the PL emission maximum is greatly influenced by the substituents. PMCYH-PV which contains strong electron-donating alkoxy groups, shows the PL emission maximum at 590 nm corresponding to orange light [19]. This PL peak position is very close to that of MEH-PPV. Generally, electron-donor substituted conjugated polymers show red-shifted PL emission due to the electron-rich polymer main chain resulting in better π-electron delocalization. The relative PL quantum efficiency of PMCYH-PV is 15% higher compared to that of MEH-PPV. This means that PMCYH-PV aggregates less each other because it has rigid cyclohexyl group in the side chain. Therefore, PMCYH-PV could have better "dilution effect", and lower non-radiative decay of the excitons caused by the interchain-exciton formation.

On the other hand, PPFPV which contains electron-withdrawing perfluorinated biphenyl group shows blue-shifted PL emission at 520 nm [20]. And when the bulky substituents such as perfluorinated biphenyl group induce a steric crowding, the PL spectrum becomes structureless because of disorder as shown in Fig. 2. Such disorder would lead to a distribution of emitting centers that smears out the vibrational structure [27].

Fluorine-substituted PPVs show a very interesting and surprising spectral properties. Recently, Shim et al. [7], Karasz et al. [28] and Benjamin et al. [21] have reported very similar optical properties of fluorine-substituted PPV derivatives.

In summary, the fluoro-PPVs have UV absorption maximum peaks that are blue-shifted relative to that of unsubstituted PPV, but have PL emissions that are substantially red-shifted relative to that of PPV. For example, PFPV [7] shows a little bit blue-shifted UV absorption but 20 nm red-shifted PL emission com-

Table 2. PL and EL emission maxima of PPV derivatives

PPV derivatives	Name	PL maximum (nm)	EL maximum (nm)
(structure)	PPV	540	540
(structure)	PMPPV	560	560
(structure)	MEH-PPV	590	590
(structure)	PMCYH-PV	590	590
(structure)	PMEHT-PV	540	540
(structure)	TMS-PPV	530	540
(structure)	PFPV	560	560
(structure)	PDFPV	590	600
(structure)	PPFPV	520	520

pared with PPV. 2,5-Difluoro-PPV which contains two fluorine atoms on the phenylene ring [28] shows over 50 nm red-shifted PL emission. These results are not well understood, but red-shifted PL emission of fluoro-PPVs clearly reflect the electronic effects of fluorine substitution. A reasonable suggestion is that the excitons generated in fluoro-PPVs may be trapped in shallow trap states associated with charge separation between polymer backbone and highly electronegative fluorine atoms. Therefore, this donor-acceptor pair recombination process exhibits a longer wavelength emission in fluoro-PPV. Table 2 summarizes the PL and EL emission maxima of PPV derivatives.

2.3
Electroluminescence (EL)

In the case of PPV polymers that have conjugation in the polymer backbone, the π-electrons are delocalized along the π-backbone and the delocalization may help making the electrons more mobile. In LED device, when electrons and holes are injected from the electrodes into the polymer, the electrons are injected into the conduction band (LUMO) from the cathode, and the holes are injected into the valence band (HOMO) from the anode. These electrons and holes generate negative polarons and positive polarons, respectively, through electron-phonon coupling. These carriers migrate along the polymer chain toward the opposite electrode under the influence of the applied electric field and combine on a segment of polymer chain to form singlet excitons as produced by photoexcitation in PL. The singlet excitons decay radiatively to give EL radiation. But the EL quantum efficiency is not so high because there are many nonradiative decay channels such as singlet-triplet crossing, formation of interchain ex-

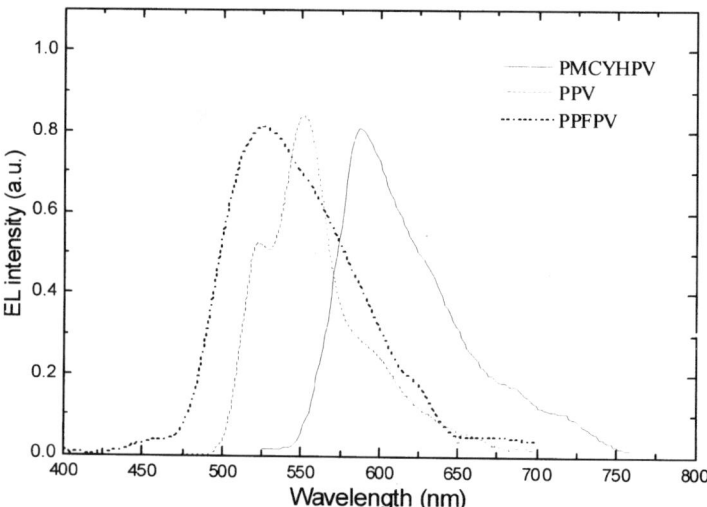

Fig. 3. EL spectra from PPV, PMCYH-PV and PPFPV

citons, exciton-exciton fusion and singlet-exciton quenching and so on. Therefore, low EL quantum efficiency indicates that only a very small fraction of singlet excitons decay radiatively and most of them decay nonradiatively. In contrast to PL in which electrons and holes are directly generated in pairs under photoexcitation and form singlet excitons, the probability of formation of singlet excitons in EL is 0.25. Figure 3 shows the EL spectra from PPV, PMCYH-PV and PPFPV.

The EL and PL emission maxima of the polymers are almost the same, indicating that the origins of luminescence are the same for both. EL peak positions are also red-shifted when strong electron-donating groups are attached to the conjugated backbone. Therefore, it is possible to tune the EL emission color by introducing different substituents in the polymer backbone. In order to obtain a blue color, it is necessary to widen the π-π^* gap by shortening the conjugation length and lowering the electron density in the conjugated backbone. PPFPV shows EL peak in the greenish blue region because the electron density of the polymer main chain is diminished by introduction of the strong electron-withdrawing perfluorobiphenyl group.

3
PPV Derivatives Bearing Anthracene, Carbazole, and Oxadiazole Pendants

3.1
Synthesis

All of the monomers needed for the synthesis of the title polymers have to be prepared via multi-step routes as reported earlier for P-1~P-5 and P-7 [29–31] and as shown in Scheme 5 and 6 for P-6 and P-8. When the final polymer is soluble in an organic solvent, it can be readily obtained by polymerizing the bis-

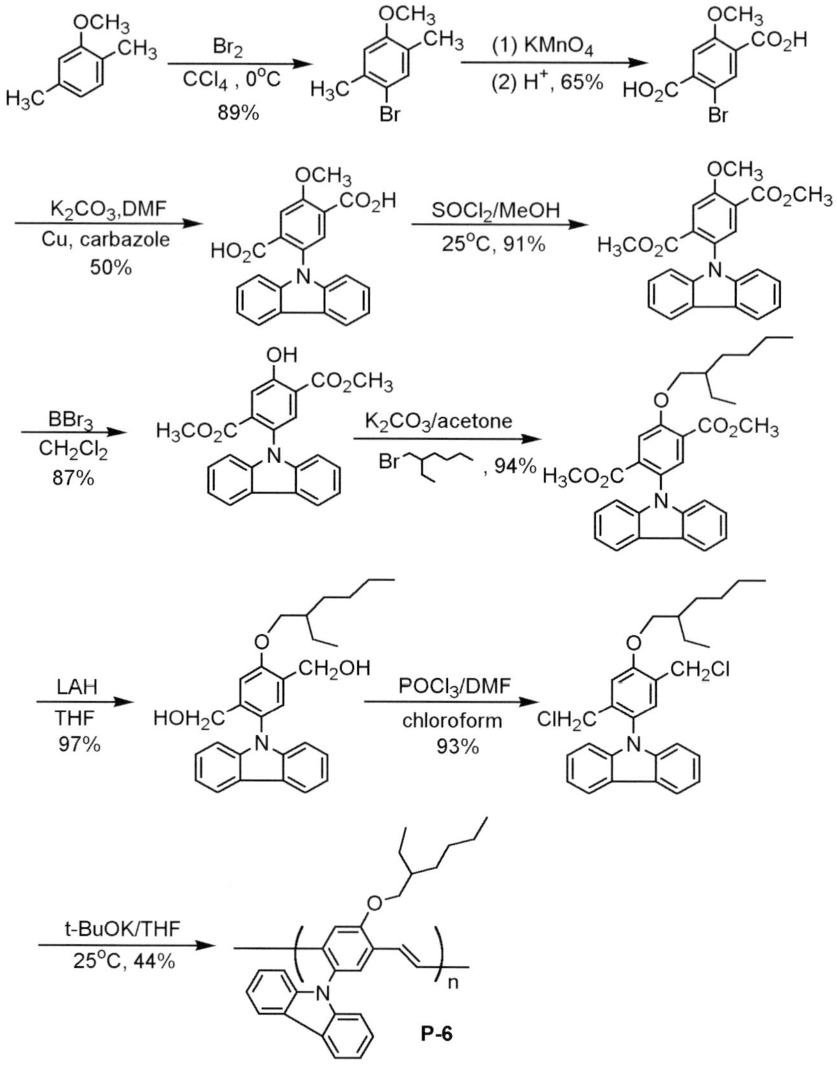

Scheme 5. Synthetic route to P-6.

halomethyl monomer in the solvent using a strong base such as potassium *t*-butoxide [32]. The strong base acts not only as a condensing agent of monomers but also as the dehydrohalogenation reagent of the formed prepolymer molecules. Purification or removal of low molecular weight fraction from the obtained polymers by Soxhlet extraction using a solvent such as acetone and methanol provides polymers of rather high molecular weight (Mn >10,000) and relatively low molecular weight distribution (PDI <2.0). Insoluble polymers have to be prepared by the conventional Wessling-Zimmerman [33] sulfonium salt pre-

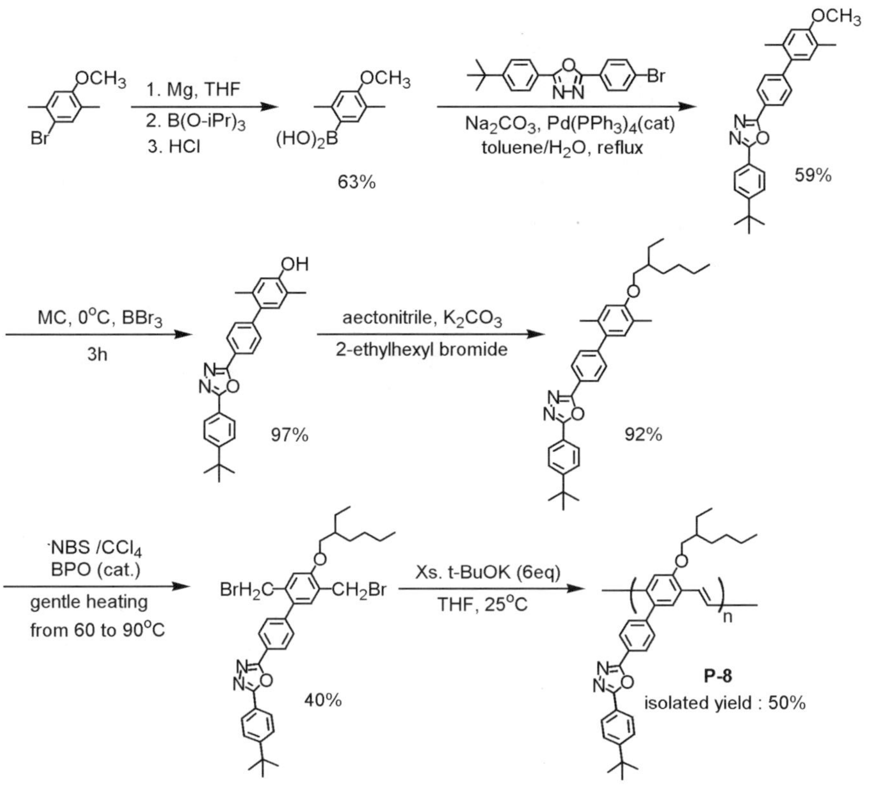

Scheme 6. Synthetic route to P-8.

cursor route. If the precursor polymer is water-soluble, it is easily purified by dialysis.

3.2
UV-vis Absorption and PL Spectra of PPV Derivatives with Anthracene Pendants

Figure 4 compares the UV-vis absorption spectra of thin films of the four polymers (P-1~P-4) bearing anthracene pendants [29,30]. The polymers absorb strongly at about 325–410 nm overlapped with much weaker, broad absorptions over the region of about 400–525 nm. The absorptions in the shorter wavelength region originate from the anthracene moieties, whereas those in the longer wavelength region arise from the π-π^* transition of the main chain π-systems. The absorption maxima of the anthracene moieties (λ_{max} at 342, 361, 378 and 402 nm) appear slightly red-shifted when compared with those of the 9,10-diphenylanthracene [34] that absorbs at 337, 355, 374 and 394 nm, which suggests the existence of a slight electronic interaction between the main chain and the pendant. The presence of the alkoxy substituent [35] attached onto one of the

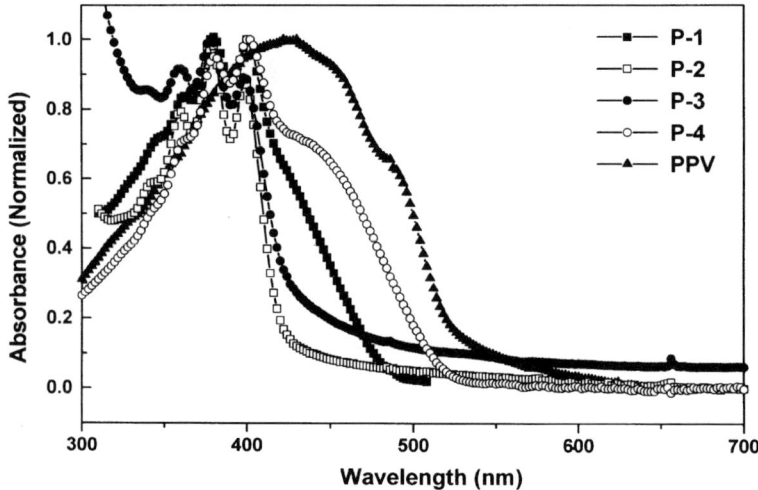

Fig. 4. UV-vis absorption spectra of P-1~P-4 and PPV

phenyl groups in diphenylanthracene moieties can also cause a red shift. Direct interaction between the anthracene pendant and the main chain in P-1 and P-4, however, appears to be minimal considering very small changes in the absorption positions for the pendants. Noncoplanar geometry of the anthracene unit relative to the main chain must be the major cause for such a low degree of electronic interaction.

It is noted that the absorption for the π-π^* transition of the main chain is significantly stronger for P-1 and P-4 in which the anthracene structure is directly attached to the backbone. Another structural information obtainable from Fig. 4 is that the absorption positions of π-π^* transition of the backbone are significantly blue-shifted when compared with that of unsubstituted PPV whose absorption extends to about 550 nm. This suggests that the presence of bulky substituents partially destruct delocalization of π-electrons along the backbone. In other words, coplanarity of the backbone π-system must be partially disrupted.

The absorption edges of the spectra for P-1 and P-4, 480 and 520 nm, respectively, correspond to the optical bandgap energies of 2.58 and 2.38 eV, both of which are larger than the bandgap energy of 2.37 eV (523 nm) for PPV. The electron-donating property of the alkoxy group in P-4 reduces the bandgap energy of P-4 relative to P-1.

Photoluminescnece (PL) spectra of the polymer thin films at an excitation wavelength of 365 nm are compared in Fig. 5 [29,30]. The PL spectra of the polymers have one common feature in that they reveal a simultaneous luminescence from the main chain and the anthracene pendants. The first part of spectra whose maximum emitting light wavelength occurs at about 500 nm, corresponds to overlapped emission from the anthracene pendant and the backbone and the second part arises mainly from the backbone. It is noted that the PL

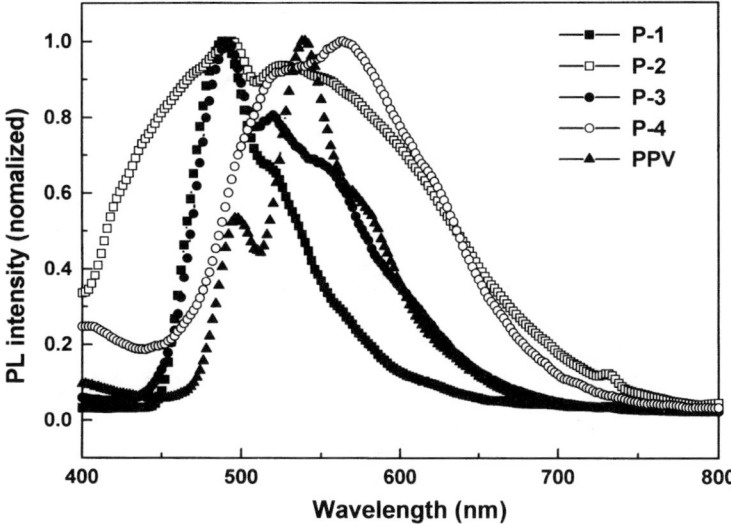

Fig. 5. PL spectra of P-1~P-4 and PPV

spectrum of P-4 covers the broadest range of wavelength from about 450 to 750 nm and most red-shifted among the PL spectra of the polymers. The presence of the anthracene pendant and the alkoxy substituent appears to be the main reason for this spectral characteristics of P-4. Emission by the pendants in P-1 and P-4 starts at about 450 nm and is significantly red-shifted when compared to the emission (400–500 nm) by 9-phenylanthracene [36] suggesting that direct attachment of the anthracene moiety to the PPV backbone causes a facile electronic interaction in the excited state including exciton migration. Such an interaction is much less expected in P-2 and P-3 where the pendants are separated from the backbone by the spacers existing between the two structural components. Therefore, the PL spectra of P-2 and P-3 appear to be nothing but superimposition of the spectra from the pendants and main chains. Their PL emission starts at about 400 nm from the pendant anthracene moieties.

3.3
EL Characteristics of PPV Derivatives with Anthracene Pendants

Electroluminescence (EL) spectra of P-1~P-4 are given in Fig. 6 [29, 30]. The EL spectra of the polymers were obtained for the light-emitting diodes (LEDs) fabricated between the indium-tin oxide (ITO) coated glass anode and the aluminum cathode. For P-1~P-3, their organic soluble precursor polymers were coated onto the ITO-coated glass and they were subjected to thermolysis at 270 °C for 12 h to convert them into the final polymers. And then aluminum cathode was vacuum deposited on the polymer films. P-4 obtained by base-catalyzed polymerization of the bis-bromomethyl monomer was organic soluble and, thus,

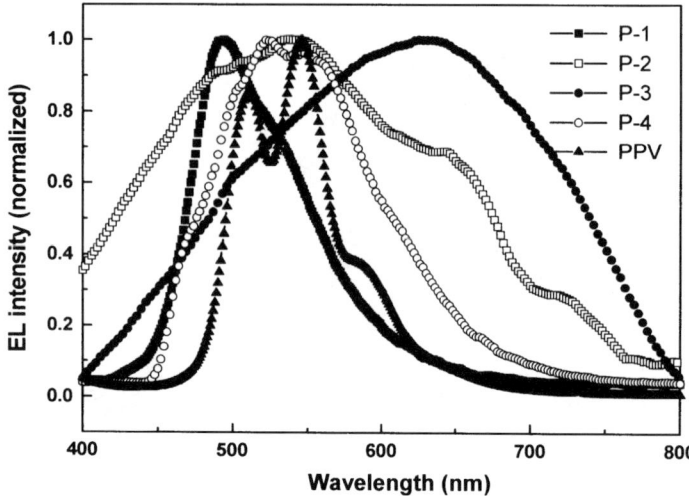

Fig. 6. EL spectra of P-1~P-4 and PPV

was directly spin-coated onto the ITO-coated glass using a solution in 1,1,2,2-tetrachloroethane.

The EL spectra (Fig. 6) of P-1, P-2, and P-4 are about the same as their PL spectra given in Fig. 5. The EL spectrum (Fig. 6) of P-3, however, is very much different from its PL spectrum (Fig. 5). In fact, P-3 emits light only from the main chain in EL, but not any from the diphenylanthracene pendants. This contrasting EL behavior can be explained by the assumption that the excited state electronic energy transfer or exciton migration between the backbone and the pendant occurs in P-1, P-2, and P-4, but not in P-3. It is rather surprising that P-2 allows for such interactions between the backbone and the pendants in spite of the presence of the σ-bonded oxyethyleneoxy spacer between them. This can be interpreted as an indication that energy transfer or excitation migration occurs through the oxyethyleneoxy spacer via the through-bond interaction mechanism. Since such an interaction diminishes rapidly as the length of the spacer increases [37, 38], the same phenomenon is not observed in the EL of P-3. Since the main chain should possess a lower ionization potential and a higher electron affinity than the anthracene moiety, it is conceivable that initial hole- and electron- injection will occur preferentially to the backbone part resulting in the formation of initial excitons only along the main chain followed by the electronic interactions with the pendants in P-1, P-2, and P-4. But the light of 365 nm utilized in PL studies is absorbed by the both parts of the polymer molecules causing PL emission from the backbone and the pendants, as described above. It also is possible, in part, in PL that a part of emitted light by the pendants is reabsorbed by the backbone followed by additional emission from the main chain.

The current density vs electric field curves and the emitted light intensity vs current density for the three LED devices fabricated from P-1 and P-4 are com-

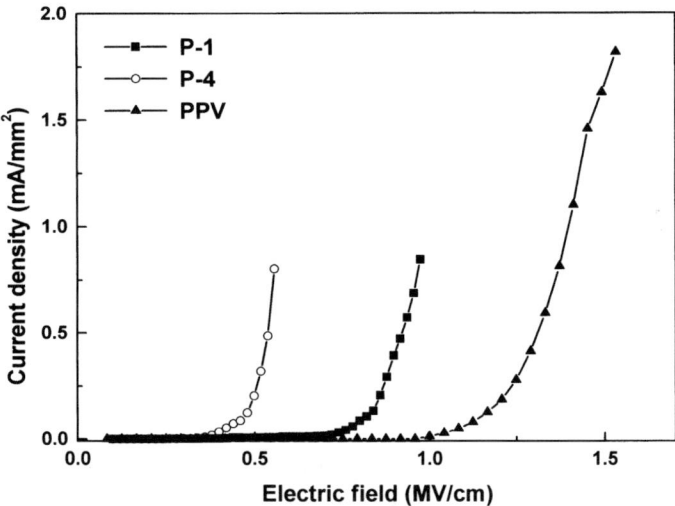

Fig. 7. Current density – electric field curves of P-1, P-4 and PPV

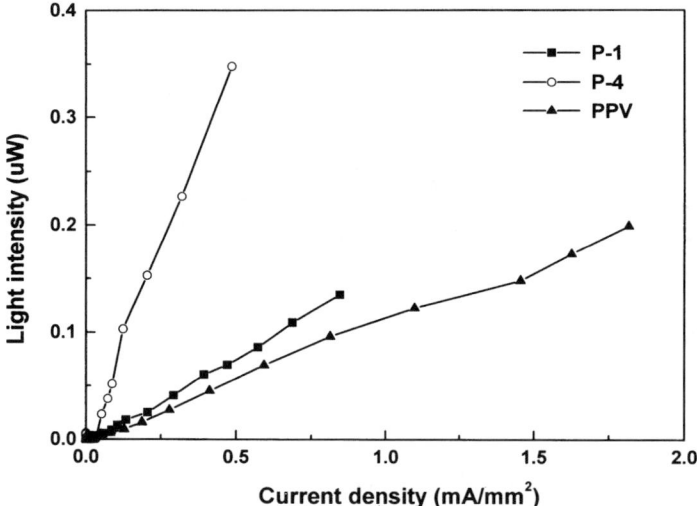

Fig. 8. Emitted light intensity – current density curves of P-1, P-4 and PPV

pared with those of PPV in Fig. 7 and 8, respectively [30]. The threshold field and emitted light intensity increase in the order of P-4<P-1<PPV and PPV <P-1<P-4, respectively. The external quantum efficiencies estimated from the data presented in Fig. 7 and 8 are represented in Fig. 9 [30]. The efficiency increases in the order of PPV <P-1<P-4. At a given current level, the EL efficiency of P-4 is more than 10 times the efficiency of PPV. The efficiency of P-1 is consistently higher than PPV, but to a small extent. Improved efficiency of P-4 can be ex-

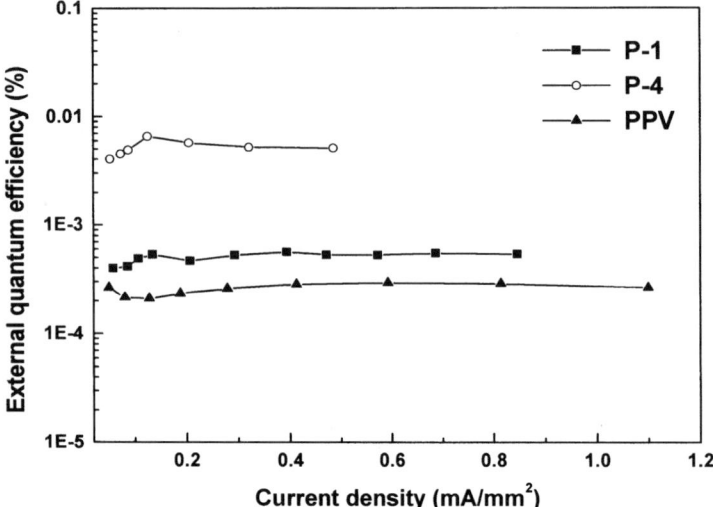

Fig. 9. External quantum efficiency of P-1, P-4 and PPV in EL

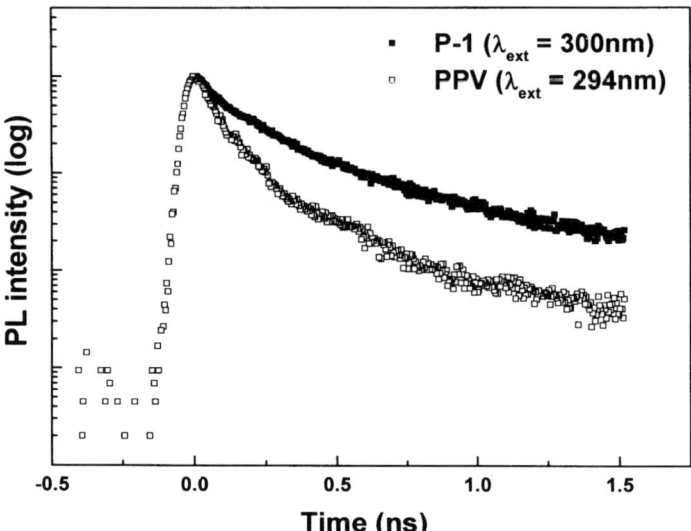

Fig. 10. Comparison of time-resolved PL spectra of P-1 and PPV films at 300 K (probed at 580 and 600 nm, respectively)

plained by its unique structure: the polymer bears the second luminescing phenylanthracene pendant and also the 2-ethylhexyloxy group that increases the interchain distance suppressing the possibility of the formation of interchain excitons. The presence of the two substituents seems to synergistically enhance the EL efficiency. In addition, the anthracene group directly attached to the main chain is able to be involved in the intrachain electronic interactions with the backbone, which increases the lifetime of excitons formed. Figure 10 clearly demonstrates the longer fluorescene lifetime for P-1 when compared with PPV. It is noted that excitation of P-1 by 300 nm lengthenes the PL lifetime probed at 580 nm that corresponds to the emission by the backbone. This strongly supports our supposition that there are strong electronic interactions between the pendant and the backbone.

3.4
UV-vis Absorption and PL Spectra of PPV Derivatives Bearing Carbazole and BPD Pendants

P-5 and P-6 have the hole-transporting carbazole (Cz) [38–40] moieties attached to the main chain, whereas P-7 and P-8 have the electron-transporting 2-(4-*t*-butylphenyl)-5-phenyl-1,3,4-oxadiazole (BPD) [39, 41, 42] pendants. Their UV-vis absorptions are compared in Fig. 11 [31]. P-5 and P-6 absorb at 310–350 nm originating from the pendant in addition to the broad absorption at about 350–540 nm arising from the π-π^* transition of the backbone. P-7 and P-8 also exhibit almost the same absorption characteristics : an absorption around 300 nm by the pendant and a broader absorption in the longer wavelength region (ca. 360–535 nm). In both cases, polymers (P-6 and P-8) carrying the electron-donating 2-ethylhexyloxy group exhibit a red-shift in absorption. It is well known that alkoxy groups cause a bathochromic shift in UV-vis absorptions.

The absorption positions of the pendants in the polymers are slightly red-shifted compared with those of corresponding low molar mass compounds [31]. The absorption positions corresponding to π-π^* transitions of the backbone of

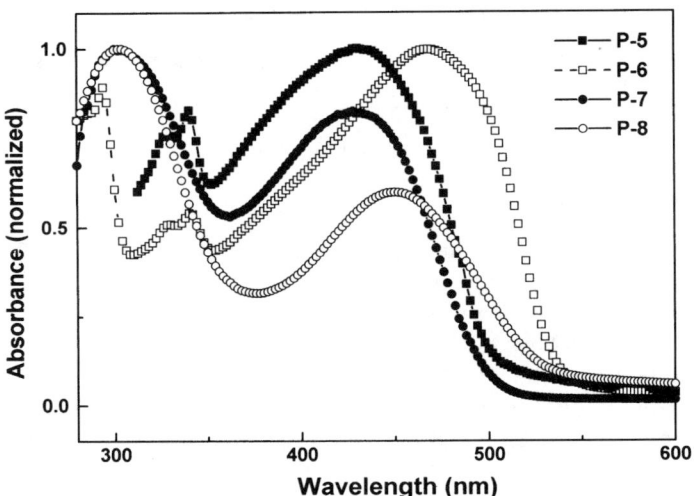

Fig. 11. UV-vis absorption spectra of P-5~P-8

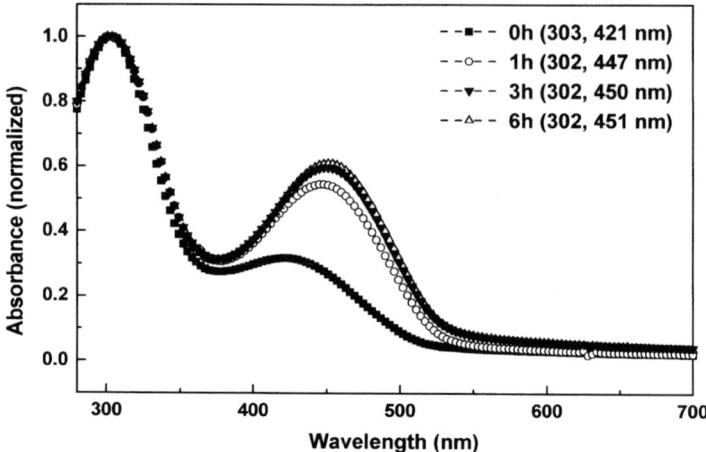

Fig. 12. The dependence of UV-vis absorption of P-8 on thermal history

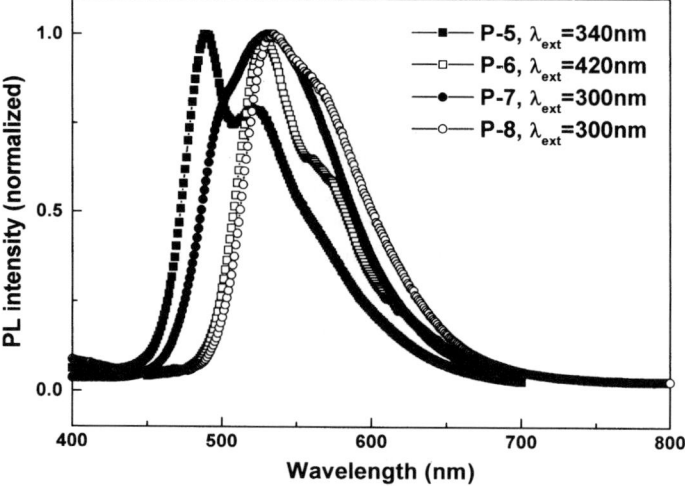

Fig. 13. PL spectra of P-5~P-8

P-5 and P-7 are significantly blue shifted when compared with that of PPV, which again suggests the occurrence of a partial disruption of the coplanarity of the main chain π-system by the presence of the bulky pendants. The same phenomenon was described above for P-1~P-4.

It should be noted that the Gilch [32] polymerization method using excess potassium t-butoxide at room temperature does not provides us with fully eliminated polymers. Figure 12 compares the UV-vis absorptions of P-8's with different thermal history and it shows that thermal treatment of the polymer at 150 °C in a vacuum not only moves the absorption position for π-π* transition to the longer wavelength side but also increases the absorption intensity for the same transition. Thermally treated samples exhibit a red-shift also in PL spectra ; the virgin P-8 shows PL maximum at 534 nm, whereas the one treated at 150 °C for 3 h at 541 nm (Fig. 13).

The PL spectra of the films of P-5~P-8 are given in Fig. 13 [31]. The PL spectra of the polymers have one common feature in that there is a luminescence only from the backbone structure. The PL intensity from the pendants may be too weak when compared with that from the backbone. It also is possible that a very facile energy transfer occurs from the pendants to backbones so that PL from the pendants is not observed. P-7 and P-8 exhibit PL spectra of the same characteristics as for P-5 and P-6, with the peak maximum positions at about 530 and 540 nm, respectively. The presence of the alkoxy substituent on the phenylene ring in P-6 and P-8 moves the absorption positions to the longer wavelength side when compared with P-5 and P-7. Here again, we can conjecture an electronic interaction between the pendant and the backbone. A study on time-resolved PL spectra demonstrates that P-7 exhibits a much longer PL decay ($\tau=750$ ps) than P-5 ($\tau=220$ ps) [43]. It also was conjectured that a charge transfer state is formed in P-7, which causes a red-shifted and structureless emission.

3.5
EL Properties of PPV Derivatives with Cz and BPD Pendants

The EL spectra of P-5~P-8 shown in Fig. 14 are very similar to their PL spectra presented in Fig. 13, but with reduced spectral details. Although the electric field-current density curves are not given, threshold electric field increases in the order of P-6 (0.63 MV/cm) <P-5 (0.75 MV/cm) <PPV (1.1 MV/cm) <P-7(2.4 MV/cm) <P-8 (2.3 MV/cm). The threshold electric field values given in the pa-

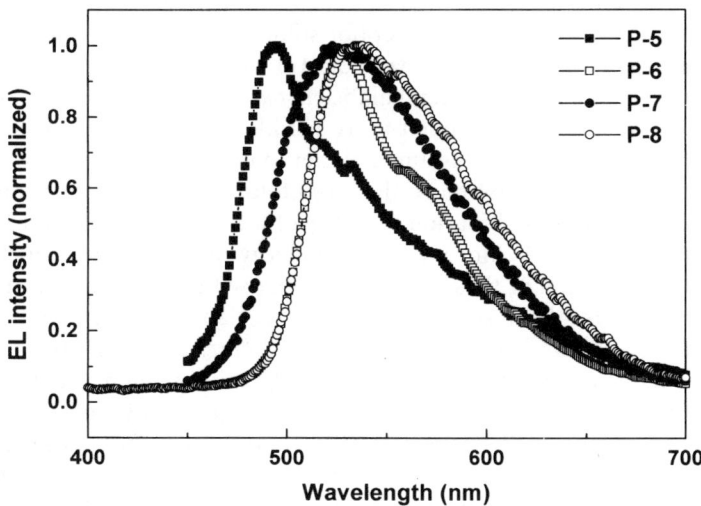

Fig. 14. EL spectra of P-5~P-8

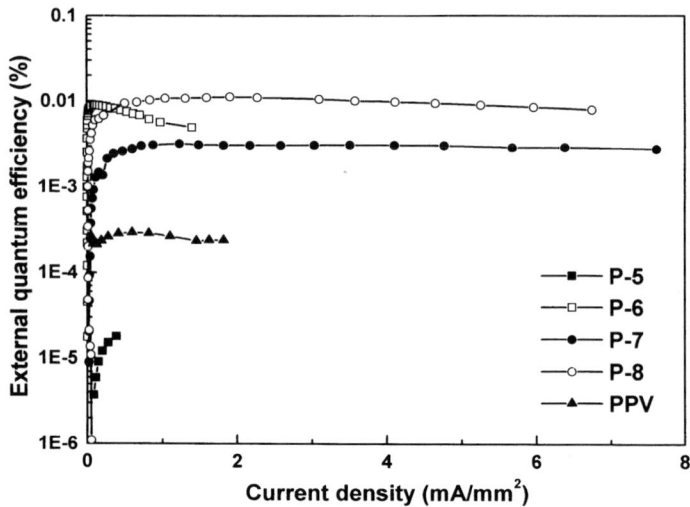

Fig. 15. External quantum efficiency of P-5~P-8 and PPV

rentheses are the values at the current density of 0.1 mA/mm^2. According to our experience the polymers bearing the Cz pendants consistently reveal a lower threshold electric field than those bearing the BPD pendants. All the data given in this discussion were obtained from the LED devices having ITO glass / polymer / Al configuration.

The external quantum efficiency of P-6 and P-8 are much higher than the rest of polymers including PPV (Fig. 15). The particularly low efficiency of P-5 is ascribed to two major different reasons: low electron affinity (3.64 eV) [31] when compared with other polymers (3.93 eV for P-7 and 3.75 eV for PPV) [31] and poor film quality originated from its poor solubility in organic solvents. In addition, the presence of the well-known hole-transporting carbazole pendants in P-5 may even cause an increased imbalance in carrier transport, since PPV itself accepts and transports holes much more readily than electrons. According to Fig. 15, the EL efficiency of P-7 is much higher than that of P-5 by more than two orders of magnitude, i.e., by more than 100-fold. Transient EL measurements [44] proved that the recombination mobility of the electron- hole pair for P-7 is much faster than for PPV. Moreover, the EL decay rate of P-7 was much slower than PPV. Combination of these factors together with improved electron injection and mobility results in superior quantum efficiency. Higher EL efficiency of P-6 and P-8 is ascribed to their better solubility that results in superior film quality and also to the presence of the additional alkoxy substituent that hinders the formation of interchain excitons that are known to cause radiationless decay resulting in low efficiency. EL properties of P-6 and P-8 are under study in detail. The EL efficiencies of P-6 and P-8 are more than 50 times the efficiency of PPV. The EL efficiency for the device prepared from P-4 is lower than those of P-6 and P-8, but higher than that of P-7.

3.6
Examples of Highly Efficient Light-Emitting Devices

Although a relationship between the structure of PPV derivatives and their EL efficiency was the subject of discussion in previous sections, the single layer EL devices fabricated from the present polymers utilizing the ITO-coated glass anode and the Al cathode exhibit rather unsatisfactory external quantum efficiency with high threshold electric field. Such an observation indicates that facilitated carrier injection and balance in carrier mobility are difficult to attain by simple modification of PPV. In order to facilitate carrier injection one can apply a conducting organic or organometallic layer on the ITO electrode and use a low work function metal cathode in stead of aluminum. An organic conducing layer on the ITO electrode not only makes the hole injection easier, but also improves the adhesion between the electrode and organic emitting layer. The use of a low work function metal cathode is expected to lower the electron injection barrier. Balance in carrier mobility appears to be much more difficult to achieve. Since PPV and their derivatives possess much higher hole mobility than electron mobility, a high portion of holes injected into the emitting layer is annihilated on the cathode, which results in low efficiency of LEDs. This undesirable phenomenon can be avoided by employing an electron-transporting layer between the metal cathode and the light emitting polymer layer.

Table 3 [45] compares EL characteristics of the devices fabricated using P-6. As alluded above, existence of a hole-transporting layer (copper phtahlocyanin ; CuPc) [46,47] and an electron-transporting layer (tris(8-hydroxyquinolino)aluminum ; Alq_3) [48,49] in the device and utilization of the low work function Li (Al:Li alloy) [50] cathode can indeed improve the LED efficiency. The Alq_3 layer, however, tends to increase the threshold electric field.

Table 3 shows that the ITO/CuPc/P-6/Alq_3/Al:Li device exhibits an external quantum efficiency of 2000 times the ITO/PPV/Al device and 40 times the ITO/P-6/Al device. The maximum emitted light intensity presented as maximum luminance in the table is 1600 cd/m^2 for the best device. Because CuPc absorbs over 500–800 nm, a part of emitted light from the emitting layers is reabsorbed by the CuPc. This is the reason why the external quantum efficiency is not improved even when the CuPc layer was coated onto the ITO electrode

Table 3. LED device characteristics of P-6 and PPV

Device structure	Φ_{EX}(%)	Relative Φ_{EX}	Electric Field(MV/cm) for 0.1 mA/mm^2	300 cd/m^2	Max. Lumin (cd/m^2)
ITO/ PPV/ Al	0.0002	1	1.1	–	–
ITO/ P-6/ Al	0.01	50	0.5	–	20
ITO/ P-6/ Alq_3 / Al	0.24	1200	1.0	1.3	360
ITO/ P-6/ Alq_3/ Al:Li	0.38	1800	1.0	1.2	1650
ITO/ CuPc/ P-6/ Alq_3/ Al:Li	0.41	2000	0.7	1.0	1600

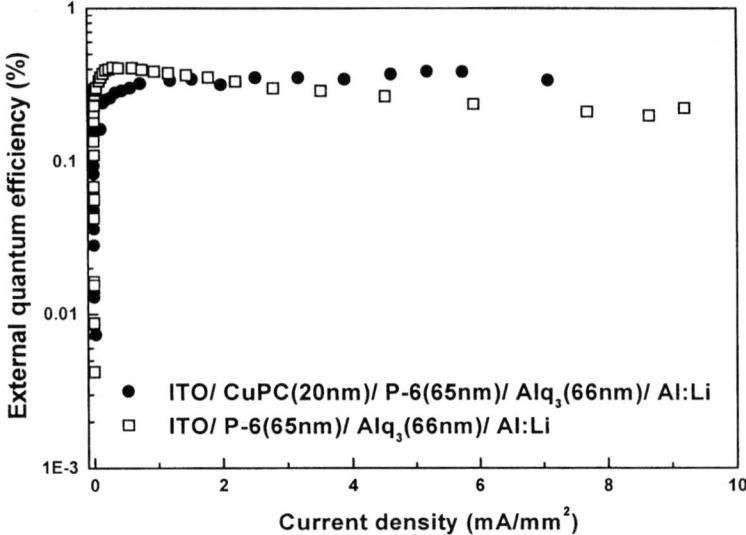

Fig. 16. LED external quantum efficiency of the ITO/ CuPc (20 nm)/ P-6 (65 nm)/ Alq$_3$ (66 nm)/ Al:Li device and the ITO/ P-6 (65 nm) /Alq$_3$ (66 nm) /Al:Li device

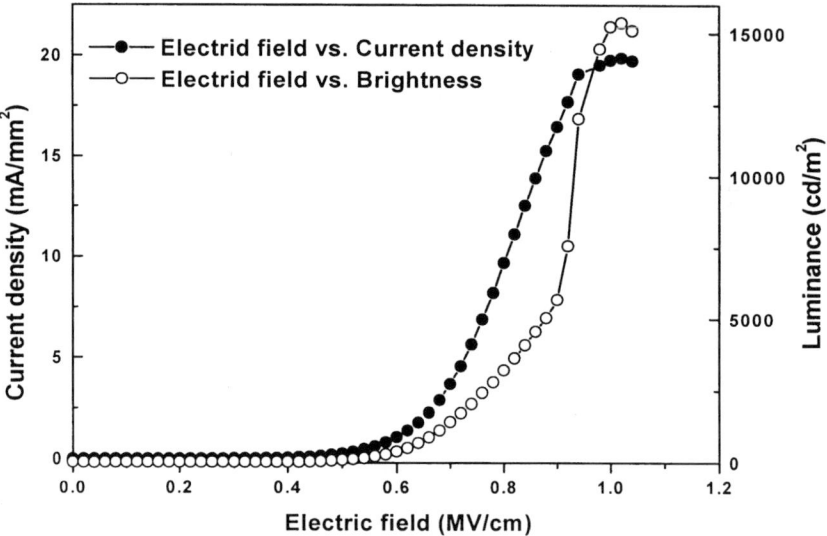

Fig. 17. The electric field – current density – luminance curves for the ITO/ PEDOT (110 nm)/ P-6 (100 nm) /Ca:Al device

(Fig. 16). But it lowers the threshold electric field and improves the stability of the LED device.

Replacing the CuPc layer with the poly(3,4-ethylenedioxythiophene) (PEDOT) layer and the Al cathode with the Ca:Al alloy electrode, we [51] construct-

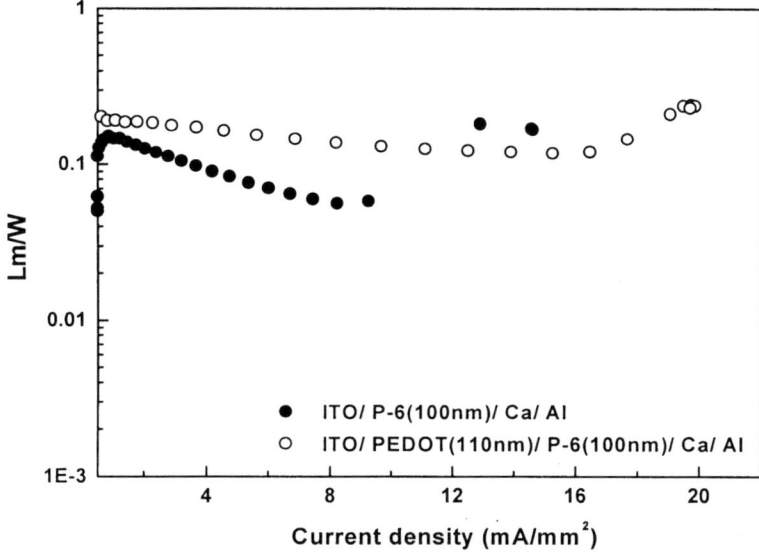

Fig. 18. Comparison of the stability of the devices with and without the PEDOT layer

ed an LED device and its EL properties were studied. The PEDOT sample (Baytron P) was purchased from Bayer and the film's electrical conductivity was measured to be 1.0 S cm^{-1}. The results are given in Fig. 17 and 18. The threshold electric field of this device was lower than 0.5 MV/cm and the maximum luminance attainable was greater than 15,000 cd/m^2 (Fig. 17). The device efficiency was 0.2 Lm/W (Fig. 18). In fact, the device without the PEDOT layer also exhibited about the same efficiency, but the stability of the device was much poorer. The maximum brightness obtained for the latter device was about 9000 cd/m^2.

4
Kinked Conjugated Polymers

Multicolor display applications require at least three basic colors; red, green and blue. Blue EL, which is difficult to achieve with inorganic semiconductors, has been sought in conjugated polymers having a high HOMO-LUMO energy gap, such as partially eliminated PPV, poly(p-phenylene), poly(alkylfluorene) and copolymers containing confined conjugated segments such as CNMBC (conjugated-nonconjugated multiblock copolymer) type polymers [52–57]. However, they show rather poor processability and mechanical properties and had high operating voltages in light-emitting diode devices. One approach to obtaining the blue-light emission is the introduction of kinked (bent) structure in the polymer main chain. Blue-shifted emissions can be obtained by incorporating meta- or ortho-phenylene units, which also improved the processability of polymers [58, 59]. Ortho-linked polymers have the bent (kinked) structures that disrupt the conjugation in polymers in the same monomer as cis linkages in PPV do [60].

4.1
Synthesis and Characterization

Kinked conjugated polymers can be synthesized by Wittig condensation reaction [61]. The synthetic routes and the polymer structures are shown in Scheme 7.

The synthesized polymers are end-capped by triphenylphosphoniummethylbenzene chloride for removing the aldehyde end group, because the carbonyl group is thought to be a quenching site in the electroluminescence mechanism [60]. All of the synthesized conjugated polymers are highly soluble in common organic solvents such as THF, chloroform, methylene chloride, 1,2-dichloroethane, DMF and so on. Especially, they could spin-cast onto various substrates to give highly transparent and homogeneous thin films without heat treatment. The weight-average molecular weights (Mw) of the polymers, determined by gel permeation chromatography using polystyrene standards, are in the range of 4000–6000 with a polydispersity index of 1.25–2.05.

The thermal properties of the polymers were evaluated by means of TGA under nitrogen atmosphere and all of the polymers exhibit very good thermal stabilities. PMEH-PPV series shows better thermal stability (up to 350 °C) than that (up to 250 °C) of the PBTMS-PPV because of the difference in bond strength of their side groups. The bond strength of the Si-C bond in PBTMS-PPV polymers is 451.5 kJ/mol which is lower than that of the C-O bond in PMEH-PPV polymers (1076.5 kJ/mol).

4.2
Optical and PL Properties

Table 4 shows the UV-vis absorption and PL emission maxima of the polymers shown in Scheme 7 were measured in the thin film state on the quartz plate. The well-known MEH-PPV shows the maximum absorption at 510 nm. But p-PMEH-PPV shows absorption maximum at 430 nm due to the presence of the

Table 4. UV-Visible absorption and PL emission maxima of the polymers

Polymers	UV-vis λ_{max} (nm)	PL λ_{max} (nm)
p-PBTMS-PPV	380	485
o-PBTMS-PPV	340	470
m-PBTMS-PPV	330	440
p-PMEH-PPV	430	550
o-PMEH-PPV	360	500
m-PMEH-PPV	390	490

*The values listed were measured in the thin-film state on the quartz plate

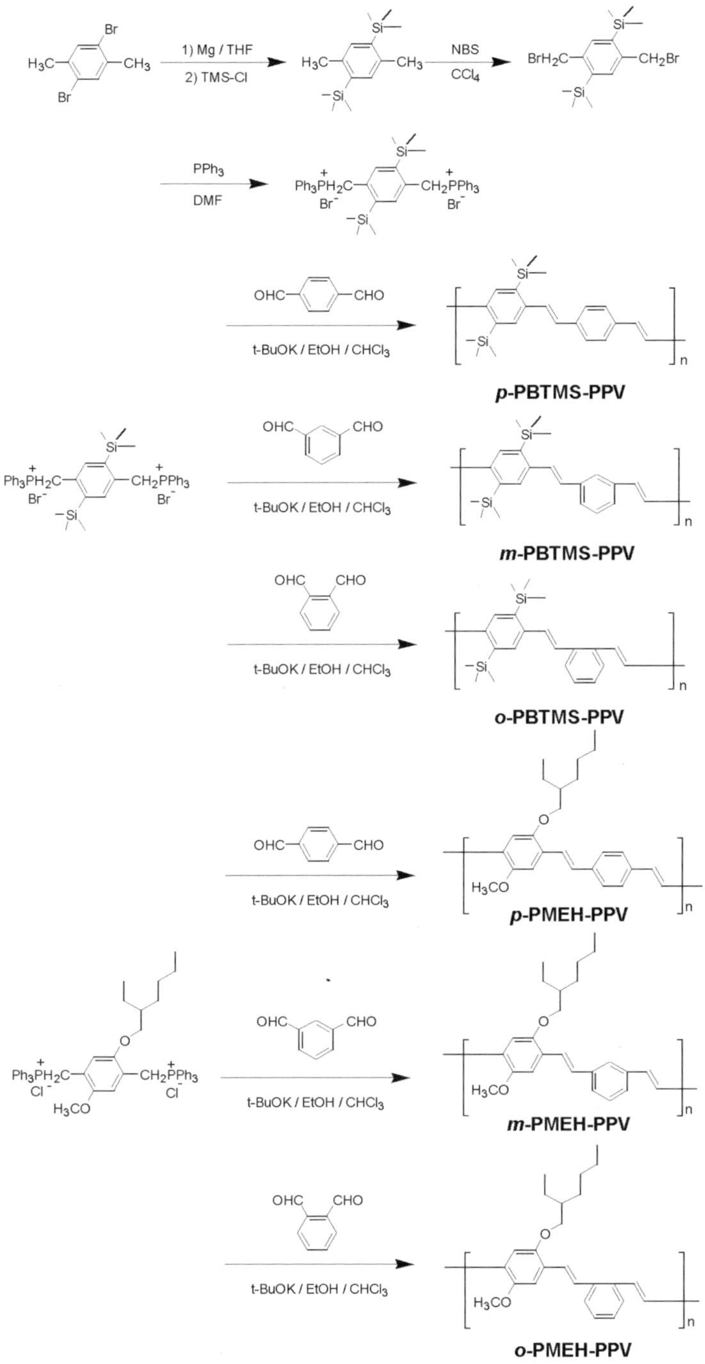

Scheme 7. Synthetic scheme of kinked conjugated polymers

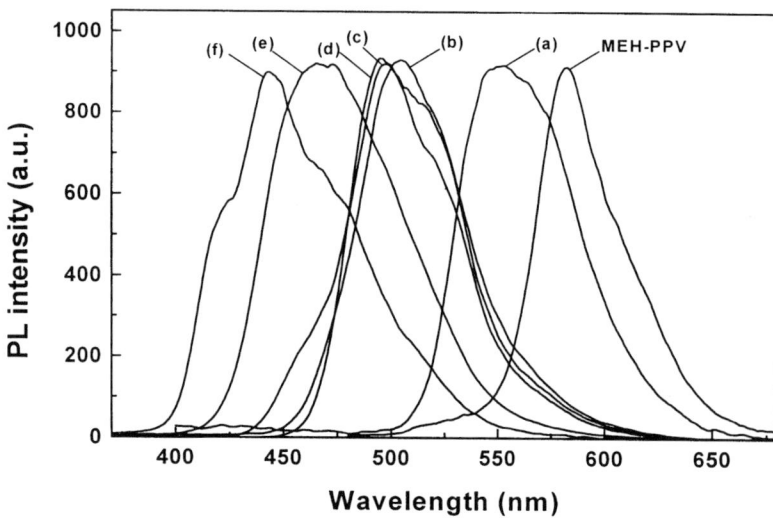

Fig. 19. Photoluminescence spectra of (a) *p*-PMEH-PPV, (b) *o*-PMEH-PPV, (c) *m*-PMEH-PPV, (d) *p*-PBTMS-PPV, (e) *o*-PBTMS-PPV, and (f) *m*-PBTMS-PPV thin films coated on a quartz plate

unsubstituted phenylene unit in the alternating copolymer. The absorption maximum bands of *o*(*m*)-PMEH-PPV are more blue-shifted compared to that of *p*-PMEH-PPV. This means that the π-electron delocalization of the polymer main chain is interrupted by ortho- and meta-linkage, resulting in a reduction of the π-conjugation length [62]. Son et al. [60] reported that cis linkages in the PPV main chain interrupt the conjugation. Trans and cis linkages have almost the same electronic structures, but cis linkage causes a bent (kinked) type structure. This deviation from the linearity of cis linkage can interrupt linear coplanarity for conjugation, which results in a diminished conjugation length compared to that of trans.

The *o*(*m*, *p*)-PBTMS-PPV series shows very similar absorption profiles with those of PMEH-PPV polymers. The maximum absorption wavelengths of *o*(*m*)-PBTMS-PPV are blue-shifted for the same reason as that of *o*(*m*)-PMEH-PPV. But PBTMS-PPV polymers show about 20–50 nm blue-shifted absorption maxima compared to those of PMEH-PPV series, because the trimethylsilyl substituent has little electron-donating property. The PL spectra show very drastic changes in emission color because of the substituents and kink effects. Figure 19 shows PL emission profiles of the polymers mentioned above.

The emission of *p*-PMEH-PPV is blue-shifted about 40 nm because the number of the alkoxy groups in the repeating unit is one half compared with MEH-PPV. As a result of the decrease in the number of the alkoxy substituent in *p*-PMEH-PPV than that of MEH-PPV, the electron density of the conjugated main chain in *p*-PMEH-PPV is diminished. Consequently, *p*-PMEH-PPV whose emission is blue-shifted, has a larger band gap than does MEH-PPV. For the case of *o*-

PMEH-PPV and *m*-PMEH-PPV, the emission peaks which are more blue-shifted compared to that of *p*-PMEH-PPV, appear at around 500 and 490 nm, respectively, due to the same reason as described for their optical absorption spectra.

On the other hand, the PL emission peaks of the PBTMS-PPV series are more blue-shifted compared to those of the counterparts of the PMEH-PPV series. This result indicates that the trimethylsilyl substituent has a smaller electron-donating effect than that of the alkoxy substituent. With an excitation wavelength of 340 nm (UV absorption maximum), the PL spectra of *o*-PBTMS-PPV and *m*-PBTMS-PPV exhibit the peaks in the blue emissive region at 470 and 440 nm, respectively. Consequently, the blue light-emitting conjugated polymers can be prepared by introducing the kink linkages and nonelectron-donating substituent without any conjugation blocking segments such as oxygen or alkyl spacers in the backbone.

4.3
EL Properties and Current-Voltage-Luminance (I-V-L) Characteristics

The *o*(*m*, *p*)-PMEH-PPV and *o*(*m*, *p*)-PBTMS-PPV polymers are deposited onto indium-tin oxide covered glass substrate by spin-coating the soluble polymers in cyclohexanone. The spin-casting technique yields uniform films with of about 50–60 nm. Figure 20 and 21 show the current-voltage (I-V) and the EL intensity-voltage (L-V) characteristics of the single-layer light-emitting diode of ITO/*o*(*m*, *p*)-PBTMS-PPV/Al. The current increases with increasing forward bias voltage, which indicates a typical rectifying characteristics. The threshold voltages of PBTMS-PPV series are in the range of 8–9V for the films 50–60 nm thick.

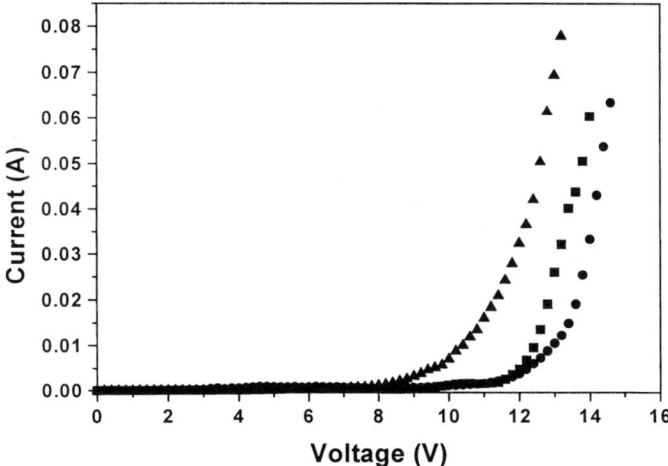

Fig. 20. Current-voltage (*I-V*) characteristics of the single-layer light-emitting diode of ITO/*o*-PBTMS-PPV/Al (■), ITO/*m*-PBTMS-PPV/Al (●), and ITO/*p*-PBTMS-PPV/Al (▲)

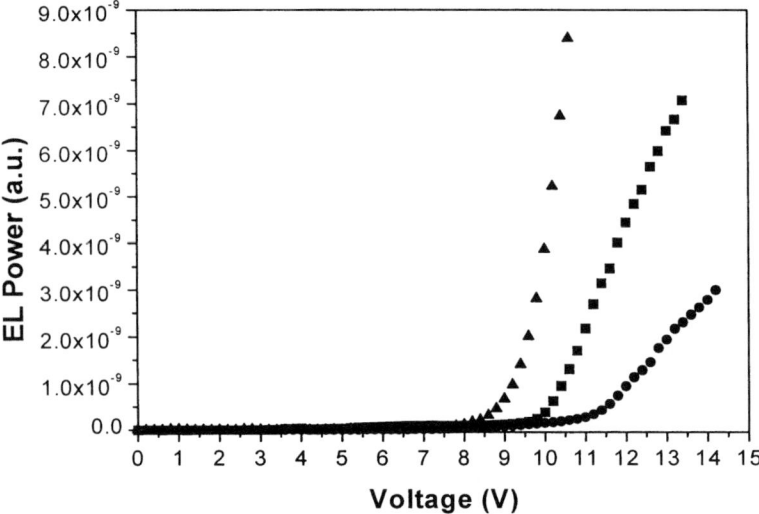

Fig. 21. Luminance-voltage (L-V) characteristics of the single-layer light-emitting diode of ITO/o-PBTMS-PPV/Al (■), ITO/m-PBTMS-PPV/Al (●), and ITO/p-PBTMS-PPV/Al (▲)

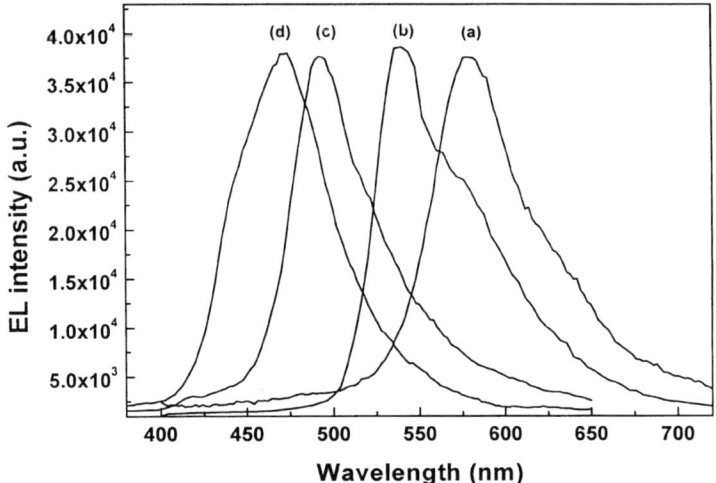

Fig. 22. Electroluminance (EL) spectra of the single-layer light-emitting diodes of (a) MEH-PPV, (b) p-PMEH-PPV, (c) p-PBTMS-PPV, and (d) o-PBTMS-PPV, which have ITO/polymer/Al configuration

As shown in Fig. 20 and 21, p-PBTMS-PPV exhibits the highest EL power at lower operating voltage because of the effective π-conjugation due to the para-linked linear chain backbone. Figure 22 shows the EL spectra of the representative polymers among the both polymer series comparing with that of MEH-PPV.

p-PMEH-PPV shows maximum EL emission at 540 nm which is similar to that of its PL spectrum, corresponding to the greenish yellow region. The EL spectrum of *p*-PBTMS-PPV shows much more blue-shifted EL emission compared to that of *p*-PMEH-PPV due to the reduction in electron-donating power of the trimethylsilyl group. Especially, kinked *o*-PBTMS-PPV shows maximum EL emission at 470 nm corresponding to pure blue region because of the reduced π-conjugation length by the nonlinear ortho linkage and non-electronic substituent. Conclusively, pure blue emission can be obtained by the introduction of nonlinear kink structure and adequate substituent, even though it is very difficult to get a pure blue emission from the fully conjugated polymers.

5
Conjugated Polymers with High Electron Affinity

In order to improve the device performance, it is necessary to balance the rate of injection of electrons and holes from opposite electrodes into the emitting polymers. Oxadiazole units are known as electron deficient groups and poor hole acceptors. Aromatic oxadiazole compounds, such as 2-(4-*tert*-butyl-phenyl)-5-biphen-yl-1,3,4-oxadiazole (PBD), have been investigated as electron-conducting/hole-blocking (ECHB) materials within multilayer devices [8, 63–65], and poly(*N*-vinylcarbazole) compounds have been used as hole-transporting materials [39, 66, 67]. PL and EL properties of PPV derivatives bearing either electron-transporting groups or hole-transporting group have been discussed above in Section 3.

5.1
Synthesis and Characterization

Scheme 8 shows the synthetic routes of the oxadiazole unit containing phosphonium salt monomers and Scheme 9 outlines the preparation of π-conjugated polymers composed of oxadiazole units [67]. The synthetic pathways to the para-linked salt monomer (A) and meta-linked salt monomer (B) are of multisteps, but the synthetic method for each step is not difficult and the product yields of intermediate compounds are very high (75–95%). Treatment of bis(4-methylphenyl)hydrazine with $POCl_3$ results in the formation of the 2,5-bis(4-methylphenyl)-1,3,4-oxadiazole.

The alternating copolymers (PPOX-CAR and PMOX-CAR) are easily prepared by the Wittig condensation polymerization between the oxadiazole containing salt monomers and the carbazole dialdehyde. The copolymer, PPOXPV, composed of oxadiazole unit and phenylenevinylene unit, can be prepared by the same way as shown in Scheme 10. But the kinked PPOXPV copolymer is prepared in different synthetic pathway. Reaction of bis(2-bromo-5-hexyloxybenzyl)-1,3,4-oxadiazole with p-divinyl benzene results in the formation of POOX-PV copolymer [69].

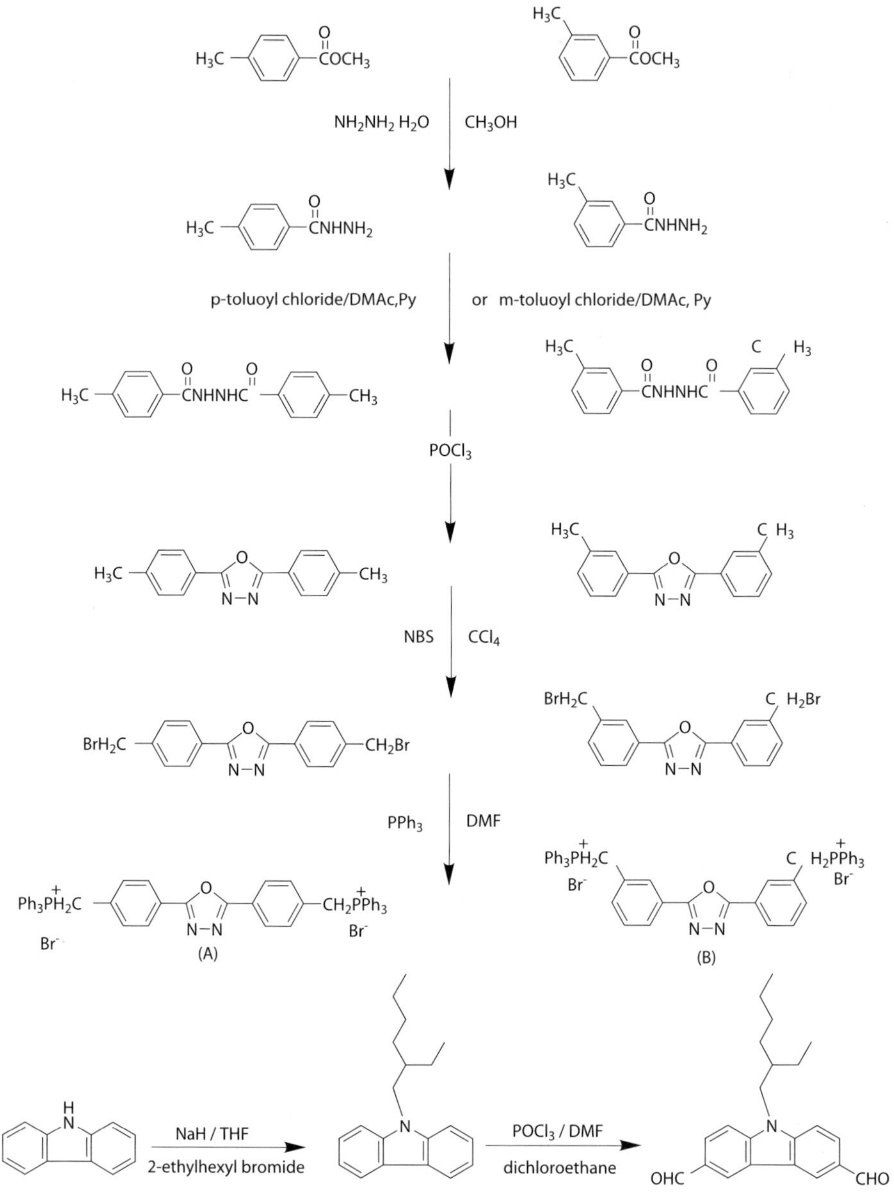

Scheme 8. Synthetic routes to the monomers

The weight-average molecular weights ($\overline{M}w$) of the polymers are in the range of 6500–15,000, with a polydispersity index of about 1.5. All of the polymers are soluble in common organic solvents such as chloroform, THF and 1,2-dichloroethane. They showed very good thermal stability in which the weight loss is less than 5% upon heating to above 350 °C under a nitrogen atmosphere. The

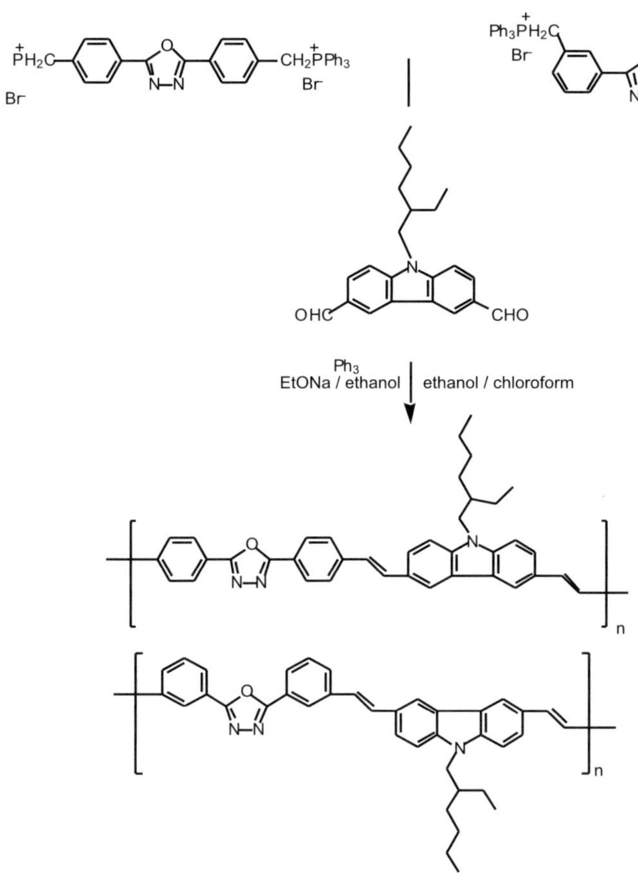

Scheme 9. Synthetic routes to PPOX-CAR and PMOX-CAR

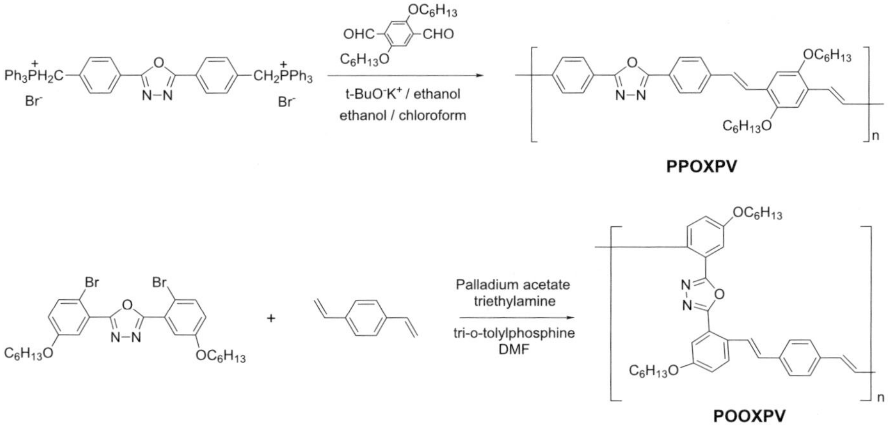

Scheme 10. The synthetic routes to PPOXPV and POOXPV

Table 5. UV-vis absorption and PL emission maxima of the polymers

Polymers	UV-Vis. λ_{max} (nm)	PL λ_{max} (nm)
PPOX-CAR	390	495
PMOX-CAR	300	450
PPOXPV	400	510
POOXPV	360	495

linear, para-linked PPOX-CAR and PPOX-PV show slightly higher decomposition temperatures (up to 400 °C) due to a better chain packing compared to those (up to 350 °C) of kinked PMOX-CAR and POOX-PV.

5.2
Optical and PL Properties

Table 5 summarizes the UV-vis absorption and PL emission maxima of the oxadiazole-based polymers. The maximum absorptions of PPOX-CAR and PMOX-CAR appear at 390 and 300 nm, respectively. The blue shift of PMOX-CAR is caused by the kink structure of meta-kinked main chain, because the meta linkage between oxadiazole and carbazole units shortens the effective conjugation length by restriction of the fully conjugated polymer structure. The same trend is observed between PPOXPV and POOXPV. The PL spectrum of PPOX-CAR exhibits a maximum peak at 495 nm, corresponding to greenish blue light emission. PPOXPV also shows similar PL emission (510 nm). PMOX-CAR and POOXPV show blue-shifted PL emission maxima compared with those of the linear counterparts. But PMOX-CAR shows much more blue-shifted PL emission at 450 nm corresponding to purple-blue light due to the lack of π-electron delocalization.

5.3
Current-Voltage-EL Power (I-V-L) Characteristics

Figure 23 shows the characteristics of the current and EL power against voltage. The PPOX-CAR polymer shows turn-on voltage of 7.5 V, but the kinked PMOX-CAR polymer shows a little bit higher turn-on voltage at 10.5 V (both polymer films have almost the same thickness, around 100 nm). It is clear that the kink structure has high turn-on voltage because of shorter conjugation length and less effective π-electron delocalization. The relatively low turn-on voltage of above polymers compared to other green and blue light emission polymers is caused by the carbazole unit, a hole transporting structure.

At a relatively low turn-on voltage, the light intensity of PPOX-CAR is much stronger than that of PMOX-CAR, as shown in Fig. 23(b). So the PPOX-CAR polymer can be used as an excellent greenish-blue light-emitting layer.

The PPOXPV and POOXPV show the very similar light-emitting properties with those of PPOX-CAR and PMOX-CAR. But the difference in PL emission

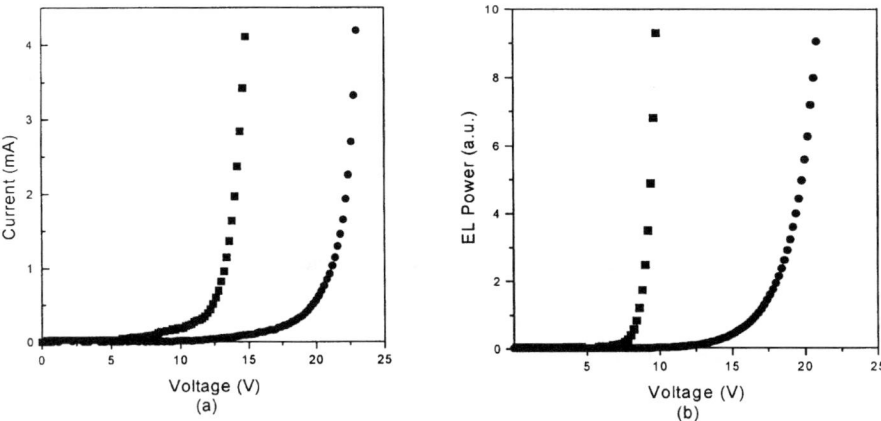

Fig. 23. EL characteristics of PPOX-CAR (■) and PMOX-CAR (●) with an ITO/polymers/Al structure. (a) Current vs. voltage and (b) EL power vs. voltage

wavelengths and also in turn-on voltages are not large compared with those between PPOX-CAR and PMOX-CAR, because ortho-linked POOXPV has a longer conjugation length and is of better π-electron delocalization compared to meta-linked PMOX-CAR.

The relative EL quantum efficiencies of PPOX-CAR and PMOX-CAR were investigated in the same device condition. The PPOX-CAR polymer exhibits 36 times higher EL quantum efficiency as compared to PPV. On the other hand, the EL quantum efficiency of PMOX-CAR is increased just 2-fold. This means that the PPOX-CAR polymer in the para linkage more easily forms balanced injections and transport of holes and electrons from opposite electrodes because the fully and linearly conjugated structure consists of the hole-transporting and the electron-tranporting units.

6
Light-Emitting Properties of Blend Polymers

The three main factors that will determine the utility of polymer LEDs as light sources are their efficiency, threshold or turn-on voltage and life time. With the appropriate choice of polymer and device design, external quantum efficiencies of up to 4% can be obtained. Therefore the research of this field is still the focus of many scientists. To increase the quantum efficiency of the polymer LEDs, Heeger et al. utilized the high work function polymer (polyaniline) as a hole injection electrode [70] and the low work function metal (Ca) as an electron injection electrode [71], resulting in highly improved efficiency of the polymer LEDs. Also multilayer structures are used to increase the quantum efficiency of the polymer LEDs. The electron-transporting layers are placed between the luminescent layer and the cathode, which enhance electroluminescent efficiency through charge carrier confinement [9, 72]. Another approach is to blend electron- or hole-transporting molecules with luminescent polymers. The efficiencies of these blend systems are also improved owing to a decrease in the energy barriers for carrier injection and transport between the electrodes and the luminescent polymers [73–76]. Recently, we have reported that the quantum efficiencies of blend system composed of high energy state polymer and low energy state MEH-PPV are highly improved due to the energy transfer effect [77, 78].

6.1
Conjugated-Nonconjugated Multiblock Copolymers (CNMBCs)

6.1.1
Preparation of Blend Polymers [77]

Nonconjugated poly[1,3-propanedioxy-1,4-phenylene-1,2-ethenylene(2,5-bis(trimethylsilyl)-1,4-phenylene)- 1,2-ethenylene-1,4-phenylene] (DSiPV) and conjugated MEH-PPV polymers are mixed by changing their weight ratios in 1,2-dichloroethane. The films of the blend polymers could be spin-cast from 1,2-dichloroethane solution with excellent reproducibility. AFM (atomic force microscopy) and SEM images show no indication of phase separation or layer formation due to the immiscibility of two polymers. The structures of DSiPV and MEH-PPV are shown in Fig. 24.

6.1.2
Optical and PL Properties of the Blend Polymers

Figure 25 shows the UV-vis absorption spectra of the polymer blends including DSiPV and MEH-PPV homopolymers. The broad absorption results from MEH-PPV appeared around 420–580 nm, which is due to π-π^* transitions of polyconjugated system. On the other hand, DSiPV shows blue-shifted absorption maximum at 340 nm due to the nonconjugated backbone structure. Blend polymer films show the separate absorption peaks corresponding to their component polymers and their intensities are varied depending on the weight fractions of two polymers.

The PL emission spectra of the blend polymers, which are normalized by the optical density at the excitation wavelength (355 nm), are shown in Fig. 26. The emission of the DSiPV film appears in the range of 400–550 nm, whereas that of

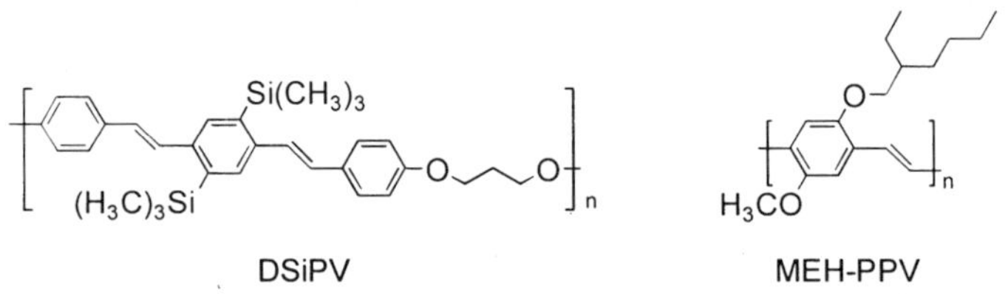

Fig. 24. The polymer structures of DSiPV and MEH-PPV

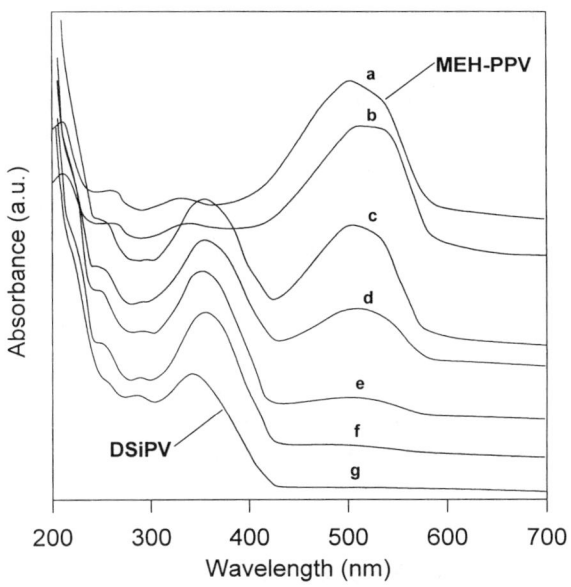

Fig. 25. Optical absorption spectra of (a) MEH-PPV, (b) 9:1, (c) 6:4, (d) 3:7, (e) 1:9, (f) 1:15 (MEH-PPV: DSiPV) blended polymers, and (g) DSiPV thin films

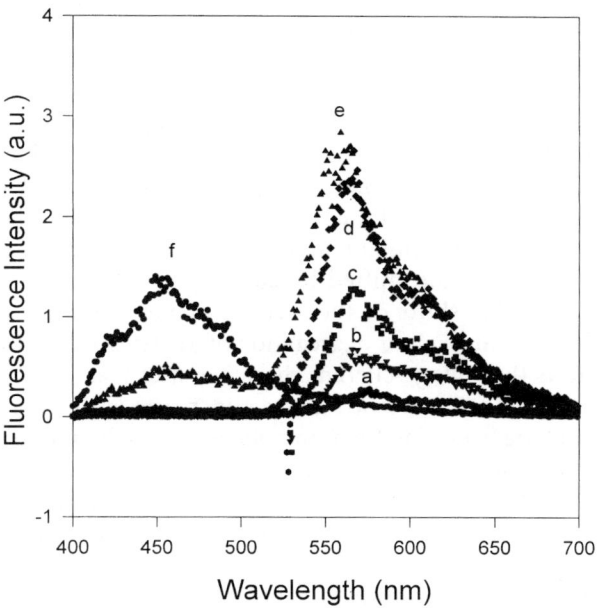

Fig. 26. Photoluminescence spectra of (a) MEH-PPV, (b) 6:4, (c) 3:7, (d) 1:9, (e) 1:15 (MEH-PPV:DSiPV) blended polymers, and (f) DSiPV thin films. All PL spectra were recorded with an excitation at 355 n m

the MEH-PPV film is weakly observed in the range 530–650 nm. Interestingly, when the blended polymers are excited by 355 nm pulses, they show strong emissions at about 560 nm which corresponds to yellowish-orange light of MEH-PPV, whereas DSiPV emission in the blue region is weakly or not observed depending on the weight fraction. These emission at 560 nm are much stronger than that of MEH-PPV homopolymer, and as the relative weight of DSiPV to MEH-PPV is increased, the emission intensities of the blend polymer films are increased. These results suggest that the energy-transfer process from the photoexcited DSiPV to low energy state MEH-PPV results in an enhancement of the yellowish-orange emission originated from MEH-PPV. This energy transfer process is probably more efficient than the fluorescence decay of DSiPV.

It has been also reported that the emission intensity is enhanced as the emitting chromophore is diluted by other inert polymers in solid state [79]. The emission intensity of the blend polymer is increased as MEH-PPV is diluted by DSiPV because the nonradiative decays, especially intermolecular quenching, are reduced by the dilution effect.

6.1.3
LED Device Characteristics of the Blend Polymers

The LED structures consist of an aluminum rectifying contact on the blend polymers which have been spin-cast onto ITO glass as a hole-injecting contact. The film thickness ranged from 60 to 80 nm. The electron-injecting aluminum contacts are deposited by vacuum evaporation at pressure below 10^{-6} torr, giving active area of 0.195 cm^2.

Figure 27 shows the EL spectra of MEH-PPV and the blend polymers. The emission maximum peaks of the blend polymers range from 560 to 580 nm depending on blending ratios, whereas the emission of DSiPV does not appear. The EL spectra of the blend polymers are very similar to their PL counterparts. The energy transfer also exists at the DSiPV and MEH-PPV interfaces, therefore the blend polymers exhibit similar emission maxima. The EL emission peaks of the blend polymers are blue shifted by 20–30 nm with respect to MEH-PPV homopolymer. The blue shifts of the blend polymers should result from the change in chain conformation or aggregation of MEH-PPV in DSiPV [80].

Figure 28 shows the voltage-current characteristics measured for MEH-PPV and the blend polymers. The forward current increases with increasing forward bias voltage for all devices. The turn-on voltages of the blend polymer devices increase as the content of DSiPV increases in the blends.

The light intensities of the blend polymers are much stronger than that of MEH-PPV homopolymer. The relative quantum efficiencies of MEH-PPV and the blend polymer devices are listed in Table 6.

The relative PL and EL quantum efficiencies of the blend polymers increase with increasing DSiPV content, and, surprisingly, the EL quantum efficiency of the 1:15 (MEH-PPV : DSiPV) blend is about 480 times that of MEH-PPV. Energy transfer or charge transfer from DSiPV to MEH-PPV exists in the EL device,

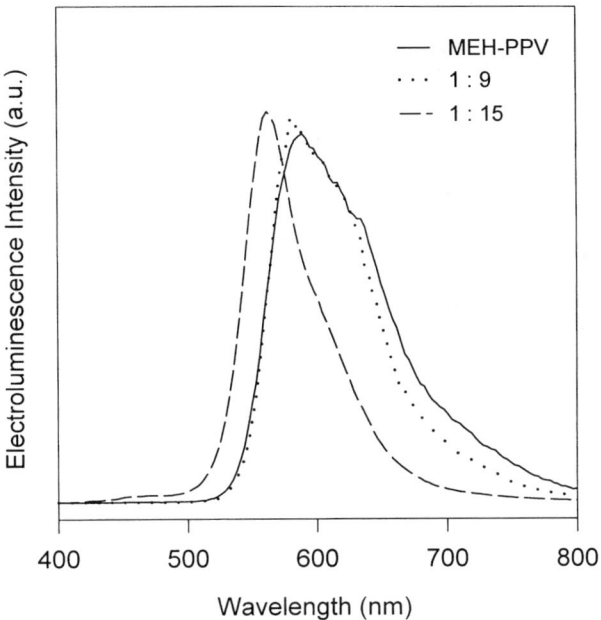

Fig. 27. Electroluminescence spectra of MEH-PPV, 1:9, and 1:15 (MEH-PPV:DSiPV) blended polymers

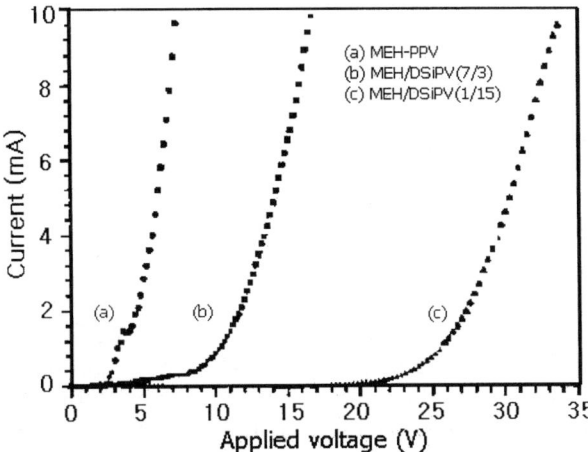

Fig. 28. Voltage-current characteristics of (a) MEH-PPV, (b) MEH-PPV/DSiPV (7/3), and (c) MEH-PPV/DSiPV (1/15)

therefore the blend polymer devices emit yellowish-orange lights at about 560–580 nm wavelength region and their EL efficiencies are also dramatically improved.

Table 6. Photoluminescence and Electroluminescence Quantum Efficiencies of MEH-PPV, DSiPV, and Blended Polymers

Polymers	Relative PL quantum efficiency		operation electric field (10^8 V/m)	relative EL quantum efficiency
	excited by 355 nm pulse	excited by 488 nm cw		
MEH-PPV	1	1	0.52	1
9:1 [a]		1.2	1.1	20
6:4 [a]	1.6	2.0	1.7	11
3:7 [a]	3.1	2.8	2.1	19
1:9 [a]	5.0	5.7	2.6	17
1:15 [a]	7.0	7.6	4.4	480
DsiPV	3.8		4.5	44

[a] The ratios are the weight ratios between MEH-PPV and DSiPV

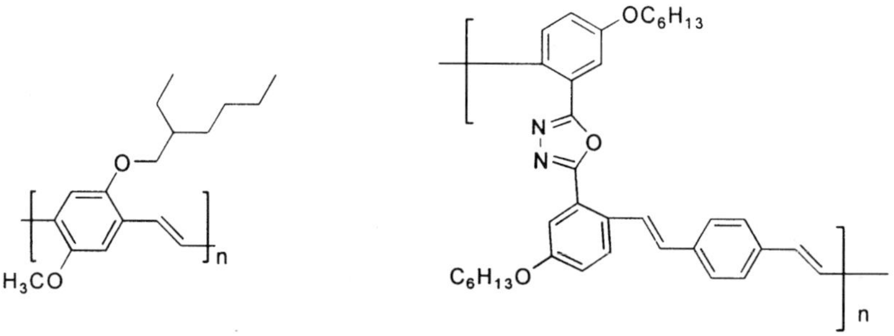

Fig. 29. The structures of MEH-PPV and POOXP V

6.2
Conjugated-Conjugated Polymer Blends

6.2.1
Preparation of Blend Polymers [81]

The kinked POOXPV polymer mentioned in Section 5 (Scheme 10) is blended with MEH-PPV in order to increase the EL intensity. The MEH-PPV/POOXPV blending ratios are 1/1, 1/5 and 1/10 by weight. The high quality films of the blend polymers are obtained from 1,2-dichloroethane solution by spin-coating technique. Figure 29 shows the structures of MEH-PPV and POOXPV.

6.2.2
Absorption and Emission Properties of the Blend Polymers

The UV-vis absorption spectra of MEH-PPV/POOXPV blend polymers are shown in Fig. 30. The π-π^* absorption transition peaks of POOXPV and MEH-PPV appear at 375 nm and 510 nm, respectively. The blend polymers show the

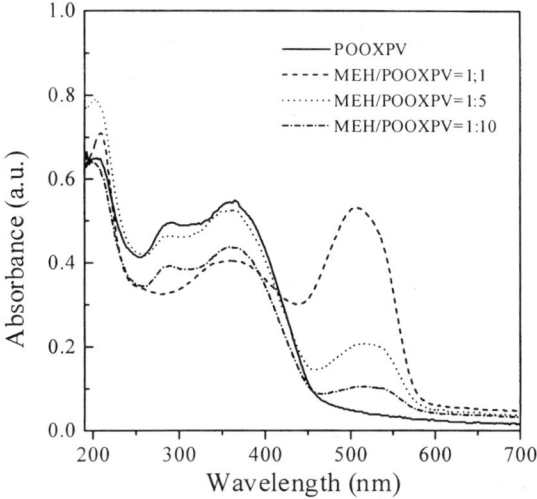

Fig. 30. UV-visible spectra of MEH-PPV/POOXPV blend polymers

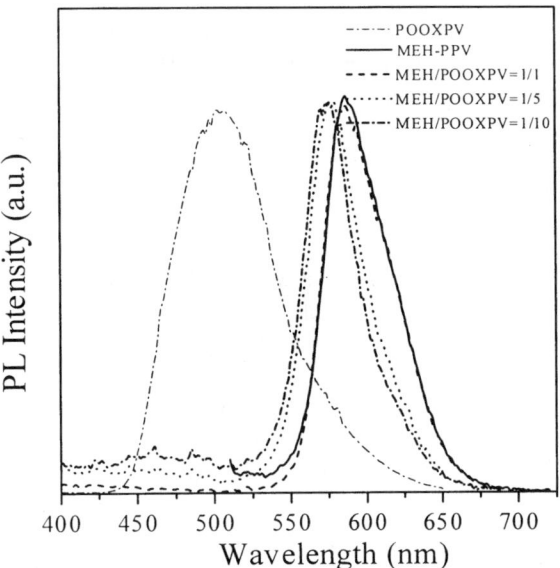

Fig. 31. PL spectra of POOXPV and MEH-PPV/POOXPV blend polymers

separated absorption peaks corresponding to the component polymers and their intensities are varied depending on the blending ratios of two polymers. This is a typical characteristics of the blend absorption spectra, and shows no interaction between two polymers in the ground state.

The photoluminescence spectra of the blend polymers are shown in Fig. 31. The blend polymers are excited at 370 nm corresponding to the absorption maximum wavelength of POOXPV. Regardless of the composition of the blend, all blend polymers exhibit very strong emissions at around 575–585 nm corresponding to that of MEH-PPV homopolymer, and the emission of POOXPV is not found because of the energy transfer from POOXPV to MEH-PPV (the PL emission maximum of POOXPV is shown to be at 495 nm). This is well coincident with the result for DSiPV/MEH-PPV blend system mentioned in Section 6.1.2.

6.2.3
LED Device Characteristics of the Blend Polymers

Figure 32 shows the normalized EL spectra of the blend polymers. The emissive wavelength is at 580 nm corresponding to that of MEH-PPV because of the energy transfer from POOXPV to MEH-PPV as mentioned in PL spectra of the blends.

Figure 33 shows the I-V-L characteristics of the blend polymers. Forward bias current is obtained when the ITO is a positive electrode and Al electrode is the ground. The forward current increases with increasing forward bias voltage for all devices. The turn-on voltage and the light-intensities of the blend polymer devices increase as the content of POOXPV increases. And light intensities of the blend polymers are much stronger than that of MEH-PPV. The emission intensity at the same driving current is markedly increased with the POOXPV content.

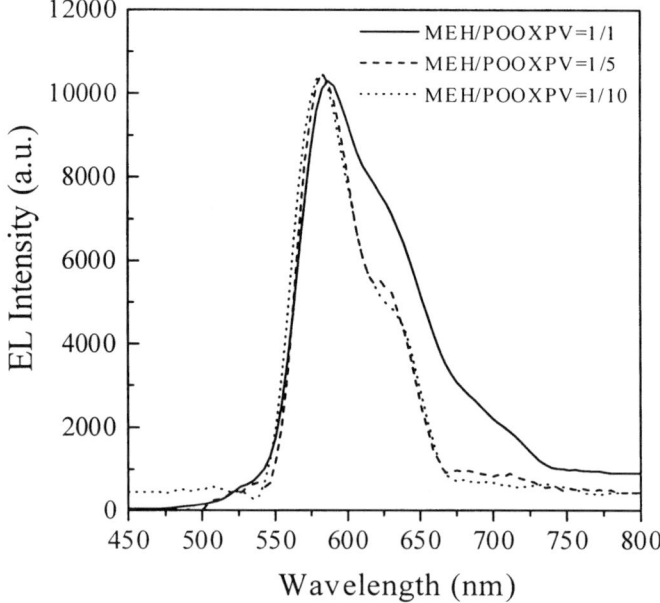

Fig. 32. EL spectra according to blending ratios

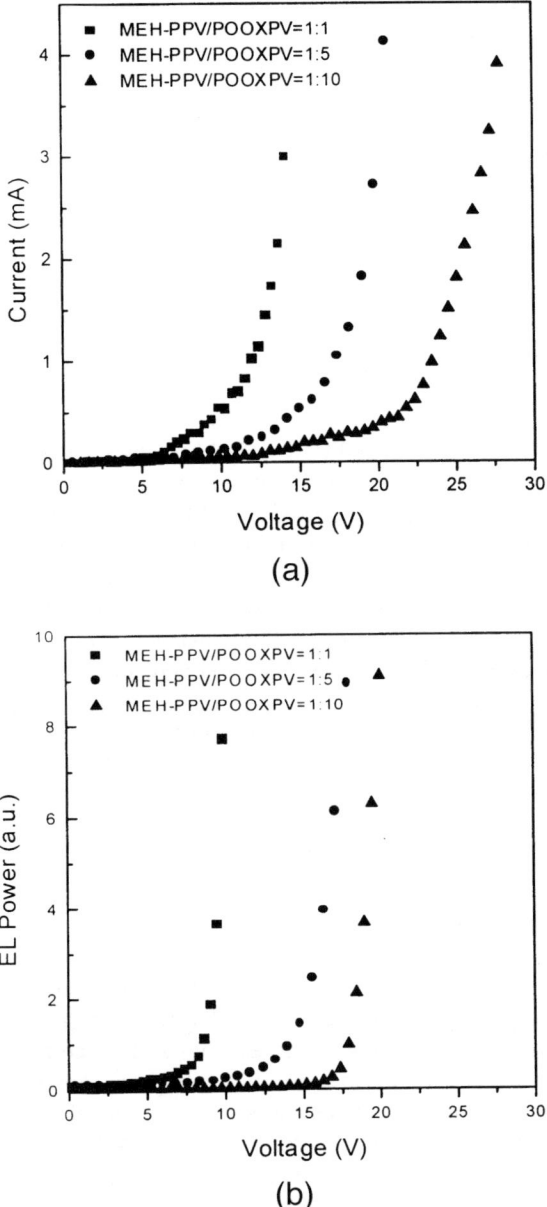

Fig. 33. EL characteristics of MEH-PPV/POOXPV blend polymers with an ITO/polymers/Al structure. (a) Current vs. voltage and (b) EL power vs. voltage

Relative quantum efficiencies of MEH-PPV and blend polymer devices are listed in Table 7. The relative quantum efficiency and brightness of the blend polymers increase with increasing POOXPV component. Interestingly, the

Table 7. Comparison of relative EL quantum efficiency, turn-on voltage, and light-intensity between POOXPV and MEH-PPV/POOXPV blends

Configuration of EL device (anode/EL polymer/cathode)	Relative EL quantum efficiency	Turn-on voltage (V)	Light intensity (cd/m^2)
ITO/MEH-PPV/Al	1	2.8	42
ITO/POOXPV/Al	8	9.5	
ITO/(MEH-PPV/POOXPV=1:1)/Al	3	6.0	1000
ITO/(MEH-PPV/POOXPV=1:5)/Al	30	9.0	2400
ITO/(MEH-PPV/POOXPV=1:10)/Al	57	11.5	

brightness of Al/[MEH-PPV/POOXPV=1:5]/ITO single layer device shows very high luminance of 2,400 cd/m^2 at 20 V. The improvement in quantum efficiency of the blend polymers can be explained by following factors. First, the excitons and/or free carriers generated from POOXPV are transferred to MEH-PPV, which, in turn, increase the concentration of excitons that enhance the EL intensity of the blend polymers. Simultaneously, as the EL active MEH-PPV is diluted by POOXPV, the intermolecular nonradiative decay is diminished by blocking of the charge carriers. Secondly, high increase in brightness is caused by the oxadiazole unit which is used as the electron conducting / hole blocking (ECHB) moiety. The POOXPV well transports electrons into the MEH-PPV, thereby enhance the probability of radiative recombination with the holes in MEH-PPV. Conclusively, POOXPV polymer is more likely to prove useful as ECHB component than as emissive material.

7
Other Conjugated Polymers

One of the simple modifications of the PPV synthesis gives access to blue-shifted light-emitting polymers with a high concentration of interruptions along the conjugated chain. Scheme 11 shows synthesis of partially conjugated PPV copolymers. Partial conversion of water-soluble salt precursor or organic-soluble alkoxy precursor polymers produces statistical partially eliminated copolymers with regions of conjugated and nonconjugated units [82, 83]. These partially eliminated copolymers emit blue-green light due to the interruption in the conjugated chain. The conversion of partially eliminated copolymers to the fully eliminated PPV may be accomplished by either thermal or base treatment [84].

The effects of silicon substitution on the luminescence properties are of interest in the field of polymer LEDs. Poly(2-dimethyloctylsilyl-1,4-phenylenevinylene) (DMOS-PPV) is completely soluble in common organic solvents. Hwang et al. reported synthesis and LED properties of DMOS-PPV [85]. Scheme 12 shows the synthetic route to DMOS-PPV by dehydrohalogenation.

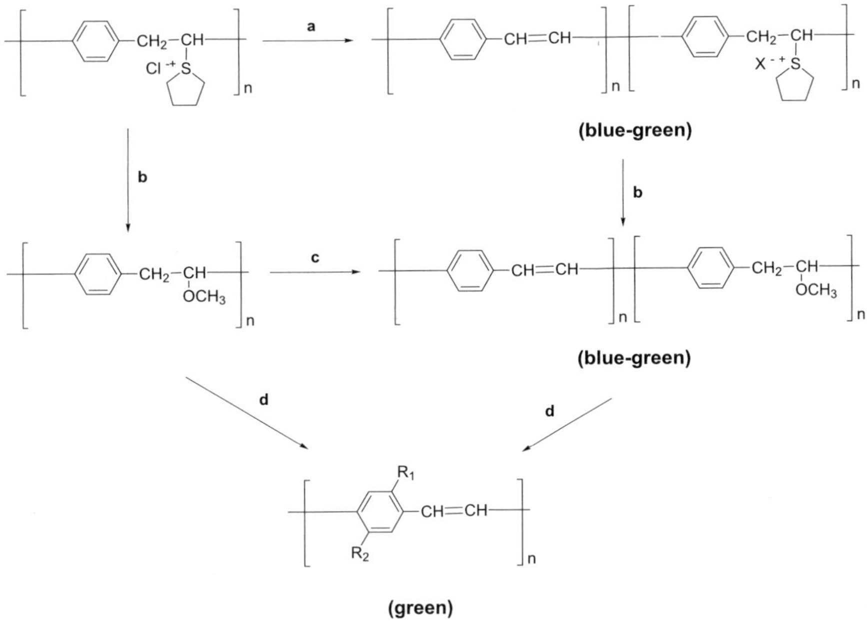

Scheme 11. Synthesis of partially conjugated PPV copolymers: a) Bu$_4$NOH, MeOH; b) MeOH, 50 °C; c) 300 °C, vacuum, 12 h; d) 220 °C, HCl (g)/Ar, 22 h.

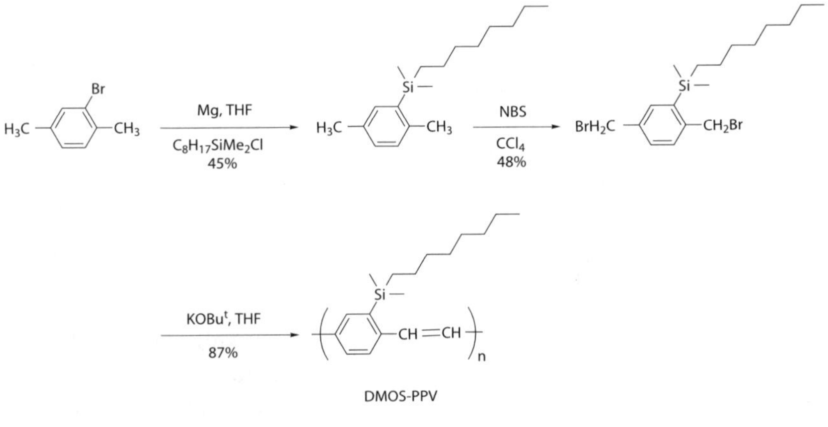

Scheme 12. The synthetic route of DMOS-PPV

DMOS-PPV shows an emission maximum at ca. 520 nm which corresponds to the green region. The absolute PL quantum efficiency for a solid film of DMOS-PPV is 60%. By comparison, the reported PL efficiencies of PPV and MEH-PPV are 27 and 15%, respectively [25]. Recently, Kim et al. have reported a series of light-emitting polymers which are interrupted by nitrogen or silicon

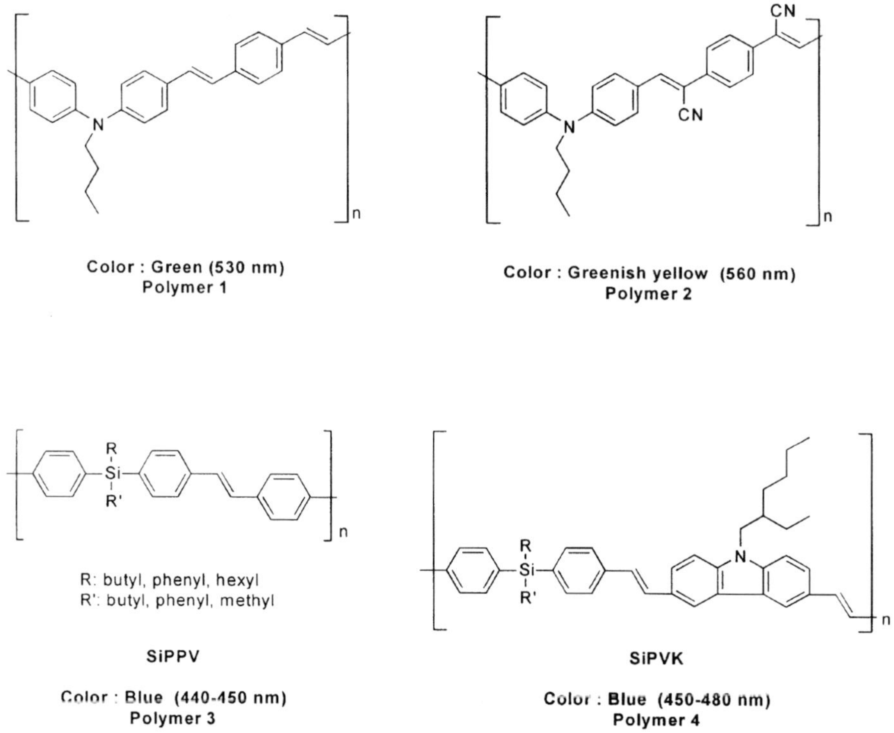

Fig. 34. Polymeric materials synthesized by Kim et al.

atoms to access blue-shifted emission [86–88]. The polymer structures are shown in Fig. 34.

The cyano substituents in polymer 2 contribute to an increase in the electron affinity of the polymer and lower the band gap. Therefore, polymer 2 shifts the emission to the longer wavelength region (560 nm) even though it is not fully π-conjugated. Polymer 3 and 4 which contain silicon atoms between conjugated main chain show blue-shifted emissions, due to the restriction of π-conjugation length.

Greenham et al. reported a red-light emitting dialkoxy-substituted PPV derivative (CN-PPV) with cyano-groups on the vinylene units [89]. Scheme 13 shows synthesis of CN-PPV. This CN-PPV is a brilliant red, fluorescent material. Cyclovoltametry studies show that the cyano groups shift the onset of reduction by about 0.6 V [90]. The low HOMO-LUMO gap of CN-PPV indicates that the nitrile groups do not introduce any significant steric torsion in the conjugated chain. The emission at 710 nm occurs exclusively from the lower energy CN-PPV.

Another interesting light-emitting moiety is a fluorene molecule. Polyfluorene derivatives have extremely high luminescence efficiencies in solution, which are largely maintained in the polymer films [91]. As a result, polyflu-

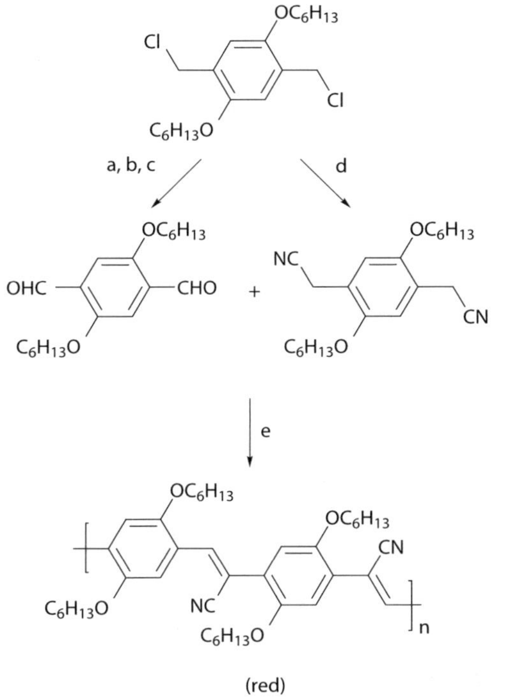

Scheme 13. Synthesis of CN-PPV: a) NaOAc; b) KOH, EtOH; c) pyridinium chlorochromate; d) NaCN; e) KOtBu or Bu$_4$NOH, tBuOH, THF, 50 °C.

orenes and conjugated polymers composed of fluorene units have been intensively studied recently as the blue light-emitting materials [92–94].

Scheme 14 shows the synthetic pathway to copoly(alkylfluorene). Linear, para-linked PDHFPV and kinked meta-linked PDHFMPV show the EL emission at 482 and 426 nm, respectively, corresponding to blue light. Promising results with fluorene contained conjugated polymers offer opportunities for stable blue devices.

8
Conclusion and Outlook

The prospect for the success of conjugated polymer-based LEDs as commercially viable display devices depends not only on the development of new materials, but also on progress in the design of devices. Strong advantages of polymers over inorganics in light emission are processibility, color tunability, low operating voltage and flexibility. Processibility is an advantage for application to large-area displays, where the device can be fabricated by casting the luminescent polymer from solution. Many π-conjugated polymers have π-π* energy gaps re-

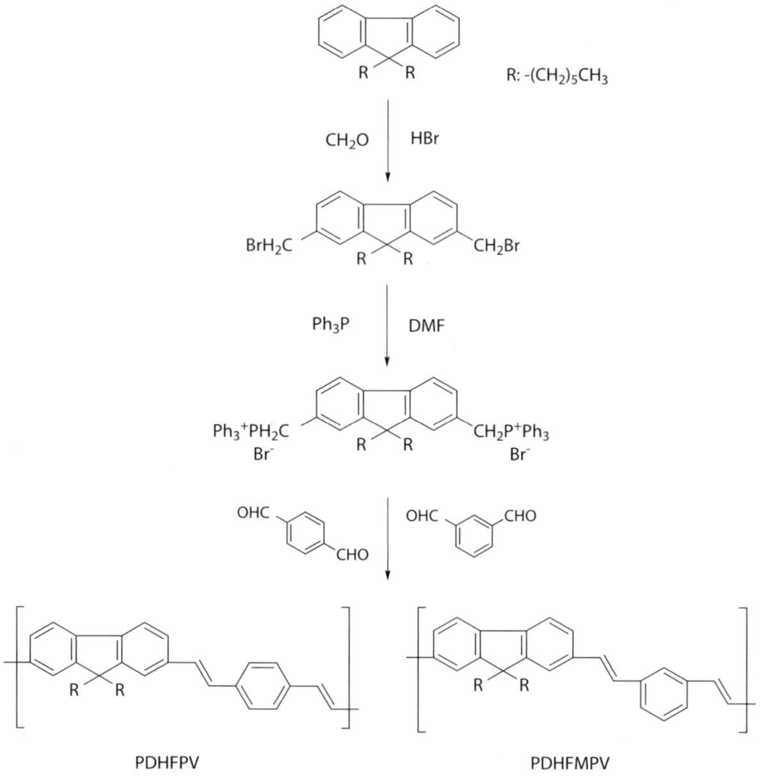

Scheme 14. Synthesis of copoly(alkylfluorene)

sponsible for EL, which can be altered through controlled changes in the molecular structure, so that color tuning is easily attainable. The possibility of device operation at low voltage allows a large-area display operable with low energy consumption. When a flexible substrate is used, LEDs or displays have a mechanica l advantage and the devices can have a variety of shapes. The available device parameters of an onset voltage of a couple of volts, luminance efficiency of 10 lm/W, and brightness of over 50,000 cd/m^2 are adequate for computer monitor, display and liquid crystal display backlight applications. Lifetimes over 10,000 h are also necessary for many commercial applications.

A number of tasks still remain to be achieved before the apparent bright prospects of conjugated polymer LEDs can be realized. The most important task is development of the best light-emitting polymer materials. And, high purity polymers should be required and high molar mass polymers are also advantageous for ensuring good film forming properties. Although the best polymer LED materials are being developed, there is still much room for research on academic and industrial aspects. The important areas for further exploration include the physics of device operation such as carrier injection, transport, and the photophysics of light emission. Polymer engineering through chemistry for higher quantum efficiency, multiple colors and device design and construction is also required. It is important to figure out the role and the control of interfaces. Process technology suitable for large-area devices, patterning within a layer and defining the pixel should also be developed. We expect that polymer LEDs will ultimately have a significant impact in the area of new display technologies.

Acknowledgement. This work was supported by Korea Science and Engineering Foundation (KOSEF) through center for advanced functional polymers (H. K. Shim) and center for photo- and electro-responsive molecules (J.-I. Jin).

References

1. Hwang DH, Kang IN, Jang MS, Shim HK (1995) Bull Korean Chem Soc 16:135
2. Burroughes JH, Bradley DDC, Brown AR, Marks RN, Friend RH, Burn PL, Holmes AB (1990) Nature 347:539
3. Gustafasson G, Cao Y, Treacy GM, Klauetter F, Colaneri N, Heeger AJ (1992) Nature 357:477
4. Bisberg J, Cumming WJ, Gaudiana RA, Hutchinson KD, Ingwall RT, Kolb ES, Mehta PG, Minns RA, . Peterswn CP (1995) Macromolecules 28:386
5. Khanna RK, Cui H (1993) Macromolecules 26:7076
6. Aguiar M, Karasz FE, Akcelrud L (1995) Macromolecules 28:4598
7. Kang IN, Shim HK, Zyung T (1997) Chem Mater 9:746
8. Jang Ms, Song SY, Lee JI, Zyung T, Shim HK (1999) Macromol Chem Phys 200:1101
9. Brown AR, Bradley DDC, Burroughes JH, Friend RH, Greenham NC, Burn PL, Holmes AB, Kraft A (1992) Appl Phys Lett 61:2793
10. Burn PL, Kraft A, Baigent DR, Bradley DDC, Brown AR, Friend RH, Gymer RW, Holmes AB, Jackson RW (1993) J Am Chem Soc 115:10117
11. Hay M, Klavetter FL (1995) J Am Chem Soc 117:7112
12. Kraft A, Grimsdale AC, Holmes AB (1998) Angew Chem Int Ed 37:402
13. Feast WJ, Tsibouklis J, Pouwer KL, Groenendaal L, Meijer EW, (1996) Polymer 37:5017
14. Reynolds JR, Karasz FE, Chien JCW, Gourley KD, Lillya CP (1983) J Phys 44:693
15. Jin J-I, Park CK, Shim HK (1993) Macromolecules 26:1779
16. Gagnon DR, Capistran JD, Karasz FE, Lenz RW (1987) Polymer 28:567
17. Halliday DA, Burn PL, Friend RH, Bradley DDC, Holmes AB (1993) Synth Met 55:902
18. Wudl F, Allemand PM, Srdanov G, Ni Z McBranch D (1991) ACS Sym Ser 455:683
19. Jang MS, Song SY, Lee JI, Shim HK, Zyung T (1999) Macromol Chem Phys 200:1101
20. Jang MS, Song SY, Shim HK (2000) Polymer in press
21. Benjamin I, Faraggi EZ, Avny Y, Davidov D, Neumann R (1996) Chem Mater 8:352
22. Zyung T, Kim JJ, Hwang WY, Hwang DH, Shim HK (1995) Synth Met 71:2167
23. Yoon CB, Kang IN, Shim HK (1997) J Polym Sci Part A Polym Chem 35:2253
24. Hwang DH, Kang IN, Jang MS, Shim HK, Zyung T (1996) Polym Bull 36:383
25. Greenham NC, Samuel IDW, Hayes GR, Philips RT, Kessener YAR, Moratti SC, Holmes AB, Friend RH (1995) Chem Phys Lett 241:89
26. Stampfl J, Tasch S, Leising G, Scherf U (1995) Synth Met 71:2125
27. Bradley DDC, Friend RH, Feast WJ (1987) Synth Met 17:645
28. Gurge RM, Sarker AM, Lahti PM, Hu B, Karasz FE (1997) Macromolecules 30:8286
29. Jin J-I, Chung S-J, Yu SH (1998) Macromol Symp 128:79
30. Chung S-J, Jin J-I, Lee CH, Lee C-I (1998) Adv Mater 10:684
31. Chung S-J, Kwon K-Y, Lee S-W, Jin J-I, Lee CH, Lee CE, Park Y (1998) Adv Mater 10:1112
32. Gilch HG, Wheelright WL (1966) J Polym Sci Part A-1 4:1337
33. Wessling RA, Zimmerman RG (1968) US Pat 3401152

34. Berlman IB (1971) Handbook of Fluorescence Spectra of Aromatic Molecules (2nd ed), Academic Press, New York, P364
35. Stern ES, Timmons CJ, 1970, Introduction to Electronic Absorption Spectroscopy in Organic Chemistry (3rd ed), Edward Arnold Publishers, London, U.K., P.185
36. Ref. 31, P363
37. Dexter DL (1953) J Chem Phys 21:836
38. Errolaev VL(1967) Sov Physics Doklady 6:600
39. Zhang C, Vonseggern H, Pakbaz K, Kraabel B, Schmidt HW, Heeger AJ (1994) Synth. Met. 62:35
40. Kido J, Hongawa K, Nagai K (1994) App Phys Lett 64:815
41. Strukelj M, Papadimitrakopoulos F, Miller TM, Rothberg LJ (1995) Science 267:1969
42. Yang Y, Pei Q (1995) J Appl Phys 77:4807
43. Kim YH, Jeoung SC, Kim D, Chung S-J, Jin J-I (2000) Chem. Mater. 12:1067
44. Jang JW, Oh DK, Lee CH, Lee CE, Lee DW, Jin J-I (2000) J Appl Phys 87:3183
45. Kim K, Hong YR, Jin J-I, (2000) Synth. Met. 121:1705
46. Yungi L, Hong M, Alex K-Y. Jen (1999) Chem Mater 11:27
47. Lee ST, Wang YM, Hou XY, Tang CW (1990) Appl Phys Lett 74:670
48. Wu CC, J. Chun KM, Burrows PE, Sturm JC (1995) Appl Phys Lett 66:653.
49. Tang CW, Vanslyke (1987) Appl Phys Lett 51:913
50. Murayama R, Kawami S, Wakimoto T, Sato H, Nakada H, Namiki T, Imai K, Nomura M (1993) Extended Abstracts (The 54th Autumn Meeting 1993) ; The Japan Society of Applied Physics 3:1127
51. Kim K, Hong Y-R, Lee S-W, Jin J-I, Park Y., Song B-H, Kim W-H, Park J-K (2001). J. Mater. Chem. 11: in press
52. Burn PL,.Holmes AB, Kraft A, Bradley DDC, Brown AR, Friend RH (1992) J Chem Soc Chem Commun 55:936.
53. Grem G, Leditzky G, Ullrich B, Leising G (1992) Synth. Met. 51(1–3):383–389.
54. Grem, G, Martin V, Meghdadi F, Paar C, Stampfl J, Sturm J, Tasch S, Leising G (1995) Synth Met 71:2193.
55. Ohmori Y, Udicha,M, Muro K, Yoshino K (1991) Jpn J Appl Phys 30:L1941.
56. Miller RD, Klaerner G (1998) Macromolecules 31:2007.
57. Sokolik I, Yang Z, Karasaz FE, Morton DCJ (1993) Appl Phys 74(5):3584.
58. Barashkov NN, Olivos HJ, Ferraris JP (1997) Synth Met 90:41
59. Pang Y, Li J, Hu B, Karasz FE (1998) Macromolecules 31:6730
60. Son S, Dodabalapur A, Lovinger AJ, Galvin ME (1995) Science 269:375–378.
61. Ahn T, Jang MS, Shim HK, Hwang DH, Zyung T (1999) Macromolecules 32:3279
62. Kang BS, Kim DH, Lim SM, Kim J, Seo ML, Bark KM, Shin SC (1997) Macromolecules 30:7196
63. Adachi C, Tokito S, Tsutsui T, Saito S (1988) Jpn J Appl Phys 27:L713.
64. Hamada Y, Adachi C, Tsutsui T, Saito S (1992) Jpn J Appl Phys 31:1812.
65. Li XC, Spencer GCW, Holmes AB, Moratti SC, Cacialli F, Friend RH (1996) Synth. Met. 76:153.
66. Wang GM, Yuan CW, Wu HW, Wei YJ (1995) Appl. Phys. 78:2679.
67. Hu B, Yang Z, Karasz FE (1994) J. Appl. Phys.76:2419.
68. Song SY, Jang MS, Shim HK, Hwang DH, Zyung T (1999) Macromolecules 32:1482
69. Shim HK (1999) The 7th workshop on Advanced Functional Polymer Materials, Taejon, Korea
70. Yang Y, Heeger AJ (1994) J Appl Phys Lett 64:1245
71. Parker ID (1994) J Appl Phys 75:1656
72. Greenham NC, Moratti SC, Bradley DDC, Friend RH, Holmes AB (1993) Nature 365:628
73. Aratani S, Zhang C, Pakbaz K, Hoger S, Wudl F, Heeger AJ (1993) J Electron Mater 22:745
74. Zhang C, Höger S, Pakbaz K, Wudl F, Heeger AJ (1994) J Electron Mater 23:453
75. Hu B, Yang Z, Karasz FE (1993) J Appl Phys 72:2419
76. Parker ID, Pei Q (1994) Appl Phys Lett 65:1272

77. Kang IN, Hwang DH, Shim HK, Zyung T, Kim JJ (1996) Macromolecules 29:165
78. Lee JI, Kang IN, Hwang DH, Shim HK, Jeoung SC, Kim D (1996) Chem Mater 8:1925
79. Lemmer U (1993) Appl Phys Lett 62:2827
80. Hotta S, Rughooputh SDDV, Heeger AJ, Wudl F (1987) Macromolecules 20:212
81. Song SY, Jang MS, Shim HK, Song IS, Kim WA (1999) Synth Met 102:1116
82. Burn PL, Holmes AB, Kraft A, Bradley DDC, Brown AR, Friend RH (1992) J Chem Soc Chem Commun 32
83. Brown AR, Burn PL, Bradley DDC, Friend RH, Kraft A, Holmes AB (1992) Mol Cryst Liq Cryst 216:111
84. Zhang C, Braun D, Heeger AJ, (1993) J Appl Phys 73:5177
85. Hwang DH, Kim ST, Shim HK, Holmes AB, Moratti SC, Friend RH (1996) J Chem Soc Chem Commun 2241
86. Kim DJ, Kim SH, Lee JH, Kang SJ, Kim HK, Zyung T, Cho I, Choi SK (1996) Mol Cryst Liq Cryst 280:391
87. Kim DJ, Kim SH, Zyung T, Kim JJ, Cho I, Choi SK (1996) Macromolecules 29:3657
88. Kim HK, Ryu MK, Lee SM (1997) Macromolecules 30:1236
89. Greenham NC, Moratti SC, Bradley DDC, Friend RH, Holmes AB (1993) Nature 365:628
90. Li XC, Kraft A, Cervini R, Spencer GCW, Cacialli F, Friend RH, Grüner J, Holmes AB, DeMello JC, Moratti SC (1996) Mat Res Soc Symp Proc 413:13
91. Miller RD, Klaerner G (1998) Macromolecules 31:2007
92. Cho HN, Kim DY, Kim YC, Lee JY, Kim CY (1997) Adv Mater 9:326
93. Ohmori Y, Uchida M, Muro K, Yoshino K (1991) J Appl Phys 30:1941
94. Grüner J, Wittmanm HF, Hamer PJ, Friend RH, Huber J, Scherf U, Mullen K, Moratti SC, Holmes AB (1994) Synth Met 67:181

Received: August 2000

Author Index Volumes 101–158

Author Index Volumes 1–100 see Volume 100

de, Abajo, J. and *de la Campa, J.G.*: Processable Aromatic Polyimides. Vol. 140, pp. 23-60.
Adolf, D. B. see Ediger, M. D.: Vol. 116, pp. 73-110.
Aharoni, S. M. and *Edwards, S. F.*: Rigid Polymer Networks. Vol. 118, pp. 1-231.
Albertsson, A.-C., Varma, I. K.: Aliphatic Polyesters: Synthesis, Properties and Applications. Vol. 157, pp. 99-138.
Albertsson, A.-C. see Edlund, U.: Vol. 157, pp. 53-98.
Albertsson, A.-C. see Söderqvist Lindblad, M.: Vol. 157, pp. 139-161.
Albertsson, A.-C. see Stridsberg, K. M.: Vol. 157, pp. 27-51.
Améduri, B., Boutevin, B. and *Gramain, P.*: Synthesis of Block Copolymers by Radical Polymerization and Telomerization. Vol. 127, pp. 87-142.
Améduri, B. and *Boutevin, B.*: Synthesis and Properties of Fluorinated Telechelic Monodispersed Compounds. Vol. 102, pp. 133-170.
Amselem, S. see Domb, A. J.: Vol. 107, pp. 93-142.
Andrady, A. L.: Wavelenght Sensitivity in Polymer Photodegradation. Vol. 128, pp. 47-94.
Andreis, M. and *Koenig, J. L.*: Application of Nitrogen-15 NMR to Polymers. Vol. 124, pp. 191-238.
Angiolini, L. see Carlini, C.: Vol. 123, pp. 127-214.
Anseth, K. S., Newman, S. M. and *Bowman, C. N.*: Polymeric Dental Composites: Properties and Reaction Behavior of Multimethacrylate Dental Restorations. Vol. 122, pp. 177-218.
Antonietti, M. see Cölfen, H.: Vol. 150, pp. 67-187.
Armitage, B. A. see O'Brien, D. F.: Vol. 126, pp. 53-58.
Arndt, M. see Kaminski, W.: Vol. 127, pp. 143-187.
Arnold Jr., F. E. and *Arnold, F. E.*: Rigid-Rod Polymers and Molecular Composites. Vol. 117, pp. 257-296.
Arshady, R.: Polymer Synthesis via Activated Esters: A New Dimension of Creativity in Macromolecular Chemistry. Vol. 111, pp. 1-42.

Bahar, I., Erman, B. and *Monnerie, L.*: Effect of Molecular Structure on Local Chain Dynamics: Analytical Approaches and Computational Methods. Vol. 116, pp. 145-206.
Ballauff, M. see Dingenouts, N.: Vol. 144, pp. 1-48.
Baltá-Calleja, F. J., González Arche, A., Ezquerra, T. A., Santa Cruz, C., Batallón, F., Frick, B. and *López Cabarcos, E.*: Structure and Properties of Ferroelectric Copolymers of Poly(vinylidene) Fluoride. Vol. 108, pp. 1-48.
Barnes, M. D. see Otaigbe, J.U.: Vol. 154, pp. 1-86.
Barshtein, G. R. and *Sabsai, O. Y.*: Compositions with Mineralorganic Fillers. Vol. 101, pp.1-28.
Baschnagel, J., Binder, K., Doruker, P., Gusev, A. A., Hahn, O., Kremer, K., Mattice, W. L., Müller-Plathe, F., Murat, M., Paul, W., Santos, S., Sutter, U. W., Tries, V.: Bridging the Gap Between Atomistic and Coarse-Grained Models of Polymers: Status and Perspectives. Vol. 152, pp. 41-156.
Batallán, F. see Baltá-Calleja, F. J.: Vol. 108, pp. 1-48.
Batog, A. E., Pet'ko, I. P., Penczek, P.: Aliphatic-Cycloaliphatic Epoxy Compounds and Polymers. Vol. 144, pp. 49-114.

Barton, J. see Hunkeler, D.: Vol. 112, pp. 115-134.
Bell, C. L. and *Peppas, N. A.*: Biomedical Membranes from Hydrogels and Interpolymer Complexes. Vol. 122, pp. 125-176.
Bellon-Maurel, A. see Calmon-Decriaud, A.: Vol. 135, pp. 207-226.
Bennett, D. E. see O'Brien, D. F.: Vol. 126, pp. 53-84.
Berry, G.C.: Static and Dynamic Light Scattering on Moderately Concentraded Solutions: Isotropic Solutions of Flexible and Rodlike Chains and Nematic Solutions of Rodlike Chains. Vol. 114, pp. 233-290.
Bershtein, V. A. and *Ryzhov, V. A.*: Far Infrared Spectroscopy of Polymers. Vol. 114, pp. 43-122.
Bigg, D. M.: Thermal Conductivity of Heterophase Polymer Compositions. Vol. 119, pp. 1-30.
Binder, K.: Phase Transitions in Polymer Blends and Block Copolymer Melts: Some Recent Developments. Vol. 112, pp. 115-134.
Binder, K.: Phase Transitions of Polymer Blends and Block Copolymer Melts in Thin Films. Vol. 138, pp. 1-90.
Binder, K. see Baschnagel, J.: Vol. 152, pp. 41-156.
Bird, R. B. see Curtiss, C. F.: Vol. 125, pp. 1-102.
Biswas, M. and *Mukherjee, A.*: Synthesis and Evaluation of Metal-Containing Polymers. Vol. 115, pp. 89-124.
Biswas, M. and *Sinha Ray, S.*: Recent Progress in Synthesis and Evaluation of Polymer-Montmorillonite Nanocomposites. Vol. 155, pp. 167-221.
Bolze, J. see Dingenouts, N.: Vol. 144, pp. 1-48.
Bosshard, C.: see Gubler, U.: Vol. 158, pp. 123-190.
Boutevin, B. and *Robin, J. J.*: Synthesis and Properties of Fluorinated Diols. Vol. 102. pp. 105-132.
Boutevin, B. see Amédouri, B.: Vol. 102, pp. 133-170.
Boutevin, B. see Améduri, B.: Vol. 127, pp. 87-142.
Bowman, C. N. see Anseth, K. S.: Vol. 122, pp. 177-218.
Boyd, R. H.: Prediction of Polymer Crystal Structures and Properties. Vol. 116, pp. 1-26.
Briber, R. M. see Hedrick, J. L.: Vol. 141, pp. 1-44.
Bronnikov, S. V., Vettegren, V. I. and *Frenkel, S. Y.*: Kinetics of Deformation and Relaxation in Highly Oriented Polymers. Vol. 125, pp. 103-146.
Brown, H. R. see Creton, C.: Vol. 156, pp. 53-135.
Bruza, K. J. see Kirchhoff, R. A.: Vol. 117, pp. 1-66.
Budkowski, A.: Interfacial Phenomena in Thin Polymer Films: Phase Coexistence and Segregation. Vol. 148, pp. 1-112.
Burban, J. H. see Cussler, E. L.: Vol. 110, pp. 67-80.
Burchard, W.: Solution Properties of Branched Macromolecules. Vol. 143, pp. 113-194.

Calmon-Decriaud, A. Bellon-Maurel, V., Silvestre, F.: Standard Methods for Testing the Aerobic Biodegradation of Polymeric Materials. Vol 135, pp. 207-226.
Cameron, N. R. and *Sherrington, D. C.*: High Internal Phase Emulsions (HIPEs)-Structure, Properties and Use in Polymer Preparation. Vol. 126, pp. 163-214.
de la Campa, J. G. see de Abajo, , J.: Vol. 140, pp. 23-60.
Candau, F. see Hunkeler, D.: Vol. 112, pp. 115-134.
Canelas, D. A. and *DeSimone, J. M.*: Polymerizations in Liquid and Supercritical Carbon Dioxide. Vol. 133, pp. 103-140.
Canva, M., Stegeman, G. I.: Quadratic Parametric Interactions in Organic Waveguides. Vol. 158, pp. 87-121.
Capek, I.: Kinetics of the Free-Radical Emulsion Polymerization of Vinyl Chloride. Vol. 120, pp. 135-206.
Capek, I.: Radical Polymerization of Polyoxyethylene Macromonomers in Disperse Systems. Vol. 145, pp. 1-56.
Capek, I.: Radical Polymerization of Polyoxyethylene Macromonomers in Disperse Systems. Vol. 146, pp. 1-56.

Capek, I. and *Chern, C.-S.*: Radical Polymerization in Direct Mini-Emulsion Systems. Vol. 155, pp. 101-166.
Carlini, C. and *Angiolini, L.*: Polymers as Free Radical Photoinitiators. Vol. 123, pp. 127-214.
Carter, K. R. see Hedrick, J. L.: Vol. 141, pp. 1-44.
Casas-Vazquez, J. see Jou, D.: Vol. 120, pp. 207-266.
Chandrasekhar, V.: Polymer Solid Electrolytes: Synthesis and Structure. Vol 135, pp. 139-206
Chang, J. Y. see Han, M. J.: Vol. 153, pp. 1-36.
Charleux, B., Faust R.: Synthesis of Branched Polymers by Cationic Polymerization. Vol. 142, pp. 1-70.
Chen, P. see Jaffe, M.: Vol. 117, pp. 297-328.
Chern, C.-S. see Capek, I.: Vol. 155, pp. 101-166.
Choe, E.-W. see Jaffe, M.: Vol. 117, pp. 297-328.
Chow, T. S.: Glassy State Relaxation and Deformation in Polymers. Vol. 103, pp. 149-190.
Chung, T.-S. see Jaffe, M.: Vol. 117, pp. 297-328.
Cölfen, H. and *Antonietti, M.*: Field-Flow Fractionation Techniques for Polymer and Colloid Analysis. Vol. 150, pp. 67-187.
Comanita, B. see Roovers, J.: Vol. 142, pp. 179-228.
Connell, J. W. see Hergenrother, P. M.: Vol. 117, pp. 67-110.
Creton, C., Kramer, E. J., Brown, H. R., Hui, C.-Y.: Adhesion and Fracture of Interfaces Between Immiscible Polymers: From the Molecular to the Continuum Scale. Vol. 156, pp. 53-135.
Criado-Sancho, M. see Jou, D.: Vol. 120, pp. 207-266.
Curro, J.G. see Schweizer, K.S.: Vol. 116, pp. 319-378.
Curtiss, C. F. and *Bird, R. B.*: Statistical Mechanics of Transport Phenomena: Polymeric Liquid Mixtures. Vol. 125, pp. 1-102.
Cussler, E. L., Wang, K. L. and *Burban, J. H.*: Hydrogels as Separation Agents. Vol. 110, pp. 67-80.

Dalton, L. Nonlinear Optical Polymeric Materials: From Chromophore Design to Commercial Applications. Vol. 158, pp. 1-86.
DeSimone, J. M. see Canelas D. A.: Vol. 133, pp. 103-140.
DiMari, S. see Prokop, A.: Vol. 136, pp. 1-52.
Dimonie, M. V. see Hunkeler, D.: Vol. 112, pp. 115-134.
Dingenouts, N., Bolze, J., Pötschke, D., Ballauf, M.: Analysis of Polymer Latexes by Small-Angle X-Ray Scattering. Vol. 144, pp. 1-48.
Dodd, L. R. and *Theodorou, D. N.*: Atomistic Monte Carlo Simulation and Continuum Mean Field Theory of the Structure and Equation of State Properties of Alkane and Polymer Melts. Vol. 116, pp. 249-282.
Doelker, E.: Cellulose Derivatives. Vol. 107, pp. 199-266.
Dolden, J. G.: Calculation of a Mesogenic Index with Emphasis Upon LC-Polyimides. Vol. 141, pp. 189-245.
Domb, A. J., Amselem, S., Shah, J. and *Maniar, M.*: Polyanhydrides: Synthesis and Characterization. Vol.107, pp. 93-142.
Doruker, P. see Baschnagel, J.: Vol. 152, pp. 41-156.
Dubois, P. see Mecerreyes, D.: Vol. 147, pp. 1-60.
Dubrovskii, S. A. see Kazanskii, K. S.: Vol. 104, pp. 97-134.
Dunkin, I. R. see Steinke, J.: Vol. 123, pp. 81-126.
Dunson, D. L. see McGrath, J. E.: Vol. 140, pp. 61-106.

Eastmond, G. C.: Poly(ε-caprolactone) Blends. Vol.149, pp. 59-223.
Economy, J. and *Goranov, K.*: Thermotropic Liquid Crystalline Polymers for High Performance Applications. Vol. 117, pp. 221-256.
Ediger, M. D. and *Adolf, D. B.*: Brownian Dynamics Simulations of Local Polymer Dynamics. Vol. 116, pp. 73-110.
Edlund, U. Albertsson, A.-C.: Degradable Polymer Microspheres for Controlled Drug Delivery. Vol. 157, pp. 53-98.
Edwards, S. F. see Aharoni, S. M.: Vol. 118, pp. 1-231.

Endo, T. see Yagci, Y.: Vol. 127, pp. 59-86.
Engelhardt, H. and *Grosche, O.*: Capillary Electrophoresis in Polymer Analysis. Vol. 150, pp. 189-217.
Erman, B. see Bahar, I.: Vol. 116, pp. 145-206.
Ewen, B, Richter, D.: Neutron Spin Echo Investigations on the Segmental Dynamics of Polymers in Melts, Networks and Solutions. Vol. 134, pp. 1-130.
Ezquerra, T. A. see Baltá-Calleja, F. J.: Vol. 108, pp. 1-48.

Faust, R. see Charleux, B: Vol. 142, pp. 1-70.
Fekete, E see Pukánszky, B: Vol. 139, pp. 109-154.
Fendler, J.H.: Membrane-Mimetic Approach to Advanced Materials. Vol. 113, pp. 1-209.
Fetters, L. J. see Xu, Z.: Vol. 120, pp. 1-50.
Förster, S. and *Schmidt, M.*: Polyelectrolytes in Solution. Vol. 120, pp. 51-134.
Freire, J. J.: Conformational Properties of Branched Polymers: Theory and Simulations. Vol. 143, pp. 35-112.
Frenkel, S. Y. see Bronnikov, S. V.: Vol. 125, pp. 103-146.
Frick, B. see Baltá-Calleja, F. J.: Vol. 108, pp. 1-48.
Fridman, M. L.: see Terent´eva, J. P.: Vol. 101, pp. 29-64.
Fukui, K. see Otaigbe, J. U.: Vol. 154, pp. 1-86.
Funke, W.: Microgels-Intramolecularly Crosslinked Macromolecules with a Globular Structure. Vol. 136, pp. 137-232.

Galina, H.: Mean-Field Kinetic Modeling of Polymerization: The Smoluchowski Coagulation Equation. Vol. 137, pp. 135-172.
Ganesh, K. see Kishore, K.: Vol. 121, pp. 81-122.
Gaw, K. O. and *Kakimoto, M.*: Polyimide-Epoxy Composites. Vol. 140, pp. 107-136.
Geckeler, K. E. see Rivas, B.: Vol. 102, pp. 171-188.
Geckeler, K. E.: Soluble Polymer Supports for Liquid-Phase Synthesis. Vol. 121, pp. 31-80.
Gehrke, S. H.: Synthesis, Equilibrium Swelling, Kinetics Permeability and Applications of Environmentally Responsive Gels. Vol. 110, pp. 81-144.
de Gennes, P.-G.: Flexible Polymers in Nanopores. Vol. 138, pp. 91-106.
Giannelis, E.P., Krishnamoorti, R., Manias, E.: Polymer-Silicate Nanocomposites: Model Systems for Confined Polymers and Polymer Brushes. Vol. 138, pp. 107-148.
Godovsky, D. Y.: Device Applications of Polymer-Nanocomposites. Vol. 153, pp. 163-205.
Godovsky, D. Y.: Electron Behavior and Magnetic Properties Polymer-Nanocomposites. Vol. 119, pp. 79-122.
González Arche, A. see Baltá-Calleja, F. J.: Vol. 108, pp. 1-48.
Goranov, K. see Economy, J.: Vol. 117, pp. 221-256.
Gramain, P. see Améduri, B.: Vol. 127, pp. 87-142.
Grest, G.S.: Normal and Shear Forces Between Polymer Brushes. Vol. 138, pp. 149-184.
Grigorescu, G, Kulicke, W.-M.: Prediction of Viscoelastic Properties and Shear Stability of Polymers in Solution. Vol. 152, p. 1-40.
Grosberg, A. and *Nechaev, S.*: Polymer Topology. Vol. 106, pp. 1-30.
Grosche, O. see Engelhardt, H.: Vol. 150, pp. 189-217.
Grubbs, R., Risse, W. and *Novac, B.*: The Development of Well-defined Catalysts for Ring-Opening Olefin Metathesis. Vol. 102, pp. 47-72.
Gubler, U., Bosshard, C.: Molecular Design for Third-Order Nonlinear Optics. Vol. 158, pp. 123-190.
van Gunsteren, W. F. see Gusev, A. A.: Vol. 116, pp. 207-248.
Gusev, A. A., Müller-Plathe, F., van Gunsteren, W. F. and *Suter, U. W.*: Dynamics of Small Molecules in Bulk Polymers. Vol. 116, pp. 207-248.
Gusev, A. A. see Baschnagel, J.: Vol. 152, pp. 41-156.
Guillot, J. see Hunkeler, D.: Vol. 112, pp. 115-134.
Guyot, A. and *Tauer, K.*: Reactive Surfactants in Emulsion Polymerization. Vol. 111, pp. 43-66.

Hadjichristidis, N., Pispas, S., Pitsikalis, M., Iatrou, H., Vlahos, C.: Asymmetric Star Polymers Synthesis and Properties. Vol. 142, pp. 71-128.
Hadjichristidis, N. see Xu, Z.: Vol. 120, pp. 1-50.
Hadjichristidis, N. see Pitsikalis, M.: Vol. 135, pp. 1-138.
Hahn, O. see Baschnagel, J.: Vol. 152, pp. 41-156.
Hakkarainen, M.: Aliphatic Polyesters: Abiotic and Biotic Degradation and Degradation Products. Vol. 157, pp. 1-26.
Hall, H. K. see *Penelle, J.:* Vol. 102, pp. 73-104.
Hamley, I. W.: Crystallization in Block Copolymers. Vol. 148, pp. 113-138.
Hammouda, B.: SANS from Homogeneous Polymer Mixtures: A Unified Overview. Vol. 106, pp. 87-134.
Han, M.J. and Chang, J.Y.: Polynucleotide Analogues. Vol. 153, pp. 1-36.
Harada, A.: Design and Construction of Supramolecular Architectures Consisting of Cyclodextrins and Polymers. Vol. 133, pp. 141-192.
Haralson, M. A. see Prokop, A.: Vol. 136, pp. 1-52.
Hassan, C.M. and Peppas, N.A.: Structure and Applications of Poly(vinyl alcohol) Hydrogels Produced by Conventional Crosslinking or by Freezing/Thawing Methods. Vol. 153, pp. 37-65.
Hawker, C. J. Dentritic and Hyperbranched Macromolecules – Precisely Controlled Macromolecular Architectures. Vol. 147, pp. 113-160.
Hawker, C. J. see Hedrick, J. L.: Vol. 141, pp. 1-44.
Hedrick, J. L., Carter, K. R., Labadie, J. W., Miller, R. D., Volksen, W., Hawker, C. J., Yoon, D. Y., Russell, T. P., McGrath, J. E., Briber, R. M.: Nanoporous Polyimides. Vol. 141, pp. 1-44.
Hedrick, J. L., Labadie, J. W., Volksen, W. and Hilborn, J. G.: Nanoscopically Engineered Polyimides. Vol. 147, pp. 61-112.
Hedrick, J. L. see Hergenrother, P. M.: Vol. 117, pp. 67-110.
Hedrick, J. L. see Kiefer, J.: Vol. 147, pp. 161-247.
Hedrick, J.L. see McGrath, J. E.: Vol. 140, pp. 61-106.
Heller, J.: Poly (Ortho Esters). Vol. 107, pp. 41-92.
Hemielec, A. A. see Hunkeler, D.: Vol. 112, pp. 115-134.
Hergenrother, P. M., Connell, J. W., Labadie, J. W. and Hedrick, J. L.: Poly(arylene ether)s Containing Heterocyclic Units. Vol. 117, pp. 67-110.
Hernández-Barajas, J. see Wandrey, C.: Vol. 145, pp. 123-182.
Hervet, H. see Léger, L.: Vol. 138, pp. 185-226.
Hilborn, J. G. see Hedrick, J. L.: Vol. 147, pp. 61-112.
Hilborn, J. G. see Kiefer, J.: Vol. 147, pp. 161-247.
Hiramatsu, N. see Matsushige, M.: Vol. 125, pp. 147-186.
Hirasa, O. see Suzuki, M.: Vol. 110, pp. 241-262.
Hirotsu, S.: Coexistence of Phases and the Nature of First-Order Transition in Poly-N-isopropylacrylamide Gels. Vol. 110, pp. 1-26.
Höcker, H. see Klee, D.: Vol. 149, pp. 1-57.
Hornsby, P.: Rheology, Compoundind and Processing of Filled Thermoplastics. Vol. 139, pp. 155-216.
Hui, C.-Y. see Creton, C.: Vol. 156, pp. 53-135
Hult, A., Johansson, M., Malmström, E.: Hyperbranched Polymers. Vol. 143, pp. 1-34.
Hunkeler, D., Candau, F., Pichot, C., Hemielec, A. E., Xie, T. Y., Barton, J., Vaskova, V., Guillot, J., Dimonie, M. V., Reichert, K. H.: Heterophase Polymerization: A Physical and Kinetic Comparision and Categorization. Vol. 112, pp. 115-134.
Hunkeler, D. see Prokop, A.: Vol. 136, pp. 1-52; 53-74.
Hunkeler, D see Wandrey, C.: Vol. 145, pp. 123-182.

Iatrou, H. see Hadjichristidis, N.: Vol. 142, pp. 71-128.
Ichikawa, T. see Yoshida, H.: Vol. 105, pp. 3-36.
Ihara, E. see Yasuda, H.: Vol. 133, pp. 53-102.
Ikada, Y. see Uyama, Y.: Vol. 137, pp. 1-40.

Ilavsky, M.: Effect on Phase Transition on Swelling and Mechanical Behavior of Synthetic Hydrogels. Vol. 109, pp. 173-206.
Imai, Y.: Rapid Synthesis of Polyimides from Nylon-Salt Monomers. Vol. 140, pp. 1-23.
Inomata, H. see Saito, S.: Vol. 106, pp. 207-232.
Inoue, S. see Sugimoto, H.: Vol. 146, pp. 39-120.
Irie, M.: Stimuli-Responsive Poly(N-isopropylacrylamide), Photo- and Chemical-Induced Phase Transitions. Vol. 110, pp. 49-66.
Ise, N. see Matsuoka, H.: Vol. 114, pp. 187-232.
*Ito, K., Kawaguchi, S,:*Poly(macronomers), Homo- and Copolymerization. Vol. 142, pp. 129-178.
Ivanov, A. E. see Zubov, V. P.: Vol. 104, pp. 135-176.

Jacob, S. and Kennedy, J.: Synthesis, Characterization and Properties of OCTA-ARM Polyisobutylene-Based Star Polymers. Vol. 146, pp. 1-38.
Jaffe, M., Chen, P., Choe, E.-W., Chung, T.-S. and *Makhija, S.*: High Performance Polymer Blends. Vol. 117, pp. 297-328.
Jancar, J.: Structure-Property Relationships in Thermoplastic Matrices. Vol. 139, pp. 1-66.
Jerôme, R.: see Mecerreyes, D.: Vol. 147, pp. 1-60.
Jiang, M., Li, M., Xiang, M. and Zhou, H.: Interpolymer Complexation and Miscibility and Enhancement by Hydrogen Bonding. Vol. 146, pp. 121-194.
Jin, J.: see Shim, H.-K.: Vol. 158, pp. 191-241.
Jo, W. H. and Yang, J. S.: Molecular Simulation Approaches for Multiphase Polymer Systems. Vol. 156, pp. 1-52.
Johansson, M. see Hult, A.: Vol. 143, pp. 1-34.
Joos-Müller, B. see Funke, W.: Vol. 136, pp. 137-232.
Jou, D., Casas-Vazquez, J. and *Criado-Sancho, M.*: Thermodynamics of Polymer Solutions under Flow: Phase Separation and Polymer Degradation. Vol. 120, pp. 207-266.

Kaetsu, I.: Radiation Synthesis of Polymeric Materials for Biomedical and Biochemical Applications. Vol. 105, pp. 81-98.
Kaji, K. see Kanaya, T.: Vol. 154, pp. 87-141.
Kakimoto, M. see Gaw, K. O.: Vol. 140, pp. 107-136.
Kaminski, W. and *Arndt, M.*: Metallocenes for Polymer Catalysis. Vol. 127, pp. 143-187.
Kammer, H. W., Kressler, H. and *Kummerloewe, C.*: Phase Behavior of Polymer Blends - Effects of Thermodynamics and Rheology. Vol. 106, pp. 31-86.
Kanaya, T. and Kaji, K.: Dynamcis in the Glassy State and Near the Glass Transition of Amorphous Polymers as Studied by Neutron Scattering. Vol. 154, pp. 87-141.
Kandyrin, L. B. and *Kuleznev, V. N.*: The Dependence of Viscosity on the Composition of Concentrated Dispersions and the Free Volume Concept of Disperse Systems. Vol. 103, pp. 103-148.
Kaneko, M. see Ramaraj, R.: Vol. 123, pp. 215-242.
Kang, E. T., Neoh, K. G. and *Tan, K. L.*: X-Ray Photoelectron Spectroscopic Studies of Electroactive Polymers. Vol. 106, pp. 135-190.
Karlsson, S. see Söderqvist Lindblad, M.: Vol. 157, pp. 139–161.
Kato, K. see Uyama, Y.: Vol. 137, pp. 1-40.
Kawaguchi, S. see Ito, K.: Vol. 142, p 129-178.
Kazanskii, K. S. and *Dubrovskii, S. A.*: Chemistry and Physics of „Agricultural" Hydrogels. Vol. 104, pp. 97-134.
Kennedy, J. P. see Jacob, S.: Vol. 146, pp. 1-38.
Kennedy, J. P. see Majoros, I.: Vol. 112, pp. 1-113.
Khokhlov, A., Starodybtzev, S. and *Vasilevskaya, V.*: Conformational Transitions of Polymer Gels: Theory and Experiment. Vol. 109, pp. 121-172.
Kiefer, J., Hedrick J. L. and *Hiborn, J. G.*: Macroporous Thermosets by Chemically Induced Phase Separation. Vol. 147, pp. 161-247.
Kilian, H. G. and *Pieper, T.*: Packing of Chain Segments. A Method for Describing X-Ray Patterns of Crystalline, Liquid Crystalline and Non-Crystalline Polymers. Vol. 108, pp. 49-90.
Kim, J. see Quirk, R.P.: Vol. 153, pp. 67-162.

Kishore, K. and *Ganesh, K.*: Polymers Containing Disulfide, Tetrasulfide, Diselenide and Ditelluride Linkages in the Main Chain. Vol. 121, pp. 81-122.
Kitamaru, R.: Phase Structure of Polyethylene and Other Crystalline Polymers by Solid-State ^{13}C/MNR. Vol. 137, pp 41-102.
Klee, D. and *Höcker, H.*: Polymers for Biomedical Applications: Improvement of the Interface Compatibility. Vol. 149, pp. 1-57.
Klier, J. see Scranton, A. B.: Vol. 122, pp. 1-54.
Kobayashi, S., Shoda, S. and *Uyama, H.*: Enzymatic Polymerization and Oligomerization. Vol. 121, pp. 1-30.
Köhler, W. and *Schäfer, R.*: Polymer Analysis by Thermal-Diffusion Forced Rayleigh Scattering. Vol. 151, pp. 1-59.
Koenig, J. L. see Andreis, M.: Vol. 124, pp. 191-238.
Koike, T.: Viscoelastic Behavior of Epoxy Resins Before Crosslinking. Vol. 148, pp. 139-188.
Kokufuta, E.: Novel Applications for Stimulus-Sensitive Polymer Gels in the Preparation of Functional Immobilized Biocatalysts. Vol. 110, pp. 157-178.
Konno, M. see Saito, S.: Vol. 109, pp. 207-232.
Kopecek, J. see Putnam, D.: Vol. 122, pp. 55-124.
Koßmehl, G. see Schopf, G.: Vol. 129, pp. 1-145.
Kramer, E. J. see Creton, C.: Vol. 156, pp. 53-135.
Kremer, K. see Baschnagel, J.: Vol. 152, pp. 41-156.
Kressler, J. see Kammer, H. W.: Vol. 106, pp. 31-86.
Kricheldorf, H. R.: Liquid-Cristalline Polyimides. Vol. 141, pp. 83-188.
Krishnamoorti, R. see Giannelis, E.P.: Vol. 138, pp. 107-148.
Kirchhoff, R. A. and *Bruza, K. J.*: Polymers from Benzocyclobutenes. Vol. 117, pp. 1-66.
Kuchanov, S. I.: Modern Aspects of Quantitative Theory of Free-Radical Copolymerization. Vol. 103, pp. 1-102.
Kuchanov, S. I.: Principles of Quantitive Description of Chemical Structure of Synthetic Polymers. Vol. 152, p. 157-202.
Kudaibergennow, S.E.: Recent Advances in Studying of Synthetic Polyampholytes in Solutions. Vol. 144, pp. 115-198.
Kuleznev, V. N. see Kandyrin, L. B.: Vol. 103, pp. 103-148.
Kulichkhin, S. G. see Malkin, A. Y.: Vol. 101, pp. 217-258.
Kulicke, W.-M. see Grigorescu, G.: Vol. 152, p. 1-40.
Kummerloewe, C. see Kammer, H. W.: Vol. 106, pp. 31-86.
Kuznetsova, N. P. see Samsonov, G. V.: Vol. 104, pp. 1-50. Labadie, J. W. see Hergenrother, P. M.: Vol. 117, pp. 67-110.

Labadie, J. W. see Hedrick, J. L.: Vol. 141, pp. 1-44.
Labadie, J. W. see Hedrick, J. L.: Vol. 147, pp. 61-112.
Lamparski, H. G. see O´Brien, D. F.: Vol. 126, pp. 53-84.
Laschewsky, A.: Molecular Concepts, Self-Organisation and Properties of Polysoaps. Vol. 124, pp. 1-86.
Laso, M. see Leontidis, E.: Vol. 116, pp. 283-318.
Lazár, M. and *RychlΩ, R.*: Oxidation of Hydrocarbon Polymers. Vol. 102, pp. 189-222.
Lechowicz, J. see Galina, H.: Vol. 137, pp. 135-172.
Léger, L., Raphaël, E., Hervet, H.: Surface-Anchored Polymer Chains: Their Role in Adhesion and Friction. Vol. 138, pp. 185-226.
Lenz, R. W.: Biodegradable Polymers. Vol. 107, pp. 1-40.
Leontidis, E., de Pablo, J. J., Laso, M. and *Suter, U. W.*: A Critical Evaluation of Novel Algorithms for the Off-Lattice Monte Carlo Simulation of Condensed Polymer Phases. Vol. 116, pp. 283-318.
Lee, B. see Quirk, R.P: Vol. 153, pp. 67-162.
Lee, Y. see Quirk, R.P: Vol. 153, pp. 67-162.
Lesec, J. see Viovy, J.-L.: Vol. 114, pp. 1-42.
Li, M. see Jiang, M.: Vol. 146, pp. 121-194.
Liang, G. L. see Sumpter, B. G.: Vol. 116, pp. 27-72.

Lienert, K.-W.: Poly(ester-imide)s for Industrial Use. Vol. 141, pp. 45-82.
Lin, J. and *Sherrington, D. C.*: Recent Developments in the Synthesis, Thermostability and Liquid Crystal Properties of Aromatic Polyamides. Vol. 111, pp. 177-220.
Liu, Y. see Söderqvist Lindblad, M.: Vol. 157, pp. 139-161
López Cabarcos, E. see Baltá-Calleja, F. J.: Vol. 108, pp. 1-48.

Majoros, I., Nagy, A. and *Kennedy, J. P.*: Conventional and Living Carbocationic Polymerizations United. I. A Comprehensive Model and New Diagnostic Method to Probe the Mechanism of Homopolymerizations. Vol. 112, pp. 1-113.
Makhija, S. see Jaffe, M.: Vol. 117, pp. 297-328.
Malmström, E. see Hult, A.: Vol. 143, pp. 1-34.
Malkin, A. Y. and *Kulichkhin, S. G.*: Rheokinetics of Curing. Vol. 101, pp. 217-258.
Maniar, M. see Domb, A. J.: Vol. 107, pp. 93-142.
Manias, E., see Giannelis, E.P.: Vol. 138, pp. 107-148.
Mashima, K., Nakayama, Y. and *Nakamura, A.*: Recent Trends in Polymerization of a-Olefins Catalyzed by Organometallic Complexes of Early Transition Metals. Vol. 133, pp. 1-52.
Mathew, D. see Reghunadhan Nair, C.P.: Vol. 155, pp. 1-99.
Matsumoto, A.: Free-Radical Crosslinking Polymerization and Copolymerization of Multivinyl Compounds. Vol. 123, pp. 41-80.
Matsumoto, A. see Otsu, T.: Vol. 136, pp. 75-138.
Matsuoka, H. and *Ise, N.*: Small-Angle and Ultra-Small Angle Scattering Study of the Ordered Structure in Polyelectrolyte Solutions and Colloidal Dispersions. Vol. 114, pp. 187-232.
Matsushige, K., Hiramatsu, N. and *Okabe, H.*: Ultrasonic Spectroscopy for Polymeric Materials. Vol. 125, pp. 147-186.
Mattice, W. L. see Rehahn, M.: Vol. 131/132, pp. 1-475.
Mattice, W. L. see Baschnagel, J.: Vol. 152, p. 41-156.
Mays, W. see Xu, Z.: Vol. 120, pp. 1-50.
Mays, J.W. see Pitsikalis, M.: Vol.135, pp. 1-138.
McGrath, J. E. see Hedrick, J. L.: Vol. 141, pp. 1-44.
McGrath, J. E., Dunson, D. L., Hedrick, J. L.: Synthesis and Characterization of Segmented Polyimide-Polyorganosiloxane Copolymers. Vol. 140, pp. 61-106.
McLeish, T.C.B., Milner, S. T.: Entangled Dynamics and Melt Flow of Branched Polymers. Vol. 143, pp. 195-256.
Mecerreyes, D., Dubois, P. and *Jerôme, R.*: Novel Macromolecular Architectures Based on Aliphatic Polyesters: Relevance of the „Coordination-Insertion" Ring-Opening Polymerization. Vol. 147, pp. 1 -60.
Mecham, S. J. see McGrath, J. E.: Vol. 140, pp. 61-106.
Mikos, A. G. see Thomson, R. C.: Vol. 122, pp. 245-274.
Milner, S. T. see McLeish, T. C. B.: Vol. 143, pp. 195-256.
Mison, P. and *Sillion, B.*: Thermosetting Oligomers Containing Maleimides and Nadiimides End-Groups. Vol. 140, pp. 137-180.
Miyasaka, K.: PVA-Iodine Complexes: Formation, Structure and Properties. Vol. 108. pp. 91-130.
Miller, R. D. see Hedrick, J. L.: Vol. 141, pp. 1-44.
Monnerie, L. see Bahar, I.: Vol. 116, pp. 145-206.
Morishima, Y.: Photoinduced Electron Transfer in Amphiphilic Polyelectrolyte Systems. Vol. 104, pp. 51-96.
Morton M. see Quirk, R.P: Vol. 153, pp. 67-162
Mours, M. see Winter, H. H.: Vol. 134, pp. 165-234.
Müllen, K. see Scherf, U.: Vol. 123, pp. 1-40.
Müller-Plathe, F. see Gusev, A. A.: Vol. 116, pp. 207-248.
Müller-Plathe, F. see Baschnagel, J.: Vol. 152, p. 41-156.
Mukerherjee, A. see Biswas, M.: Vol. 115, pp. 89-124.
Murat, M. see Baschnagel, J.: Vol. 152, p. 41-156.
Mylnikov, V.: Photoconducting Polymers. Vol. 115, pp. 1-88.

Nagy, A. see Majoros, I.: Vol. 112, pp. 1-11.
Nakamura, A. see Mashima, K.: Vol. 133, pp. 1-52.
Nakayama, Y. see Mashima, K.: Vol. 133, pp. 1-52.
Narasinham, B., Peppas, N. A.: The Physics of Polymer Dissolution: Modeling Approaches and Experimental Behavior. Vol. 128, pp. 157-208.
Nechaev, S. see Grosberg, A.: Vol. 106, pp. 1-30.
Neoh, K. G. see Kang, E. T.: Vol. 106, pp. 135-190.
Newman, S. M. see Anseth, K. S.: Vol. 122, pp. 177-218.
Nijenhuis, K. te: Thermoreversible Networks. Vol. 130, pp. 1-252.
Ninan, K.N. see Reghunadhan Nair, C. P.: Vol. 155, pp. 1-99.
Noid, D. W. see Otaigbe, J.U.: Vol. 154, pp. 1-86.
Noid, D. W. see Sumpter, B. G.: Vol. 116, pp. 27-72.
Novac, B. see Grubbs, R.: Vol. 102, pp. 47-72.
Novikov, V. V. see Privalko, V. P.: Vol. 119, pp. 31-78.

O'Brien, D. F., Armitage, B. A., Bennett, D. E. and *Lamparski, H. G.:* Polymerization and Domain Formation in Lipid Assemblies. Vol. 126, pp. 53-84.
Ogasawara, M.: Application of Pulse Radiolysis to the Study of Polymers and Polymerizations. Vol.105, pp. 37-80.
Okabe, H. see Matsushige, K.: Vol. 125, pp. 147-186.
Okada, M.: Ring-Opening Polymerization of Bicyclic and Spiro Compounds. Reactivities and Polymerization Mechanisms. Vol. 102, pp. 1-46.
Okano, T.: Molecular Design of Temperature-Responsive Polymers as Intelligent Materials. Vol. 110, pp. 179-198.
Okay, O. see Funke, W.: Vol. 136, pp. 137-232.
Onuki, A.: Theory of Phase Transition in Polymer Gels. Vol. 109, pp. 63-120.
Osad'ko, I.S.: Selective Spectroscopy of Chromophore Doped Polymers and Glasses. Vol. 114, pp. 123-186.
Otaigbe, J. U., Barnes, M. D., Fukui, K., Sumpter, B. G., Noid, D. W.: Generation, Characterization, and Modeling of Polymer Micro- and Nano-Particles. Vol. 154, pp. 1-86.
Otsu, T., Matsumoto, A.: Controlled Synthesis of Polymers Using the Iniferter Technique: Developments in Living Radical Polymerization. Vol. 136, pp. 75-138.

de Pablo, J. J. see Leontidis, E.: Vol. 116, pp. 283-318.
Padias, A. B. see Penelle, J.: Vol. 102, pp. 73-104.
Pascault, J.-P. see Williams, R. J. J.: Vol. 128, pp. 95-156.
Pasch, H.: Analysis of Complex Polymers by Interaction Chromatography. Vol. 128, pp. 1-46.
Pasch, H.: Hyphenated Techniques in Liquid Chromatography of Polymers. Vol. 150, pp. 1-66.
Paul, W. see Baschnagel, J.: Vol. 152, p. 41-156.
Penczek, P. see Batog, A. E.: Vol. 144, pp. 49-114.
Penelle, J., Hall, H. K., Padias, A. B. and *Tanaka, H.:* Captodative Olefins in Polymer Chemistry. Vol. 102, pp. 73-104.
Peppas, N. A. see Bell, C. L.: Vol. 122, pp. 125-176.
Peppas, N.A. see Hassan, C.M.: Vol. 153, pp. 37-65
Peppas, N. A. see Narasimhan, B.: Vol. 128, pp. 157-208.
Pet'ko, I. P. see Batog, A. E.: Vol. 144, pp. 49-114.
Pichot, C. see Hunkeler, D.: Vol. 112, pp. 115-134.
Pieper, T. see Kilian, H. G.: Vol. 108, pp. 49-90.
Pispas, S. see Pitsikalis, M.: Vol. 135, pp. 1-138.
Pispas, S. see Hadjichristidis: Vol. 142, pp. 71-128.
Pitsikalis, M., Pispas, S., Mays, J. W., Hadjichristidis, N.: Nonlinear Block Copolymer Architectures. Vol. 135, pp. 1-138.
Pitsikalis, M. see Hadjichristidis: Vol. 142, pp. 71-128.
Pötschke, D. see Dingenouts, N.: Vol 144, pp. 1-48.

Pokrovskii, V. N.: The Mesoscopic Theory of the Slow Relaxation of Linear Macromolecules. Vol. 154, pp. 143-219.
Pospíšil, J.: Functionalized Oligomers and Polymers as Stabilizers for Conventional Polymers. Vol. 101, pp. 65-168.
Pospíšil, J.: Aromatic and Heterocyclic Amines in Polymer Stabilization. Vol. 124, pp. 87-190.
Powers, A. C. see Prokop, A.: Vol. 136, pp. 53-74.
Priddy, D. B.: Recent Advances in Styrene Polymerization. Vol. 111, pp. 67-114.
Priddy, D. B.: Thermal Discoloration Chemistry of Styrene-co-Acrylonitrile. Vol. 121, pp. 123-154.
Privalko, V. P. and *Novikov, V. V.:* Model Treatments of the Heat Conductivity of Heterogeneous Polymers. Vol. 119, pp 31-78.
Prokop, A., Hunkeler, D., Powers, A. C., Whitesell, R. R., Wang, T. G.: Water Soluble Polymers for Immunoisolation II: Evaluation of Multicomponent Microencapsulation Systems. Vol. 136, pp. 53-74.
Prokop, A., Hunkeler, D., DiMari, S., Haralson, M. A., Wang, T. G.: Water Soluble Polymers for Immunoisolation I: Complex Coacervation and Cytotoxicity. Vol. 136, pp. 1-52.
Pukánszky, B. and *Fekete, E.:* Adhesion and Surface Modification. Vol. 139, pp. 109-154.
Putnam, D. and *Kopecek, J.:* Polymer Conjugates with Anticancer Acitivity. Vol. 122, pp. 55- 124.

Quirk, R.P. and Yoo, T., Lee, Y., M., Kim, J. and Lee, B.: Applications of 1,1-Diphenylethylene Chemistry in Anionic Synthesis of Polymers with Controlled Structures. Vol. 153, pp. 67-162.

Ramaraj, R. and *Kaneko, M.:* Metal Complex in Polymer Membrane as a Model for Photosynthetic Oxygen Evolving Center. Vol. 123, pp. 215-242.
Rangarajan, B. see Scranton, A. B.: Vol. 122, pp. 1-54.
Ranucci, E. see Söderqvist Lindblad, M.: Vol. 157, pp. 139–161.
Raphaël, E. see Léger, L.: Vol. 138, pp. 185-226.
Reddinger, J. L. and *Reynolds, J. R.:* Molecular Engineering of π-Conjugated Polymers. Vol. 145, pp. 57-122.
Reghunadhan Nair, C.P., Mathew, D. and *Ninan, K.N.,* : Cyanate Ester Resins, Recent Developments. Vol. 155, pp. 1-99.
Reichert, K. H. see Hunkeler, D.: Vol. 112, pp. 115-134.
Rehahn, M., Mattice, W. L., Suter, U. W.: Rotational Isomeric State Models in Macromolecular Systems. Vol. 131/132, pp. 1-475.
Reynolds, J.R. see Reddinger, J. L.: Vol. 145, pp. 57-122.
Richter, D. see Ewen, B.: Vol. 134, pp.1-130.
Risse, W. see Grubbs, R.: Vol. 102, pp. 47-72.
Rivas, B. L. and *Geckeler, K. E.:* Synthesis and Metal Complexation of Poly(ethyleneimine) and Derivatives. Vol. 102, pp. 171-188.
Robin, J. J. see Boutevin, B.: Vol. 102, pp. 105-132.
Roe, R.-J.: MD Simulation Study of Glass Transition and Short Time Dynamics in Polymer Liquids. Vol. 116, pp. 111-114.
Roovers, J., Comanita, B.: Dendrimers and Dendrimer-Polymer Hybrids. Vol. 142, pp 179-228.
Rothon, R. N.: Mineral Fillers in Thermoplastics: Filler Manufacture and Characterisation. Vol. 139, pp. 67-108.
Rozenberg, B. A. see Williams, R. J. J.: Vol. 128, pp. 95-156.
Ruckenstein, E.: Concentrated Emulsion Polymerization. Vol. 127, pp. 1-58.
Rusanov, A. L.: Novel Bis (Naphtalic Anhydrides) and Their Polyheteroarylenes with Improved Processability. Vol. 111, pp. 115-176.
Russel, T. P. see Hedrick, J. L.: Vol. 141, pp. 1-44.
Rychlý, J. see Lazár, M.: Vol. 102, pp. 189-222.
Ryner, M. see Stridsberg, K. M.: Vol. 157, pp. 27–51.
Ryzhov, V. A. see Bershtein, V. A.: Vol. 114, pp. 43-122.

Sabsai, O. Y. see Barshtein, G. R.: Vol. 101, pp. 1-28.
Saburov, V. V. see Zubov, V. P.: Vol. 104, pp. 135-176.

Saito, S., Konno, M. and *Inomata, H.:* Volume Phase Transition of N-Alkylacrylamide Gels. Vol. 109, pp. 207-232.
Samsonov, G. V. and *Kuznetsova, N. P.:* Crosslinked Polyelectrolytes in Biology. Vol. 104, pp. 1-50.
Santa Cruz, C. see Baltá-Calleja, F. J.: Vol. 108, pp. 1-48.
Santos, S. see Baschnagel, J.: Vol. 152, p. 41-156.
Sato, T. and *Teramoto, A.:* Concentrated Solutions of Liquid-Christalline Polymers. Vol. 126, pp. 85-162.
Schäfer R. see Köhler, W.: Vol. 151, pp. 1-59.
Scherf, U. and *Müllen, K.:* The Synthesis of Ladder Polymers. Vol. 123, pp. 1-40.
Schmidt, M. see Förster, S.: Vol. 120, pp. 51-134.
Schopf, G. and *Koßmehl, G.:* Polythiophenes - Electrically Conductive Polymers. Vol. 129, pp. 1-145.
Schweizer, K. S.: Prism Theory of the Structure, Thermodynamics, and Phase Transitions of Polymer Liquids and Alloys. Vol. 116, pp. 319-378.
Scranton, A. B., Rangarajan, B. and *Klier, J.:* Biomedical Applications of Polyelectrolytes. Vol. 122, pp. 1-54.
Sefton, M. V. and *Stevenson, W. T. K.:* Microencapsulation of Live Animal Cells Using Polycrylates. Vol. 107, pp. 143-198.
Shamanin, V. V.: Bases of the Axiomatic Theory of Addition Polymerization. Vol. 112, pp. 135-180.
Sheiko, S. S.: Imaging of Polymers Using Scanning Force Microscopy: From Superstructures to Individual Molecules. Vol. 151, pp. 61-174.
Sherrington, D. C. see Cameron, N. R., Vol. 126, pp. 163-214.
Sherrington, D. C. see Lin, J.: Vol. 111, pp. 177-220.
Sherrington, D. C. see Steinke, J.: Vol. 123, pp. 81-126.
Shibayama, M. see Tanaka, T.: Vol. 109, pp. 1-62.
Shiga, T.: Deformation and Viscoelastic Behavior of Polymer Gels in Electric Fields. Vol. 134, pp. 131-164.
Shim, H.-K., Jin, J.: Light-Emitting Characteristics of Conjugated Polymers. Vol. 158, pp. 191-241.
Shoda, S. see Kobayashi, S.: Vol. 121, pp. 1-30.
Siegel, R. A.: Hydrophobic Weak Polyelectrolyte Gels: Studies of Swelling Equilibria and Kinetics. Vol. 109, pp. 233-268.
Silvestre, F. see Calmon-Decriaud, A.: Vol. 207, pp. 207-226.
Sillion, B. see Mison, P.: Vol. 140, pp. 137-180.
Singh, R. P. see Sivaram, S.: Vol. 101, pp. 169-216.
Sinha Ray, S. see Biswas, M: Vol. 155, pp. 167-221.
Sivaram, S. and *Singh, R. P.:* Degradation and Stabilization of Ethylene-Propylene Copolymers and Their Blends: A Critical Review. Vol. 101, pp. 169-216.
Söderqvist Lindblad, M., Liu, Y., Albertsson, A.-C., Ranucci, E., Karlsson, S.: Polymer from Renewable Resources. Vol. 157, pp. 139-161
Starodybtzev, S. see Khokhlov, A.: Vol. 109, pp. 121-172.
Stegeman, G. I.: see Canva, M.: Vol. 158, pp. 87-121.
Steinke, J., Sherrington, D. C. and *Dunkin, I. R.:* Imprinting of Synthetic Polymers Using Molecular Templates. Vol. 123, pp. 81-126.
Stenzenberger, H. D.: Addition Polyimides. Vol. 117, pp. 165-220.
Stevenson, W. T. K. see Sefton, M. V.: Vol. 107, pp. 143-198.
Stridsberg, K. M., Ryner, M., Albertsson, A.-C.: Controlled Ring-Opening Polymerization: Polymers with Designed Macromoleculars Architecture. Vol. 157, pp. 27-51.
Suematsu, K.: Recent Progress of Gel Theory: Ring, Excluded Volume, and Dimension. Vol. 156, pp. 136-214.
Sumpter, B. G., Noid, D. W., Liang, G. L. and *Wunderlich, B.:* Atomistic Dynamics of Macromolecular Crystals. Vol. 116, pp. 27-72.
Sumpter, B. G. see Otaigbe, J.U.: Vol. 154, pp. 1-86.

Sugimoto, H. and *Inoue, S.*: Polymerization by Metalloporphyrin and Related Complexes. Vol. 146, pp. 39-120.
Suter, U. W. see Gusev, A. A.: Vol. 116, pp. 207-248.
Suter, U. W. see Leontidis, E.: Vol. 116, pp. 283-318.
Suter, U. W. see Rehahn, M.: Vol. 131/132, pp. 1-475.
Suter, U. W. see Baschnagel, J.: Vol. 152, p. 41-156.
Suzuki, A.: Phase Transition in Gels of Sub-Millimeter Size Induced by Interaction with Stimuli. Vol. 110, pp. 199-240.
Suzuki, A. and *Hirasa, O.*: An Approach to Artifical Muscle by Polymer Gels due to Micro-Phase Separation. Vol. 110, pp. 241-262.

Tagawa, S.: Radiation Effects on Ion Beams on Polymers. Vol. 105, pp. 99-116.
Tan, K. L. see Kang, E. T.: Vol. 106, pp. 135-190.
Tanaka, H. and *Shibayama, M.*: Phase Transition and Related Phenomena of Polymer Gels. Vol. 109, pp. 1-62.
Tanaka, T. see Penelle, J.: Vol. 102, pp. 73-104.
Tauer, K. see Guyot, A.: Vol. 111, pp. 43-66.
Teramoto, A. see Sato, T.: Vol. 126, pp. 85-162.
Terent'eva, J. P. and *Fridman, M. L.*: Compositions Based on Aminoresins. Vol. 101, pp. 29-64.
Theodorou, D. N. see Dodd, L. R.: Vol. 116, pp. 249-282.
Thomson, R. C., Wake, M. C., Yaszemski, M. J. and *Mikos, A. G.*: Biodegradable Polymer Scaffolds to Regenerate Organs. Vol. 122, pp. 245-274.
Tokita, M.: Friction Between Polymer Networks of Gels and Solvent. Vol. 110, pp. 27-48.
Tries, V. see Baschnagel, J:. Vol. 152, p. 41-156.
Tsuruta, T.: Contemporary Topics in Polymeric Materials for Biomedical Applications. Vol. 126, pp. 1-52.

Uyama, H. see Kobayashi, S.: Vol. 121, pp. 1-30.
Uyama, Y: Surface Modification of Polymers by Grafting. Vol. 137, pp. 1-40.

Varma, I. K. see Albertsson, A.-C.: Vol. 157, pp. 99-138.
Vasilevskaya, V. see Khokhlov, A.: Vol. 109, pp. 121-172.
Vaskova, V. see Hunkeler, D.: Vol.:112, pp. 115-134.
Verdugo, P.: Polymer Gel Phase Transition in Condensation-Decondensation of Secretory Products. Vol. 110, pp. 145-156.
Vettegren, V. I.: see Bronnikov, S. V.: Vol. 125, pp. 103-146.
Viovy, J.-L. and *Lesec, J.*: Separation of Macromolecules in Gels: Permeation Chromatography and Electrophoresis. Vol. 114, pp. 1-42.
Vlahos, C. see Hadjichristidis, N.: Vol. 142, pp. 71-128.
Volksen, W.: Condensation Polyimides: Synthesis, Solution Behavior, and Imidization Characteristics. Vol. 117, pp. 111-164.
Volksen, W. see Hedrick, J. L.: Vol. 141, pp. 1-44.
Volksen, W. see Hedrick, J. L.: Vol. 147, pp. 61-112.

Wake, M. C. see Thomson, R. C.: Vol. 122, pp. 245-274.
Wandrey C., Hernández-Barajas, J. and *Hunkeler, D.*: Diallyldimethylammonium Chloride and its Polymers. Vol. 145, pp. 123-182.
Wang, K. L. see Cussler, E. L.: Vol. 110, pp. 67-80.
Wang, S.-Q.: Molecular Transitions and Dynamics at Polymer/Wall Interfaces: Origins of Flow Instabilities and Wall Slip. Vol. 138, pp. 227-276.
Wang, T. G. see Prokop, A.: Vol. 136, pp.1-52; 53-74.
Whitesell, R. R. see Prokop, A.: Vol. 136, pp. 53-74.
Williams, R. J. J., Rozenberg, B. A., Pascault, J.-P.: Reaction Induced Phase Separation in Modified Thermosetting Polymers. Vol. 128, pp. 95-156.
Winter, H. H., Mours, M.: Rheology of Polymers Near Liquid-Solid Transitions. Vol. 134, pp. 165-234.

Wu, C.: Laser Light Scattering Characterization of Special Intractable Macromolecules in Solution. Vol 137, pp. 103-134.
Wunderlich, B. see Sumpter, B. G.: Vol. 116, pp. 27-72.

Xiang, M. see Jiang, M.: Vol. 146, pp. 121-194.
Xie, T. Y. see Hunkeler, D.: Vol. 112, pp. 115-134.
Xu, Z., Hadjichristidis, N., Fetters, L. J. and *Mays, J. W.*: Structure/Chain-Flexibility Relationships of Polymers. Vol. 120, pp. 1-50.

Yagci, Y. and *Endo, T.*: N-Benzyl and N-Alkoxy Pyridium Salts as Thermal and Photochemical Initiators for Cationic Polymerization. Vol. 127, pp. 59-86.
Yannas, I. V.: Tissue Regeneration Templates Based on Collagen-Glycosaminoglycan Copolymers. Vol. 122, pp. 219-244.
Yang, J. S. see Jo, W. H.: Vol. 156, pp. 1-52.
Yamaoka, H.: Polymer Materials for Fusion Reactors. Vol. 105, pp. 117-144.
Yasuda, H. and *Ihara, E.*: Rare Earth Metal-Initiated Living Polymerizations of Polar and Nonpolar Monomers. Vol. 133, pp. 53-102.
Yaszemski, M. J. see Thomson, R. C.: Vol. 122, pp. 245-274.
Yoo, T. see Quirk, R.P.: Vol. 153, pp. 67-162.
Yoon, D. Y. see Hedrick, J. L.: Vol. 141, pp. 1-44.
Yoshida, H. and *Ichikawa, T.*: Electron Spin Studies of Free Radicals in Irradiated Polymers. Vol. 105, pp. 3-36.

Zhou, H. see Jiang, M.: Vol. 146, pp. 121-194.
Zubov, V. P., Ivanov, A. E. and *Saburov, V. V.*: Polymer-Coated Adsorbents for the Separation of Biopolymers and Particles. Vol. 104, pp. 135-176.

Subject Index

Absorption spectrum, red tail 99
Acceptor 156-157, 164-173, 177-182
Aggregation 13
All-optical signal processing 125, 130-131, 136-140, 182-186
Alternating copolymer 218, 221
Analog-to-digital conversion 65
Anomalous dispersion phase-matching 88, 93, 104
Anthracene pendant 201-208
Asymmetry 160-173, 182
Attenuated total reflection (ATR) 16

Backplane interconnection 65
Bandwidth 60
Beta site testing 63
Bias voltage stability 63
Birefringence phase-matching 88, 105
Birefringent modulator 6
Blend polymers 225-234
Blue light emission 210, 219, 221, 224, 237
Blue-shift 197, 198, 204, 215, 218-226, 234, 236
– Bond 155-157
– Bond 155
Bond length alternation 163-164
Buried channel waveguides 50

Cable television 65
Calibration of nonlinearity 134-136, 147
Carbazole 201, 209, 212, 221, 224
Carbazole pendant 201, 209, 212, 221, 224
Carbon tetrafluoride etching 55
Carotenoids 168
Cascaded prism structure 68
Centrosymmetry 160-173, 182
Cerenkov geometries 97
Cerenkov phase-matching 87, 94, 100
Chalcogenide glasses 183
Charge transfer molecules 5
Claddings 53

CLD chromophore 27
Commercialization 65
Conducting claddings 53
1D-Conjugation 133, 157, 161-162, 173-177, 182
2D-Conjugation 134, 157, 161-163, 168-173, 182
Convention, electric field 128-, nonlinearities 126-134
Copper phthalocyanin (CuPc) 213
Counter-propagating phase-matching 88, 103
Critical conjugation length 173-177
Cross-phase modulation (XPM) 128-132
Crosslinking 47
Crystal growth 73
CS_2 135, 147
Cut-off method 20
CWC chromophore 29
Cyanofuran (CF) 24

DANS 62, 184-185
DAST 9
Degenerate four wave mixing (DFWM) 128-129, 135, 145-149
Delocalization 155-159
Dendrimers 71
Device performance 60
Differential scanning calorimetry 12
Dilinking chromophores 50
Dilution effect 198, 228
Dimensional tolerances 91
Dipolar term 161-162, 165-168, 171-172
Dipole moment 22, 127, 155, 158-163, 168-173
– -hyperpolarizability product 17
Directional coupler 6
Disperse red 62
Donor 156-157, 164-173, 177-182
Double bond 156-157, 163, 177
Drive voltage 60

EFISH 15, 141, 176
Einstein convention 126
Electric field sensors 65
Electrical conductivity 12
Electrical field poling 41
Electro-absorptive modulators 8
Electro-optic coefficient 5
Electro-optic modulators 8
Electroluminescence (EL) 191, 200, 205, 216
Electron affinity 206, 212, 221, 236
Electron microscopy 20
– Electrons 155-157
Electrostatic interactions 31
Ellipsometry 16
Energy transfer 191, 206, 211, 226, 228, 232
Excited state 132, 136, 146, 154-155, 159-161, 173-174, 177-181
Excited state absorption 152, 159, 177-181
Expansion of polarization 126-128
External quantum efficiency 207, 212, 213, 225

Ferroelectric order 5
Fiber optic communication links 8
Fiber optic telecommunications 65
Figures of merit 96, 136-139, 182-185
Finline structures 56
Flip chip bonding 8
FTC chromophore 27
Futerrex PC3-6000 54

Gallium arsenide 9
Gamma radiation 63
Gilch polymerization 210
Glass transition temperature 13
Glasses 182-183
GLD chromophore 28
Gray scale mask processing 55
Ground state 146, 158-163, 173-174, 177-181
Growth anisotropy 10
Guidelines, optimization 182
Gyroscopes, optical 65

Heck's coupling 193
Heterodyne detection 60
Hole blocking 221, 234
Hole-transporting 209-213, 221, 225, 226
Horizontal integration 54
Hyper Rayleigh scattering (HRS) 15
Hyperpolarizability 6
Hyrazones 166-167

Immersion method 20
In-line modulators 59
Insertion loss 59

Interaction length 60
Interference modulation 16
Interferometer, DFWM 149
Ising model 30

Kerr gating 141
Kinked polymers 215-217, 221, 237, 243
Kleinman symmetry 129-130, 133

Land mine detection 65
Langmuir-Blodgett fabrication 11
Langmuir-Blodgett media 101
Lattice hardening 47
Light-emitting diode (LED) 192, 200, 205, 213, 215, 219, 220
Link gain 8
Lithium niobate 9
Local field correction 127, 132
Lorentz approximation 132

Mach Zehnder interferometer 6
Maximum luminance 213, 215
MDPM 112, 116
Methylnitroaniline 9
Microwave conductivity 13
Modal-dispersion phase-matching 88, 93, 94, 105
Mode mismatch 52
Modulation efficiency 8
Molecular orbitals 156
Molecular states 159-161
Monte Carlo calculations 34
Multiphoton-induced fluorescence 15

Negative polaron 200
Negative term 161-162, 166, 172
Noise figure 8
Non-collinear wavevector matching 88, 111
Nonlinear optical waveguide 139-140, 182-185
Nonlinear refractive index 130-131, 135-140, 149-151, 182-185
Nonlinear transmission 152-153
Nonlinearity requirements 136-140

Octopolar symmetry 5
Offset lithography 57
OLED technology 64
Oligothienyleneethynylenes 176
Optical absorption 12
Optical limiting 131, 151-153, 181
Optical loss 62
Opto-chip 54
pz Orbitals 156
Order parameter 5

Subject Index

Orientational average 132-134
Oxadiazole 201, 209, 221, 224, 234
- pendant 201, 209, 221, 224, 234

Permutation symmetry 129-130, 162
Perturbation theory 158, 162
Phase retardation 6
Phase separation 40
Phase-matching 87, 89, 92, 113
Phase-matching length/temporal overlap 100
Phased array radar 65
Photochemical poling 10
Photochemical stability 21
Photolithography 51
Photoluminescence (PL) 191, 197-209, 211, 216-221, 224-235
Photonic bandgap structure 69
Photonic RF phase shifter 67
Photostability 140-141
Photothermal deflection spectroscopy 19
Polarization 4
- splitters 58
Polarization-insensitive modulation 66
Poling-induced damage 62
Poling-induced optical loss 45
Poly(2-dimethyloctylsilyl-1,4-phenylenevinylene) (DMOS-PPV) 234, 235
Poly(3,4-ethylenedioxythiophene) (PEDOT) 214
Poly[2-methoxy-5-(2-ethylhexyloxy)-1,4-phenylenevinylene] (MEH-PPV) 194-198, 216, 218-221, 226-235
Poly(1,4-phenylenevinylene) (PPV) 192-197, 204, 207-213
Polycarbonate, amorphous (APC) 47
Polydiacetylene (PDA) 157, 184-185
Polyenes 157, 160, 163, 168, 176
Polyfluorenes 236
Polymer LED 192, 225, 234, 238, 239
Polyphenylenevinylene (PPV) 157, 178-181, 184-185
Polythiophene 157, 176
Polytriacetylene (PTA) 157, 164-166, 173-176, 185
Positive polaron 200
Power splitters 58
PPV derivatives 191-201, 205, 209-213, 221
Precursor 193, 195, 203
Price issues 141
PTS 184-185
Pulsed operation 140
Pump-probe 141
Push-pull poling 44

QPM 112
Quadratic parametric interactions 87
Quantum efficiency 154, 178
Quasi-epitaxy 11
Quasi-phase-matching 88, 92, 93, 108, 115

Radiofrequency interference 68
Reactive ion etching 51
Red tail, absorption spectrum 99
Red-light emission 236
Red-shift 197, 198, 201, 203-211
Reference values 134-136, 147
Relative quantum efficiency 233
Resolution 132, 152 3D-
Resonance enhancement 130, 136-139, 169-171, 176, 182
Rotational average 132-134, 142, 146, 171

Satellite telecommunications 65
Scaling law 173-177
Second-harmonic generation (SHG) 15, 128-129
Selection rules 159-161
Self light organization 88, 111
Self-assembly 73
Self-phase modulation (SPM) 128-132, 150
Semiconductors 183-184
Sequential synthesis 71
Shadow mask processing 55
Silica, fused 135-136, 142-144, 182-183
Singlet exciton 197, 200, 201
Singlet state 159-160
Slab waveguide 20
Solvent damage 14
Spatial light modulation (SLM) 69
Spatial overlap integral 99
Spectrum analyzers 65
Spin casting 41
Spin-on glass technology 55
Stripline electrodes 56
Sublimation 13, 161, 164-173, 177-182
Sum-over-states 158-163, 165, 171

Tapered transitions 52
Temporal overlap 100
Tetraethynylethene (TEE) 157, 168-173, 185
Thermal decomposition 12
Thermal effects 141, 151
Thermal gravimetric analysis 12
Third-harmonic generation (THG) 128-129, 142-145
Three level model 161-163, 165, 171
Three-photon absorption 137-139
Threshold electric field 207, 211-215, 219, 225

Time stretching 65
Time-resolved PL 208, 211
Transient EL 212
Transition dipole moment 158-163
Transparent communication links 8
Trilinking chromophores 50
Triple bond 157
Triplet state 152, 159-160, 178
Tris(8-hydroxyquinolino)aluminum (Alq3) 213
Turn-on voltage 224, 225, 228, 232
Two level model 159
Two-photon absorption 131-132, 134, 137-139, 151-155, 177-182
Two-photon fluorescence 154-155
Two-photon term 161-162, 166, 172
Two-state model 23

UV curable epoxy 14
UV-VIS absorption 195-198, 203, 209, 210, 216, 219, 224, 226, 230

Vertical integration 54
Vertical slopes 57
Vibrational overtone absorption 19
VLSI electronics 54

Waveguide, segmented 117
Waveguide birefringence 93
Wavelength division multiplexing (WDM) 66
Wavevector mismatch 95
Wavevector matching 87-94
Wessling-Zimmerman 202
Wittig condensation 194, 216, 221
Work function 213, 225

Z-scan 149-151

You are one **click** *away from a* **world of chemistry** *information!*

Come and visit Springer's
Chemistry Online Library

Books
- Search the Springer website catalogue
- Subscribe to our free alerting service for new books
- Look through the book series profiles

You want to order?   Email to: orders@springer.de

Journals
- Get abstracts, ToC´s free of charge to everyone
- Use our powerful search engine LINK Search
- Subscribe to our free alerting service LINK *Alert*
- Read full-text articles (available only to subscribers of the paper version of a journal)

You want to subscribe?   Email to: subscriptions@springer.de

Electronic Media
- Get more information on our software and CD-ROMs

You have a question on an electronic product?   Email to: helpdesk-em@springer.de

••••••••••••• Bookmark now:

www.springer.de/chem/

Springer · Customer Service
Haberstr. 7 · D-69126 Heidelberg, Germany
Tel: +49 6221 345-217/218 · Fax: +49 6221 345-229
d&p · 006756_001x_1c

Printing (Computer to Film): Saladruck Berlin
Binding: Stürtz AG, Würzburg